Encyclopedia of Animal Rights and Animal Welfare

Encyclopedia of Animal Rights and Animal Welfare

SECOND EDITION

VOLUME 2: I–Z

Edited by Marc Bekoff

Foreword by Jane Goodall

GREENWOOD PRESS
An Imprint of ABC-CLIO, LLC

A B C ☰ C L I O

Santa Barbara, California • Denver, Colorado • Oxford, England

Library of Congress Cataloging-in-Publication Data

Encyclopedia of animal rights and animal welfare / edited by Marc Bekoff ; foreword by
Jane Goodall.—2nd ed.

    v. cm.

  Includes bibliographical references and index.

  ISBN 978-0-313-35255-3 (set : alk. paper) — ISBN 978-0-313-35257-7 (vol. 1 : alk.
paper) — ISBN 978-0-313-35259-1 (vol. 2 : alk. paper) — ISBN 978-0-313-35256-0
(ebook) — ISBN 978-0-313-35258-4 (vol. 1 : ebook) — ISBN 978-0-313-35260-7
(vol. 2 : ebook)

1. Animal rights—Encyclopedias. 2. Animal welfare—Encyclopedias. I. Bekoff, Marc.
  HV4708.E53   2009
   179′.3—dc22   2009022275

14   13   12   11   10   1   2   3   4   5

This book is also available on the World Wide Web as an eBook.
Visit www.abc-clio.com for details.

ABC-CLIO, LLC
130 Cremona Drive, P.O. Box 1911
Santa Barbara, California 93116-1911

This book is printed on acid-free paper (∞)
Manufactured in the United States of America

Contents

Alphabetical List of Entries

Guide to Related Topics

Below are the headwords for all entries in *The Encyclopedia of Animal Rights and Animal Welfare,* arranged under broad topics.

Activism (*See also* Global Efforts for Animal Protection)

Abolitionist Approach to Animal Rights

American Society for the Prevention of Cruelty to Animals (ASPCA), The

Animal Protection: The Future of Activism

Antivivisectionism

People for the Ethical Treatment of Animals (PETA)

Rescue Groups

Royal Society for the Prevention of Cruelty to Animals (RSPCA) Reform Group

Scholarship and Advocacy

Silver Spring Monkeys, The

Student Objections to Dissection

Student Rights and the First Amendment

Alternatives to Animal Use (*See also* Experimentation and Models)

Alternatives to Animal Experiments in the Life Sciences

Alternatives to Animal Experiments: Reduction, Refinement, and Replacement

The Animal Body

Animal Body, Alteration of

Docking

Domestication

Animal Reproduction, Human Control of

Animal Reproduction: Human Control

Wildlife Contraception

Animal Welfare

Animal Welfare

Animal Welfare: Assessment

Cognition and Sentience

Affective Ethology

Animal Subjectivity

Consciousness, Animal

Deep Ethology

Great Apes and Language Research

Pleasure and Animal Welfare

Sentience and Animal Protection

Sentientism

Whales and Dolphins: Sentience and Suffering in

Companion Animals

Companion Animals

Companion Animals, Welfare, and the Human-Animal Bond

Pet Renting

Puppy Mills

Shelters, No-Kill

Conservation Ethics. *See* **Wildlife Ethics**

Cruelty (*See also*** Law)**

Cruelty to Animals and Human Violence

Cruelty to Animals: Enforcement of Anti-Cruelty Laws

Cruelty to Animals: Prosecuting Anti-Cruelty Laws

Deviance and Animals

Disasters

Disasters and Animals

Disasters and Animals: Legal Treatment in the United States

Dissection

Dissection in Science and Health Education

Student Objections to Dissection

Education

Humane Education

Humane Education, Animal Welfare, and Conservation

Speciesism

Teleology and Telos

Utilitarianism

Utilitarianism and Assessment of Animal Experimentation

Virtue Ethics

Political Rights of Animals

Political Subjectivity of Animals, The

Religion

Religion and Animals

Religion and Animals: Animal Theology

Religion and Animals: Buddhism

Religion and Animals: Christianity

Religion and Animals: Daoism

Religion and Animals: Disensoulment

Religion and Animals: Hinduism

Religion and Animals: Islam

Religion and Animals: Jainism

Religion and Animals: Judaism

Religion and Animals: Judaism and Animal Sacrifice

Religion and Animals: Pantheism and Panentheism

Religion and Animals: Reverence for Life

Religion and Animals: Saints

Religion and Animals: Theodicy

Religion and Animals: Theos Rights

Religion and Animals: Veganism and the Bible

Religion, History, and the Animal Protection Movement

Representation of Animals

Animal Body, Alteration of

Art, Animals, and Ethics

Disneyfication

Docking

Entertainment and Amusement: Animals in the Performing Arts

Museums and Representation of Animals

Objectification of Animals

Poetry and the Representation of Animals

I

INDIA: ANIMAL EXPERIMENTATION

The history of animal experimentation in India in the 20th century parallels that of the United Kingdom in the same period. However, after independence from Britain in 1945, there was a sudden rise in the number of animals used, with a much sharper decline in the conditions under which the animals were housed and experimented upon. India's first Prime Minister, Jawaharlal Nehru, gave a very high degree of importance to building scientific institutions, which led to several animal-using laboratories being expanded.

Since the late 1950s, there was some amount of protest against the use of animals under conditions that many considered shameful. The initial cry was from Rukmini Devi Arundel, who, as an independent Member of the Upper House of Parliament, submitted a bill called the Prevention of Cruelty to Animals Bill. The government, realizing the importance of the measure, requested she withdraw her bill and introduced an identical bill which, in 1960, became the Prevention of Cruelty to Animals (PCA) Act.

From 1959 onward, the group which was registered in 1964 as the Blue Cross of India pursued the issue very seriously. In March 1965, the Blue Cross held an international conference on the subject. The National Anti-Vivisection Society of the U.K. and the British Union for the Abolition of Vivisection sent a delegate, and the Scottish Society for the Prevention of Vivisection provided a large amount of literature and films for the conference. Rukmini Devi, who had been nominated by the Government of India as the first chairperson of the Animal Welfare Board of India, a statutory body set up in 1962 under the PCA Act of 1960, presided over the conference. Under this pressure, the Government of India set up the first Committee for the Purpose of Control and Supervision of Experiments on Animals (CPCSEA), whose members visited various institutions that used animals and met with various stakeholders on this issue. The author and three other members of the Blue Cross were the first to depose before the CPCSEA, and their claims were met with disbelief, if not ridicule. It is pertinent to note that the chairman of the CPCSEA was Kamal Nayan Bajaj, a Member of Parliament, and most of the other members of the CPCSEA were the heads of government laboratories that used animals.

Yet, after visiting all the major labs over a period of a year, the CPCSEA issued a paper on the issue, which began by saying:

Vivisection, or animal experimentation, is one of the most inhuman cruelties against animals, which are being perpetrated in the world to-day. The object of these experiments is said to be in order to advance scientific knowledge, and to undertake research to save or prolong human or animal life and alleviate suffering. In the name of science, however, animals are made to endure the most barbaric tortures ever invented by the human brain, often lasting over long periods and without any sort of anesthetic.

Animals are frozen, boiled, have electric currents passed through their brains, or are driven insane, all in an insatiable "quest for knowledge," which can do nothing whatever to benefit the human race. Many experiments which are successful with animals are a complete failure when applied to human beings. Vested interests, however, make it necessary for the experiments to continue, although what they are showing may be completely useless or already known.

Not to be out done by the rest of the world in cruelties, we perform them here in India on a smaller scale, but no less horrible, as can be seen in the photograph on the facing page. In this Calcutta experiment, several monkeys were given experimental dropsy, leading to their death in one to four months. In the various research laboratories and institutions of India, experiments are performed on certain animals like buffaloes, horses, sheep, goats, dogs, cats, monkeys, rabbits, guinea-pigs, mice, frogs, and fowl. Apart from experimenting upon them here, the export of monkeys from India for experimentation in other countries provides a

handsome source of income, especially in dollars, as a monkey which fetches about Rs.10 in India costs more like 7 pounds or 8 pounds in the U.K.

Rhesus monkeys in large numbers in the Northern Indian State of Uttar Pradesh are trapped, stuffed into cages, and carried on shoulder-poles to Lucknow. A train journey of 260 miles takes them to New Delhi, whence a transport plane carries them the 4,000 miles to London Airport. From London they may be flown another 3,000 miles across the Atlantic to New York, from where they travel on for a further 700 miles in trucks to Okatia Farms in South Carolina. Here the Rhesus monkeys of India are caged with other hordes of "Java" monkeys from the Philippines. After twenty-one days of rigorous health checks, they are dispatched to laboratories in Toronto, Pittsburgh, Detroit, and Berkeley, California. It is disturbing to note that although the Okatia Farms may receive 5,000 monkeys a month, the supply never catches up with the demand, but is rather on the increase all the time. The above gives some small idea upon this vast subject which to deal with adequately is far beyond the scope of this small hand-out. ("Animal Experimentation," 1965)

With typical bureaucratic efficiency, the CPCSEA then abdicated its responsibility by saying:

Much might be done towards alleviating the suffering of animals held for experimental purposes, through personal contacts with

Medical Students and the Staff of Medical Colleges. It should be made more generally known that a very large proportion of painful experiments serve no useful purpose, and instances are known in which treatment successfully made on animals, when applied to humans proved disastrous. Seminars organized to discuss the subject with doctors and students might well prove profitable. Failing all else, there is nothing to prevent airing our views wherever possible and denouncing the cowardice of gaining human benefits at the expense of animals which are entirely at our mercy and unable to protect themselves. The Government of India have now set up a committee for the purpose of regulating animal experimentations. ("Animal Experimentation")

In the meantime, the Blue Cross continued with its campaign. In 1966, the World Coalition Against Vivisection, headquartered in Geneva, started its India office with Captain V. Sundaram as its president and S. Chinny Krishna as honorary secretary.

The attitude of the scientists who constituted the bulk of the members, with an occasional civil servant thrown in, can be judged by the fact that, over the next three decades, the CPCSEA was reconstituted three times and met only twice. Even the reconstitution was only done after numerous letters by the Blue Cross and the Animal Welfare Board. In 1990, the Indian National Science Academy (INSA) brought out its *Guidelines for Animal Experimentation*. These guidelines remained totally ignored by the scientists.

In 1996, Maneka Gandhi became the Minister for Environment and Forests. Her first move was to reconstitute the CPCSEA with herself as chairperson. Author S. Chinny Krishna was nominated as a member. Many of the other members were from animal-using laboratories. With a great deal of difficulty, rules were formulated based on the guidelines issued in 1990 by the Indian National Science Academy. Realizing that these rules would finally be effectively implemented, the scientists literally took to the streets in protest. Dozens of identical letters of protest were sent to the Prime Minister, alleging that implementation of these rules would set back India's scientific progress. A small committee of five was set up by the Prime Minister, with the condition that Mrs. Gandhi could not be on this Committee. Three scientists who used animals in their labs, the Director of the Animal Welfare Division of the Ministry of Environment and Forests, and Dr. Krishna were given the task of addressing the objections raised by the scientists and drafting suitable rules. A fresh set of rules was then drawn up, considered by the CPCSEA, and passed after following all required procedures.

For the next two years, over 500 laboratories and research institutions were inspected. Conditions were found to be deplorable in most of the places. Photographs were posted on the CPCSEA web site. Realizing that the government meant business, government and private institutions began to implement the rules regarding registration, housing, and treatment of animals.

By 2002, serum and vaccine manufacturers, and breeders and animal users were not averse to following the rules, since they began to realize that they got better and more dependable results. The vast majority of the larger institutions were registered. Things were looking much better for the millions of animals bred and used in experiments each year (CPCSEA.org).

In mid-2002, some disgruntled scientists were able to have Maneka Gandhi removed, first as Minister and then as chair of the CPCSEA. The CPCSEA was purged of Dr. Sultan Ismail, Dr. P.Y. Guru, and Dr. S. Chinny Krishna, all of whom were strongly in favor of good science and proper regulations. The Ministry of Health took control of the CPCSEA, but the reforms had been set in motion and much of the progress was irreversible. Implementation is not as effective as it could have been, but the law is there now to protect the animals.

In 2006, thanks to the efforts of Norma Alvarez, the Government of India introduced the Fourth R—rehabilitation—to the existing 3 Rs concept. In addition to legally requiring reduction, refinement, and replacement in all experimentation, the rehabilitation of all animals has been added, meaning that those animals that do not need to be killed during the experimentation must be rehabilitated. Funds for this must be provided by the institution or person carrying out the study. The cost must be budgeted for while applying for grants for the experiments. Unfortunately, however, with the CPCSEA firmly under the control of the Ministry of Health, no worthwhile efforts are being made to implement the many safeguards in the laws in force.

Further Reading

"Animal Experimentation." 1965. Published by the Committee for the Purpose of Controlling and Supervision of Experiments on Animals, Government of India, Ministry of Agriculture.

CPCSEA.org—the official website of the CPCSEA, Ministry of Environment & Forests.

S. Chinny Krishna

INSTITUTIONAL ANIMAL CARE AND USE COMMITTEES (IACUCs)

Beginning in 1985, with extensive revision of the Federal Animal Welfare Act and the adoption of new policies by the National Institute of Health, most institutions that conduct animal research rely on an Institutional Animal Care and Use Committee standard or IACUC to determine whether research meets generally accepted ethical standards for the use of animals. Before 1985, such committees were generally called Animal Care Committees, and while they had some oversight of the care and housing of laboratory animals, they did not review the actual research procedures. Now, however, any organization which receives federal funds must follow Public Health Service (PHS) policies on animal research. Institutions engaged in interstate commerce in covered species of animals (mammals, with the exception of mice, rats, and animals used in agricultural practice) fall under U. S. Department of Agriculture (USDA) regulations, particularly the Animal Welfare Act, first passed in 1966, substantially revised in 1985, and amended several times since then. Both sets of regulations require an IACUC to ensure that the institution follows all applicable regulations, and that any proposal to use animals in research has been reviewed.

Although the USDA regulations and PHS policy differ in important ways (e.g.,

PHS policy applies to all vertebrates, while the USDA exempts many species), an IACUC must include a veterinarian, someone who does not use animals for research, typically referred to as the non-scientist, and someone who does not work for the institution. The two main duties of an IACUC are to review all proposals or protocols for use of covered species of animals, and to ensure compliance with all government regulations. Policies are written in a way that allows great latitude in how these duties are discharged, for example, how many of the protocols are reviewed and discussed by the full committee, as opposed to going through an expedited review process, and how facilities are inspected by the IACUC. As a result, practices vary widely depending on the size of the institution, the amount and range of animal research, and the policies set up by the individual IACUC.

From an ethical perspective, it must first be recognized that the whole system of IACUCs is based on the starting point that animal research is justified as long as it is carried out as well as possible, given the research goals. The questions they consider are almost never of the form should we be doing research on animals but rather, given that Dr. Smith is investigating X, has she shown that the study requires the use of this many animals of this species, and that she has designed the procedure to use appropriate care of the animals, including anesthetics and analgesics. The two questions are not entirely separable, because Dr. Smith will have to give some explanation for why animals must be used; however, the general presumption is in favor of animal research. Given that starting point, there are still at least two other ethical issues raised by the practice of using IACUCs

to regulate research: the scope of an IACUCs authority, and the assumption that self-regulation is the best way to bring institutions into compliance with appropriate standards for ethical research.

With regard to scope, it was noted above that many animals are not covered by the relevant regulations. Most notably, rats and mice are not currently covered by USDA regulations, and farm animals used for production-oriented research also fall into an ambiguous category. No cold-blooded species is covered by USDA regulations, and no invertebrate is covered by PHS policy. Moreover, many IACUCs have adopted the policy that issues of scientific merit fall outside the scope of their decision-making process. This has the effect of restricting, sometimes in significant ways, the nature of the deliberation process when trying to decide whether a particular proposal should be approved. Few attempts have been made to evaluate or ground these scope restrictions with a well-formulated ethical theory.

The second ethical issue focuses on the fact that IACUCs are a way in which research institutions regulate themselves. Some countries, for example, Sweden, have adopted systems of outside regulation. Arguments that have been advanced in favor of outside regulation include a higher probability of impartial and consistent standards that might also better reflect the standards of the general public. Arguments in favor of institution-based systems such as IACUCs include increased flexibility, and the fact that outside review, while feasible in localized areas with a small amount of research, would not be practical in the United States. A broader perspective on the inside/outside issue might ask whether the review process should be carried out primarily by those inside the research community, or primarily by ordinary citizens who do not themselves carry out research. In most review systems today, including the U.S. system, the majority of decision-makers (on a typical IACUC, the proportion may be six or eight to one) are people who themselves are or have been engaged in animal research.

Lilly-Marlene Russow

INSTITUTIONAL ANIMAL CARE AND USE COMMITTEES: NONAFFILIATED MEMBERS

Laws stipulate that Institutional Animal Care and Use Committees (IACUCs) should include a person or persons who are not affiliated with the research facility, to represent the concerns of the community about animal care and use. These members are referred to as non-affiliated members (NAM).

NAMs review research proposals submitted to the IACUC and participate in meetings of the committee. Questions about the proposals can be raised, and the researcher has the opportunity to answer these questions. While some committees require unanimous approval for passage of a proposal, most committees require a simple majority vote. Thus, in most research facilities, a NAM cannot block a proposal.

Only anecdotal information is available concerning the views of individuals being selected as NAMs. Nonetheless, Barbara Orlans states that individuals who are selected are typically not known within their communities as animal advocates. In fact, people with possible biases, for example, practicing scientists or staff of pro-vivisectionist organizations, have reportedly sat on these committees.

Levin and Stephens have proposed that NAMs should be community members known for their advocacy of animal protection. They propose that these people should be neither mouthpieces for the facility nor spies for local activists. Rather, they should be advocates for the research animals operating within an imperfect oversight mechanism.

Understandably, some feel uncomfortable if the NAM is or was a practicing scientist, for they believe that such a person cannot be an advocate for the animals. However, this issue should be resolvable if NAMs are chosen after careful deliberation. In fact, as we learn more about the effectiveness of NAMs in the past, for example, background and records, we will be able to make recommendations for the future. People of all views should be represented in these deliberations.

Further Reading

Levin, L. H., and Stephens, M. L. 1994/1995. Appointing Animal Protectionists to Institutional Animal Care and Use Committees. *Animal Welfare Information Center Newsletter* 5 (4):1–10.

Orlans, F. B. 1993. *In the name of science, issues in responsible animal experimentation.* New York: Oxford University Press.

Orlans, F. B., Simmonds, R. C., and Dodds, W. J., eds. 1987. Effective animal care and use committees. *Laboratory Animal Science,* Special Issue, January.

U.S. Congress 1985. *Health research extension act of 1985.* Public Law 99–158, November 20,1985.

U.S. Congress 1985. Improved Standards for Laboratory Animals Act. *Congressional Record* 131 (175):H12335-H12336.

Marjorie Bekoff

INSTITUTIONAL ANIMAL CARE AND USE COMMITTEES (IACUCs): REGULATORY REQUIREMENTS

Since 1985, U. S. regulatory agencies have required most institutions and facilities involved in animal research to establish Institutional Animal Care and Use Committees (IACUCs). These committees are responsible for reviewing animal research proposals, inspecting animal housing and laboratory areas, and monitoring programs related to scientific uses of animals.

The two major U.S. regulatory systems governing laboratory animal use, the Animal Welfare Act, and the Public Health Service Policy on Humane Care and Use of Laboratory Animals, require IACUCs. Both systems have similar requirements for IACUC membership, duties, and authority.

Committees must have at least three members. At least one doctor of veterinary medicine must serve on the committee. In addition, at least one person on the committee must have no other affiliation with the research institution. According to the Animal Welfare Act, this person should "provide representation for general community interests in the proper care and treatment of animals." Before a research project involving animals can go forward, it must be reviewed by the IACUC. The Animal Welfare Act and the Public Health Service Policy direct committees to ensure that laboratory animal pain, discomfort, and distress are reduced through the use of anesthetics, analgesics, tranquilizers, and humane killing methods. Committees must also determine that no scientifically acceptable alternatives to painful or distressing procedures on animals are available. Committees must consider whether a scientist has chosen an appropriate species and number of animals for the project as well. Finally, committees are to apply the following principle: "procedures involving animals should be designed and performed with due consideration of their relevance to human or animal health, the advancement of knowledge, or the good of society."

There are advantages and disadvantages of IACUCs. The federal government has often adopted a system of institutional committee oversight to address ethical issues in research. Institutional committees were first adopted in the 1970s as a means of monitoring research involving human subjects. Institutional committees are also used to address problems involving scientific misconduct and financial conflicts of interest affecting researchers.

Committee oversight systems reduce government expenses by assigning most of the monitoring responsibilities to research institutions, rather than to government officials. Researchers are also more likely to respect and cooperate with a committee of their colleagues than with a group of government outsiders.

Although committees must comply with certain general rules, they have a great deal of flexibility and freedom to tailor the rules to their specific institution's situation. The committee's mixed membership is intended to allow diverse values to shape ethical decision making. The hope is that this approach will produce reasonable positions on a variety of controversial bioethical issues.

Yet the committee system has its critics as well. Institutions bear financial and other burdens of administering the oversight system; faculty and staff must put aside their other duties to serve on the committees. Because the federal rules are somewhat general, different individual committees can reach different decisions on proposed research. For example, an experiment found unacceptable by one institutional committee could be labeled acceptable by a committee in a different institution.

Animal advocates also question whether the inclusion of one public member can prevent the scientific viewpoint from dominating in IACUC deliberations. They argue that committees would be more effective if one member were assigned to represent the interests of animals against pro-research interests. Thus far, however, these advocates have not persuaded Congress to revise the rules governing IACUCs.

Committees also face challenges in developing an effective approach to working with the scientists whose projects they evaluate, and in establishing meaningful programs for training on humane approaches to animal care and experimentation. They must also develop a defensible approach to recruiting and selecting new committee members, particularly the persons chosen from outside the institution.

Many of the issues facing IACUCs reflect general uncertainty over the appropriate use of animals in science. Persons favoring the elimination of or drastic reduction in laboratory animal use are unlikely to see IACUCs as providing meaningful oversight of animal research. On the other hand, persons who believe that scientists should have complete control over their experiments are likely to label IACUC activities an unjustified invasion of scientific freedom.

The IACUC system was designed to implement a third ethical perspective. This view is that animal research is ethical if conducted to advance important social goals, and if harm to laboratory animals is reduced to the minimum necessary to achieve those goals. IACUCs will continue to operate within this ethical framework unless advocates of another view successfully persuade Congress to alter the current regulatory approach.

Further Reading

Animal Welfare Act, available at http://www.aphis.usda.gov/animal_welfare/awa.shtml.

Institutional Animal Care and Use Committee Guidebook (2d ed., 2002), available at ftp://ftp.grants.nih.gov/IACUC/Guidebook.pdf.

Plous, Scott, and Herzog, Harold. 2001. Reliability of Protocol Reviews for Animal Research. *Science* 293:608–609.

Public Health Service Policy on Humane Care and Use of Laboratory Animals, available at http://grants.nih.gov/grants/olaw/references/phspol.htm.

Rebecca Dresser

ISRAEL: ANIMAL PROTECTION

While enormous progress has been made in the Western world in raising awareness about the human-animal bond and its importance to human and nonhuman species alike, in other countries this work has just begun. When Concern for Helping Animals in Israel (CHAI), www.chai-online.org, was founded in 1984, animal advocacy in Israel barely existed. There was no Animal Protection Law, no veterinary school, and only two very small animal shelters able to do little to promote spaying and neutering. Animal overpopulation control consisted exclusively of mass poisoning of cats and dogs using slow-acting, painful poisons such as strychnine and alpha chlorolose. Abused work animals were a common sight, and humane education was unknown.

For more than two decades, CHAI's desire to raise consciousness in teachers, veterinarians, and government officials, as well as in the general public, about the need to help animals has motivated their efforts and projects. CHAI's mission is to prevent and relieve animal suffering in Israel and to elevate consciousness about animals through education. Its projects foster empathy, respect, and responsibility toward all living beings, and inspire and empower Jews, Arabs, and Christians alike to recognize the interconnectedness of all living beings and to make compassionate choices for the good of all.

For the first two decades, CHAI participated in the process of drafting Israel's first Animal Protection Law. It provided funds, veterinary supplies, and equipment, including the first animal ambulance, to help start shelters in areas where there were none and to assist existing shelters; promoted spaying and neutering, and sent the first mobile spay/neuter clinic in the Middle East to Israel; successfully pressed veterinary services to switch to the use of humane oral rabies vaccine to replace mass strychnine poisonings; co-sponsored educational projects, including a Jewish/Arab program, and national and international educational conferences with Israel's Ministry of Education, on topics such as the connection between violence toward people and animals and integrating humane education in the classroom; co-sponsored, with Israel's Ministries of Agriculture, Health, and the Environment, training in animal shelter management and humane overpopulation control for municipal and shelter vets; and successfully campaigned to end various cruelties, including the Army's use of dogs as live bombs.

Today, CHAI works through its sister charity in Israel, Hakol Chai (Everything Lives), founded in 2001. To prevent and reduce the overpopulation that results in so much suffering, CHAI/Hakol Chai's state-of-the-art mobile spay/neuter clinic provides low-cost operations and education on responsible animal care throughout the country. During the evacuation of settlements in Gaza and the West Bank, the clinic's professional veterinary staff and volunteers played a major role in rescuing and finding new homes for companion and farm animals abandoned in the territories.

The organization has also rescued and rehabilitated abused horses, actively promotes legislation to prevent their abuse, and is raising funds to construct a horse/donkey sanctuary. CHAI campaigns against specific cruelties, including filing an appeal with Israel's Supreme Court to prevent gambling on horse racing from gaining a foothold there. CHAI's Alternatives Fund offers grants to promote alternatives to the use of animals in laboratories.

All CHAI/Hakol Chai's projects have an educational component, as these organizations believe planting seeds of respect, empathy, and responsibility in future generations is essential for positive change. CHAI/Hakol Chai created educational materials and videos for secular, as well as Jewish schools, and provides education on animal-related issues in schools and community centers. Only when the importance of the human-animal bond is understood worldwide will all living beings share a compassionate planet.

Further Reading
Concern for Helping Animals in Israel (CHAI): www.chai-online.org, www.hakolchai.org.il.

Nina Natelson

K

KENYA: CONSERVATION AND ETHICS

On a continent where the struggle for survival for most people is stark, there is a tendency to forget that the welfare of animals is intricately linked to the welfare of people. More often than not, development activities carried out by local people and governments, as well as international development partners, tend to overlook this basic fact.

Kenya, for example, has a variety of animals ranging from farm, to working, companion, and free-ranging wildlife. It is a wonderful diversity, yet the animals are more often than not poorly treated. A number of animals such as dogs, cats, donkeys, cows, sheep, and goats are not properly cared for by their owners. Wild animals, however, face the greatest danger, mainly from poaching, snaring, and encroachment on their natural habitat. The explanation for their welfare status could be well captured through the recognized five animal freedoms approach.

Almost two-thirds of Kenya is classified as Arid/Semi Arid Land (ASAL) area. These areas experience long spells of dry seasons, meaning no rains and therefore insufficient water and pasture. Ironically, most pastoral communities, that is, the Turkana, Pokot, Samburu, Masai, and several other communities in

the northeastern province inhabit these ASAL-classified areas. As pastoralists, the inhabitants' economic mainstay is livestock farming. The animals rarely feed to their full stomachs, and rarely quench their thirst even after moving long distances in search of these two precious commodities.

In 2006, hundreds of head of livestock succumbed to famine as dry spells ravaged the ASAL regions. The camel is understood to be the most enduring animal in the desert but, just to underline the hardship the animals face, several of them died. In other parts of the country, livestock did not suffer much, but dogs bear the brunt of hunger. In the recent past, stray dogs and cats have increased, especially in urban centers. Their owners no longer take care of them; therefore, they have to scavenge for and fight over the remnants of bones around slaughterhouses and dumping sites. They lack access to clean water, so they drink stagnant water during rainfall, and from sewers during dry seasons, or simply burn with thirst. Africa Network for Animal Welfare, together with its partners the Department of Veterinary Services, Worldwide Veterinary Service, University of Nairobi, and Kenya Veterinary Association, held a dog population and rabies control program for six days in September 2008, vaccinating a total of 1,384 animals, and spaying and neutering 152 cats and dogs in Nairobi's informal settlement areas. During that time, the team observed many stray dogs.

On the other hand, there is ample widespread and concrete evidence that many countries on the continent are losing numerous wild animals to people who snare them for the commercial bushmeat trade. And though wildlife in countries like Kenya and elsewhere in Africa face many threats, killing wildlife for bushmeat is probably one of the most potent. If unchecked, this might reverse all the gains made by conservationists over the years.

There are many ways by which hapless wild animals meet their deaths. However, poaching and snaring are the two most potent ways. and this has been going on not just away from protected areas, but also in parks and sanctuaries that are protected by armed rangers. The scale at which animals are been snared for meat is quite worrisome. For instance, during a desnaring project that took place between July 23 and August 6, 2008, a desnaring team was able to remove and destroy 156 snares in addition to arresting two poachers, one with a giraffe carcass and the other with poaching tools. They ended up getting jail sentences of five and two years respectively.

In many cases, animals are trapped using wire snares. Powerful torches and the blowing of horns are used to blind, confuse, then kill the animals on dark nights. There are different types of snares. Wire snares of different strengths, ranging from simple telephone wires, to tow ropes and unbreakable winch and break cables which are tied to trees, and others that are timed to trigger in the event that an animal steps on them. This results in the animal hanging up in the air, which makes it easy for the poacher to land a deathblow. Then there are the wire snares that are tied to tree stumps and placed on plastic sheet that is itself placed expertly on top of a dung hole. Once the animal unknowingly steps on the plastic sheet, its foot pierces a hole in it, essentially making it difficulty for the snare to slip out. As the animal pulls in an attempt to free itself, the noose tightens, making it impossible for the animal to escape.

Death might be the ultimate price animals end up paying under the hands of poachers, but in many cases, the targeted animal might end up escaping with the snare either hanging from its neck, or piercing different parts of its body. This often results in extensive suffering for the animal which, more often than not, might develop a gangrenous, pus-oozing wound around its neck and finally die after days of intense pain. Snares are also indiscriminate, and oftentimes catch non-target animals such as elephants, who although they may break away, die later due to deep wounds resulting from wire cuts.

In Kenya, poaching is an illegal activity punishable by jail sentences and/or fines. However, poaching is not the only threat affecting the survival of animals, and particularly wild animals. Owing to the rise in human population, which translates into an increase in demand for land, water, shelter, and other environmental services, animals in Kenya and elsewhere in Africa have lost, and continue to lose, much of the habitat that has sustained them since time immemorial. The critical issue here is the fact that tens of millions of people have continued to rely directly on such environmental goods as forests for timber and charcoal, salt licks, and water in animal habitats, and even pastures that wild herbivores depend on. In essence, this has continued to deplete the lands that the animals use for procreation, feeding, and general survival, as well as the water catchment areas so crucial for the country's river systems.

With increasing population, there is pressure for more land for human settlement and other economic activities such as farming. The glaring effect is currently being felt around Lake Naivasha in Kenya, where horticultural farming has taken root. Due to high water demands, the flower farmers have fenced off the lake, barring wild animals from accessing the fresh water body. More perturbing is that the lake is highly polluted, and several studies in the area have documented many deaths of both wild and domestic animals that drink the water. Marine species such as hippos, for example, have declined by more than 25 percent, according to Food & Water Watch report (Food & Water Watch, 2008,

p. 2).There were 1,500 hippos in 2004 and 1,100 in 2006.

Kenya's pastoral communities mostly survive on animal blood, meat and milk. The cows are pricked by sharp spears on the neck to draw blood, causing untold pain. They are severely injured and left weak, since this is done on weekly basis. Some communities such as the Masai use clubs with which they knock the animal on the head until it dies, while the Turkana spear the animal's heart through the ribs. These are very inhumane ways of killing animals for food. In slaughterhouses, animals witness the slaughter of others while they wait in a line for the same. These animals are also inhumanely slaughtered.

The Africa Network for Animal Welfare and its partners produced an

Rangers stack elephant ivory at the Kenya Wildlife Headquarters. (AP Photo/Khalil Senosi)

emergency response to displaced animals in 2008 in Rift Valley province, following post-election violence in Kenya, where the group vaccinated, dewormed and treated a total of 10,439 farm, working, and companion animals belonging to the Internally Displaced Persons, who had lost virtually everything they owned except the few animals the lucky ones had left. The team also came across dry cattle dips, which left the animals to the mercies of tick-borne diseases. Dogs have notably suffered the most. The government has been employing very inhumane and brutal means of controlling the stray dog population. They are baited and poisoned using strychnine poison, thus dying an agonizing death. Donkeys and camels are the mode of transport especially in ASAL areas. Despite ravaging famine at times, they are overloaded with household goods and building materials as the pastoralists move around in such of pasture and water.

The transportation of livestock and poultry destined for slaughter is seriously wanting. Kenya has only one meat processing plant in Athi River while most of the animals slaughtered are from pastoralist communities averaging 400km (approximately 250 miles) from the meat processing plant. The animals are cruelly stuffed into lorries during transportation. On the other hand, poultry farmers stuff their chicken in crates, as well as tying their legs and loading them onto the carriers of public service vehicles to drive them to urban centers for sale to consumers. By the time they get to the market, most of them are featherless or have broken limbs due to congestion and heavy winds atop the vehicle. In the villages, oxen and donkeys are whipped to force them pull heavy loads as well as perform strenuous jobs like plowing. They end up

with wounds on their necks and severe injuries on their backs.

The soaring human population is negatively affecting animals, as space becomes more limited for both wild and domestic animals. Animals, especially young ones, naturally need to run around as well as play. Migration corridors for wildlife have been interfered with or encroached on, locking the wildlife in one area. Currently there is frequent human wildlife conflict as the animals, especially elephants, leopards, and lions stray past their natural habitat due to very limited space in the game reserves and national parks. Farming in forests has also interfered with the jungle lifestyle, as the forest grounds are cleared and natural feedstuff for primates such as baboons and monkeys are uprooted.

In a nutshell, there is a need for policies and legislation that safeguard the welfare of animals. These need to be enacted and implemented. There is also a need for humane education geared towards valuing and appreciating animals as the sentient beings they are. Indeed we appreciate what we love, we love what we understand, and we understand what we are taught.

Further Reading

Food and Water Watch and The Council of Canadians. 2008. Lake Naivasha withering under the assault of international flower vendors. Available at http://www.foodandwaterwatch.org/world/africa/water-for-flowers/NaivashaReport.pdf.

Josphat Ngonyo Kisui

KROGH PRINCIPLE

The Krogh principle is one of the guiding principles of animal investigations. In a lecture delivered in 1929, Danish physiologist August Krogh (1874–1949) said: "For a large number of problems there will be some animal of choice, or a few such animals, on which it can be most conveniently studied" (quoted in Krebs, 1975, p. 221). While there is no nonhuman animal upon which all problems can be conveniently studied, for most problems there exists a convenient animal model.

Animal researchers have generally adopted the Krogh principle. They seek out species whose members have, for any problem of interest, anatomical structures of useful size or arrangement, or physiological and biochemical processes that make it easy to conduct their experiment. This principle is primarily applicable in the context of basic research. It is less clear how it is to be applied in the context of applied research, especially where the aim is to make predictions about humans. Even if an animal provides a convenient subject, we cannot automatically assume that the findings in such an animal will be applicable to humans. This problem is especially acute in the context of risk assessment, for example, predictive toxicology and teratology. Moreover, many nonhuman primates, our close phylogenic relatives, have not proven uniformly useful as predictors for human disease or genetic disorders. So there are substantial risks involved in using them as substitutes for humans. The Krogh principle, although perhaps useful as a methodological guide to basic animal research, is less useful in applied, predictive contexts.

Further Reading

Bernard, C. 1865/1949. *An introduction to the study of experimental medicine.* Paris: Henry Schuman, Inc.

Gold, L., Slone, T., Manley, N., and Bernstein, L. 1991. Target organs in chronic bioassays of 533 chemical carcinogens. *Environmental Health Perspectives,* 233–46.

Krebs, H. 1975. The August Krogh principle. *Journal of Experimental Zoology,* 194: 309–344.

Lave, L. B., Ennever, F. K., Rosencrantz, H. S., and Omenn, G. S. 1988. Information value of the rodent bioassay. *Nature,* 336, 631–633.

LaFollette, H., and Shanks, N. 1996. *Brute science: The dilemmas of animal experimentation.* London: Routledge.

Nishimura, H., and Shiota, K. 1978. Summary of comparative embryology and teratology. In J. Wilson and F. Fraser, eds. *Handbook of teratology* (vol. 3), 119–54. New York: Plenum Press.

Hugh LaFollette and Niall Shanks

L

LABORATORY ANIMAL USE—SACRIFICE

Different language is used to refer to the killing of different categories of animals. Companion animals are euthanized, farm animals are slaughtered, and research animals are sacrificed. Unlike the first two terms, however, use of the term sacrifice has been particularly controversial.

Spokespersons from the scientific community have called upon its members not to use the term sacrifice because it is unnecessary, too regularly used, and meaningless, and because it has religious and unscientific connotations. In recent years there has been a serious effort to delete the term from biological journals and grant proposals as part of a trend in this century to remove subjectivity and personalization from science. Some individuals critical of animal experimentation have also challenged its use because it makes it easier for researchers to kill animals and glorifies a practice that, in their opinion, should be seriously questioned if not stopped.

Despite official efforts to ban the term, it can still be overheard in the laboratory conversations of scientists and technicians as well as in the presentations of scientific papers at professional meetings. Direct observation of scientists and technicians has led sociologists to conclude that sacrifice is not used in the religious sense, but rather in a broader sacred sense within the scientific community. According to sociologists, sacrifice means more than simply killing laboratory specimens; it is part of a sequence of procedures that transforms animals into tools having a clear and valuable place in laboratories. Although sociologists agree that this transformation enables researchers to use animals in experiments, they disagree about the processes that create this transformation.

On the one hand, Michael Lynch argues that the transformation entails a single social process where the naturalistic animal found in nature is redefined as an analytic object signifying data and having only research value. The animal's death has meaning only to the extent that it assists research. On the other hand, Arnold Arluke maintains that the transformation involves two opposing social processes. Like Lynch, Arluke argues that laboratory animal sacrifice involves the stripping away of the everyday or nonscientific identity of animals so that they can be regarded as instruments or data. Arluke also contends that sacrifice involves a process of identification with lab animals. Some researchers, especially those who have routine contact with nonhuman primates or domestic animals, attribute human qualities to them. For these researchers, the animal's death has personal meaning. The concept of sacrifice embraces both of these tendencies by acknowledging the simultaneous distancing from and identification with laboratory animals that occur in research settings.

Rather than getting rid of the term sacrifice, the metaphor can be institutionalized by creating and openly acknowledging group rituals commemorating the death of laboratory animals. Rituals link individuals and culture by pulling together, in a personally meaningful way, the paradoxes of existence into something sensible, and the fragmentation of reality into something whole.

See also Euthanasia

Further Reading

Arluke, Arnold. 1988. Sacrificial symbolism in animal experimentation: Object or pet? *Anthrozoös* 2: 98–117.

Birke, Lynda, Arluke, Arnold, and Michael, Mike. 2007. *The sacrifice: How scientific experiments transform animals and people.* West Lafayette, IN: Purdue University Press.

Douglas, Mary. 1970. *Natural symbols.* New York: Pantheon Books.

Hubert, H., and Mauss, M. 1964. *Sacrifice: Its nature and function.* Chicago: University of Chicago Press.

Lynch, Michael. 1988. Sacrifice and the transformation of the animal body into a scientific object: Laboratory culture and ritual practice in the neurosciences. *Social Studies of Science* 18: 265–289.

Arnold Arluke

LABORATORY ANIMAL WELFARE

Millions of animals are used in laboratories around the world. Scientists may use animals to test toxic chemicals or to develop new surgery techniques. They may cause cancers and infections in animals to study them and develop cures. They may kill the animals to collect tissues and study their cells. Despite the death and pain that these acts can bring, there are ways to limit animal suffering in the laboratory.

The principle that underlies most regulation of laboratory animal use is that it can be justifiable to harm animals for science, but that pain and distress must be limited to that which is unavoidable to accomplish the scientific goal.

England led the way, with its 1868 Animals Act, in placing some government restrictions on how animals are used. Since then, other countries and jurisdictions have enacted laws. Along with these laws and regulations, scientists and veterinarians have developed standards for self-regulation of animal use over the years.

In the United States, the first national law was the Laboratory Animal Welfare Act of 1966. In 1966, Congress sought to regulate some peripheral aspects of laboratory animal welfare without actually interfering with how scientific experiments were performed. The law dictated how animals, especially dogs and cats, may be obtained for research, and how a dog vendor or a laboratory must document that they were not trafficking in stolen animals. The law specified how animals should be housed in a laboratory, and created a team of inspectors in the U.S. Department of Agriculture (USDA) to visit laboratories. The law also required that adequate veterinary care be provided for laboratory animals. But the law stopped its coverage as soon as the animal left the animal housing area and went down the hall into the laboratory.

The Laboratory Animal Welfare Act regulated animal care but not animal use, with some curious results. Although inspectors scrupulously enforced any departure from strict hygiene that might result in animal infections, no welfare rules covered intentionally infecting animals as part of an experiment. Though the law required veterinary care of animals, scientists had no oversight in conducting experiments that might intentionally make an animal sick.

This exclusion of laboratory practices from laboratory animal welfare laws was not a stable arrangement and it did not stand. The U.S. Congress has amended the Laboratory Animal Welfare Act (now called simply the Animal Welfare Act) several times. They expanded the requirement for adequate veterinary care, adding the use of painkillers and anesthetics for many experiments, with a veterinarian, not the scientist, prescribing the pain medications. Similarly, they expanded the regulations about housing animals, to include providing exercise for caged dogs and psychological wellbeing programs for caged monkeys and apes.

The most important innovation in the regulation of laboratory animal welfare has been the requirement that most institutions that conduct experiments on animals have some sort of animal care and use committee that reviews every planned use of animals. These Institutional Animal Care and Use Committees (IACUC) have been required in most American laboratories since 1986, and many other countries require similar committees. A scientist who wishes to use animals must apply to the committee for approval. She or he must describe why animals are necessary, must consult with a veterinarian on pain management and anesthesia, must describe the qualifications of everyone who will work with the animals and, in general, must assure that everything is being done to minimize animal pain and distress to that which is unavoidable. She or he must document a search for alternatives

to the painful use of animals. The IACUC must also have a system in place that allows concerned individuals and whistle-blowers to anonymously report their concerns about animal care and use.

Animal pain and distress are the overriding focus of animal welfare policies. One framework for reducing pain and distress is to think in terms of the Three Rs of alternatives: replace, reduce, and refine. Replacement alternatives are conceptually the most straightforward: researchers must find ways to generate research data without using sentient animals at all. Candidates for consideration include studying cells in tissue culture (in vitro techniques), developing computer simulations, making better use of human epidemiological data and human volunteers, or using inanimate models in teaching. Reduction is just what it sounds like—efforts to lower the numbers of animals used. This often means rethinking statistical tests and using just the number necessary for statistically valid results. Refinement alternatives are the most varied, because they comprise all the myriad ways to rethink animal care and use to reduce the potential for pain or distress. Refinements can include more aggressive use of painkillers, using noninvasive techniques such as X-rays instead of invasive dissections or surgeries to see inside the animal's body, or housing animals in compatible groups instead of all alone in steel cages.

Scientists are not barred from causing pain and disease in animals. There are some studies where this is unavoidable. Many studies of cancer, for example, call for inducing cancer in the test animals. Developing a new painkiller for human use often involves causing pain to animals to see if the new medication is effective. The job of the scientists, the veterinarian, and the committee is to limit the pain and illness. If the project is studying cancer prevention or biological processes that happen early in cancer, then there may be no reason to allow animals to progress to advanced disease. Thus, the scientist and the committee refine their experimental endpoints, either treating the cancer at its first appearance, or humanely euthanizing the animals. In studies of pain and the development of painkillers, the scientist may mostly use pain stimuli that are mild and that the animal can opt to end. For example, a scientist may time how long a rat who has received a particular painkiller will tolerate having her foot on a hot plate before she withdraws it on her own.

Critics of current animal welfare regulation raise several issues. For one, scientists are still allowed to hurt and kill animals, even though they should try to minimize any suffering. Second, in the American system, some animals are not covered by the regulations. The Animal Welfare Act only covers warm-blooded animals and, even at that, excludes mice and rats, overwhelmingly the most numerous mammals in laboratories. Another law (the Health Research Extension Act of 1985) covers all vertebrate animals if they are involved in projects or on campuses that receive federal research grants. This leaves the potential that animals at private companies and small schools that receive no federal grants may not have government oversight at all. Invertebrate animals, even sensitive species such as octopuses, are not covered by American laws at all.

Another criticism of the current system is that IACUCs are a form of self-regulation, and that this has the potential for abuse. Scientists on the committee assess the work of their peers, coworkers, and department-mates. To limit the potential for abuse, the laws require that a public or unaffiliated member or members be appointed to the committee. USDA inspectors review the work of the IACUC during their inspections, and institutions and their IACUC report annually to the USDA and to the National Institutes of Health's Office of Laboratory Animal Welfare, which enforces the Health Research Extension Act. Additionally, many institutions voluntarily seek to have their animal facilities and their IACUC program accredited by the Association for the Assessment and Accreditation of Laboratory Animal Care International, with its teams of scientists and veterinarians who conduct site visits. Still, self-regulation, with its strengths and weaknesses, remains the core of welfare oversight for the majority of laboratory animals.

Working in the animals' favor is the realization that to a great extent good science and good animal care are intertwined. Though animals may get ill or may suffer during the course of an experiment, the vast majority of experiments require that animals enter the study in uniformly good health and that pain and distress are minimized throughout. If animals are carrying various infections as they start an experiment, the scientist may never know whether she or he is seeing the results of the experiment or simply the results of the illness. If animals are stressed during an experiment, their biology is affected, and again, interpretation of data is muddied. Although it is best that scientists feel a moral responsibility to treat their animals well, there is also self-interest in keeping their animals free of disease and distress. To this end, the Institute of Laboratory Animal Resources published its *Guide for the Care and Use of Laboratory Animals,* a combination of ethical standards and expert guidance on ways to minimize pain and distress.

Ultimately, no law and no committee can see everything everywhere at all times, and so the personal, ethical responsibility of the scientists, veterinarians, and students working with animals is the main determinant of animal welfare.

Further Reading

Carbone, L. (2004). *What animals want: Expertise and advocacy in laboratory animal welfare policy.* New York: Oxford University Press.

Institute of Laboratory Animal Resources. (1996). *Guide for the care and use of laboratory animals.* Washington, DC: National Academy Press.

Orlans, F. B. (1993). *In the Name of science: Issues in responsible animal experimentation.* New York: Oxford University Press.

Rowan, A. N. (1984). *Of mice, models, and men: A critical evaluation of animal research.* Albany: State University of New York Press.

Russell, W.M.S., & R. L. Burch (1959). *The principles of humane experimental technique.* London: Methuen & Co., Ltd.

Stevens, C. (1990). "Laboratory animal welfare." In *Animals and their legal rights,* 66–105. Washington, DC: Animal Welfare Institute.

Larry Carbone

LAW AND ANIMALS

Until the 19th century, at least as far as the Western tradition is concerned, non-humans were excluded completely from the moral and legal community. Humans could use animals for whatever purpose we wanted and inflict pain and suffering on them pursuant to those uses without violating any obligations we owed to them. That is, nonhumans were regarded as things that were indistinguishable from inanimate objects, and toward which we thus could have no moral or legal obligations.

To the extent that the cruel treatment of animals was thought to raise a moral issue, it was only because of a concern that humans who abused animals were more likely to ill-treat other humans. The moral obligation concerned animals, but was really owed to other humans. Similarly, to the extent that the law provided any protection for animals, that protection was almost exclusively incidental to the animal being the property of another. Judicial condemnation of animal cruelty, with rare exception, reflected the moral concern that gratuitous cruelty to animals would translate into cruelty to other humans, or that acts of cruelty to animals might offend public decency and cause a breach of the peace.

This exclusion of animals from the moral community and denial of direct legal protection was justified on the ground that nonhumans were the spiritual inferiors of humans, were not made in God's image, and lacked a soul, or on the ground that animals were natural inferiors and lacked certain cognitive characteristics thought to be uniquely human, such as the ability to use symbolic communication or abstract concepts, or engage in reasoning or reciprocal moral relationships, or some combination of spiritual or natural inferiority. The paradigm ostensibly shifted in the 19th century as social progressives, many of whom also opposed human slavery and supported greater equality for women, maintained that any differences between humans and nonhumans did not serve to justify the treatment of animals as things. For example, moral philosopher and legal reformer Jeremy Bentham (1748–1832) maintained that animals had been degraded into the class of things, and he observed that although animals shared the characteristics regarded as unique to humans to some degree and that, in any event, the absence of these characteristics did not grant humans a license to treat animals in any way that they wished. As Bentham put it, "The question is not, Can they reason? nor, Can they talk? but, Can they suffer?"

As a direct result of the influence and efforts of Bentham and other reformers, the legal systems of Great Britain, the United States, and other nations enacted animal welfare laws that purported to provide legal protection for animals. These laws were of two kinds: general and specific. General animal welfare laws, such as anticruelty laws, prohibit cruelty or the infliction of unnecessary or unjustified suffering, or require the humane treatment of animal, without regard to particular use. Specific animal welfare laws purport to require the protection of animal interests in particular contexts, such as the use of animals in experiments or the slaughter of animals for food.

The emergence of animal welfare laws recognized that humans owed legal obligations to animals. This is not to say that these laws did not also reflect the concern that the cruel treatment of animals would have the effect of making humans treat one another badly. But it is also clear that, for the first time, animals were seen not merely as things, but as members of the moral community who were inherently deserving of some legal protection. Anticruelty laws are often explicit in applying to all animals, whether owned or unowned.

Animal welfare laws are based on the principle that animals are morally inferior to humans and that it is acceptable for humans to use animals for human purposes as long as any pain or suffering the animal incurs are considered necessary and the treatment is regarded as humane. It is often suggested that the animal welfare approach requires that we balance the interests of animals against our interests as humans in order to determine whether animal suffering is necessary. To balance interests means to assess the relative strengths of conflicting interests. If the benefits that will accrue to humans from using animals outweigh the animal interest in not suffering, then our interests prevail, and the animal suffering is regarded as necessary. If no justifiable human interests are at stake, then the infliction of suffering on animals must be regarded as unnecessary.

Many animal welfare laws, such as anticruelty statutes, are criminal laws. For the most part, only those moral rules that are widely accepted, such as prohibitions against killing other humans, inflicting physical harm on them, or taking or destroying their property, are enshrined in criminal laws. That many animal welfare laws are criminal laws suggests that we take animal interests seriously enough to punish violations of the humane treatment principle with the social stigma of a criminal penalty.

Although the emergence of animal welfare laws ostensibly represented a dramatic departure from the view that animals are merely things, the laws that were enacted in Britain, the United States, and other nations have, for the most part, failed to provide any significant level of protection for animal interests. Animals are property; they are economic commodities that have no value except that which we accord them. Under the law, the owner of an animal is entitled to exclusive physical possession of the animal, the use of the animal for economic and other gain, and the right to make contracts with respect to the animal, or to use the animal as collateral for a loan. The owner is under a duty to ensure that her animal property does not harm other humans or their property, but she can sell or bequeath the animal, give the animal away, or have the animal taken from her as part of the execution of a legal judgment against her. She can also kill the animal. Wild animals are generally regarded as owned by the state and held in trust for the benefit of the people, but they can be made the property of particular humans through hunting, or by taming and confining them.

The property status of animals renders meaningless any balancing that is supposedly required under the humane treatment principle or animal welfare laws, because what we really balance are the interests of property owners against the interests of their animal property. It is, of course, absurd to suggest that we can balance human interests, which are protected by claims of right in general and of a right to own property in particular, against the interests of property, which exists only as a means to the ends of human property owners. Although we claim to recognize that we may prefer animal interests over human interests only when there is a conflict of interests, there is always a conflict between the interests of property owners who want to use their property and the interests of their animal property. The human property interest will almost always prevail. The animal in question is always a pet or a laboratory animal, or a game animal, or a food animal, or a rodeo animal, or some other form of animal property that exists solely for our use and has no value except that which we give it. There is really no choice to be made between the human and the animal interest, because the

choice has already been predetermined by the property status of the animal. The suffering of property owners who cannot use their property as they wish counts more than animal suffering.

There are several specific ways in which the property status of animals renders animal welfare laws ineffective. First, it costs money to protect animal interests. We generally spend money to protect animal interests only when it is justified as an economic matter; that is, only when we derive an economic benefit from doing so. In most cases, animal welfare laws are limited to practices that are economically inefficient. For example, in the United States, federal law requires large animals to be stunned before being shackled, hoisted, and butchered. But this requirement merely recognizes that if animals are not stunned, carcasses will be damaged and workers will be injured. As a general matter, animal welfare laws do little more than ensure that animal exploitation is economically efficient.

Second, many of these laws explicitly exempt most forms of institutionalized property use, which account for the largest number of animals that we use. The most frequent exemptions from state anticruelty statutes involve animal agriculture, the use of animals in scientific experiments, and hunting. In some cases, specific animal welfare statutes exempt certain species of animals widely used in the practice that is supposedly regulated.

Third, even if anticruelty statutes do not contain explicit exemptions, courts have effectively exempted our common uses of animals from scrutiny by interpreting these statutes as not prohibiting the infliction of even extreme suffering if it is incidental to an accepted use of animals and a customary practice on the part of animal owners. For example, courts have consistently held that animals used for food may be mutilated in ways that unquestionably cause severe pain and suffering, and that would normally be regarded as cruel or even as torture. These practices are permitted, however, because animal agriculture is an accepted institutionalized animal use, and those in the meat industry regard these practices as normal and necessary to facilitate that use. Courts often presume that animal owners will act in their best economic interests and will not intentionally inflict more suffering than is necessary on an animal, because to do so would diminish the monetary value of the animal.

Fourth, anticruelty laws are generally criminal laws, and the state must prove beyond a reasonable doubt that a defendant engaged in an unlawful act with a culpable state of mind. The problem is that if a defendant is inflicting pain or suffering on an animal as part of an accepted institutionalized use of animals, it is difficult to prove that she acted with the requisite mental state to justify criminal liability.

Fifth, many animal welfare laws have inadequate penalty provisions, and we are reluctant, in any event, to impose the stigma of criminal liability on animal owners for what they do with their own property. Moreover, those without an ownership interest generally do not have the requisite interest, otherwise known as standing, to bring legal challenges to the use or treatment of animals by their owners.

In certain respects, the regulation of animal exploitation is similar to the regulation of human slavery in North America. Although many laws supposedly required the humane treatment of slaves and prohibited the infliction of unnecessary punishment, these laws offered almost no protection for slaves. In conflicts between slave owners and slaves, the latter almost always lost. Slave welfare laws, like animal welfare laws, generally required that slave owners merely act as rational property owners, but did not recognize the inherent value of the slaves. Slave owners were, of course, free to treat their slaves, or particular slaves, better. But as far as the law was concerned, slaves were merely economic commodities with only extrinsic or conditional value, and slave owners were essentially free to value the interests of their slaves as they chose, just as we are free to value the interests of our dogs and cats, and treat them as members of our families, or to abandon them at a shelter or have them killed because we no longer want them.

In recent years, animal lawyers have developed a practice that focuses on veterinary malpractice cases, pet trust cases, pet custody cases, and similar cases. These sorts of cases do not move animals away from the property paradigm; they enmesh them further into it.

See also Utilitarianism

Further Reading

Francione, Gary L. 1995. *Animals, property, and the law*. Philadelphia: Temple University Press.

Francione, Gary L. 2000. *Introduction to animal rights: Your child or the dog?* Philadelphia: Temple University Press.

Francione, Gary L. 2008. *Animals as persons: Essays on the abolition of animal exploitation*. New York: Columbia University Press.

Francione, Gary L., and Charlton, Anna E. 2008. Animal advocacy in the 21st Century: The abolition of the property status of nonhumans. In T. L. Bryant, R. J. Huss, and D. N. Cassuto, eds., *Animal Law in the Courts: A Reader*, 7–35. St. Paul, MN.: Thomson/West.

See also Animal Rights: The Abolitionist Approach, www.AbolitionistApproach.com

Gary L. Francione

LAW AND ANIMALS: AUSTRALIA

Over the last few decades, society's understanding of animals has changed enormously. While there is still a range of views about how animals should be treated, people everywhere are increasingly willing to accept that animals can have highly developed cognitive abilities and that they can experience a multitude of emotions. This increased awareness of the complexity of animals has led to a proliferation of animal welfare laws, which seek to regulate interactions between humans and animals. These laws exist in many Western nations, including Australia.

Australia is a large and diverse island continent situated in the Asia Pacific. It is a land of wide-ranging climates and terrains, which is home to some of the world's most unique and complex animals. While it is perhaps best known for its native animals such as the kangaroo, koala, platypus, and emu, Australia is also home to countless other native and introduced species. These include wild animals such as crocodiles, camels, buffalo, goats, rabbits, and domesticated animals such as pigs, chickens, cows, and sheep. Companion animals, such as dogs and cats, also play an important role in Australian society, with billions of dollars being spent annually to ensure the health and wellbeing of family pets.

As in most countries throughout the world, animals in Australia have no fundamental legal rights. They are considered to be the property of their owner and therefore cannot rely on the law to protect many of their basic needs and interests. In reality, because the law does not protect animals in a meaningful way, animal-based industries are free to use staggering numbers of animals for commercial gain. For example, each year in Australia:

- close to half a billion pigs, cows, sheep, and chickens are used for food and food production
- millions of animals, including rats, mice, birds, and guinea pigs are used for scientific research
- millions of kangaroos are killed for their meat, fur, and skin, because they are viewed by some members of the community as pests that compete with farming interests, and
- countless other introduced wild species such as foxes, rabbits, and wild dogs are shot and poisoned in the name of conservation

The use of animals in entertainment is also widespread, in sport and gaming events such as horse and greyhound racing, and in zoos, circuses, and rodeos.

Overview of Animal Law in Australia

Although animals in Australia do not have fundamental legal rights, a large number of laws have been enacted which claim to protect their health and wellbeing. These laws are based on the assumption that most animals are resources, and that some harm to them is justified in order to satisfy human wants and needs.

Australia has a federal political system with three tiers of government, federal, state, and local. Although there is no national animal welfare law, the federal government (Commonwealth) plays an important role in relation to the international wildlife trade and the live export of animals such as sheep, cattle, and goats, particularly as Australia is one of the largest exporters of live animals in the world.

State and territory governments have enacted animal welfare laws which regulate most other aspects of animal use. Broadly speaking, these laws, which are also referred to as anticruelty or animal protection laws, apply to all animals. They protect companion animals against cruelty, and regulate the use of animals in educational and research institutions, in zoos and circuses, in food production, in the wild, and in urban and rural communities. Some issues that do not relate directly to animal welfare, such as responsible pet ownership and unwanted animals, are addressed in separate laws made by local government.

While the federal government is in the process of coordinating a national animal welfare strategy with a key aim of establishing nationally consistent animal welfare laws, at the present time there are still numerous inconsistencies in state and territory laws, both in relation to the treatment of animals and to law enforcement. Some examples of these inconsistencies are set out below.

- The definition of animal is not the same in all states and territories, as some laws exclude crustaceans, cephalopods, and fish. This means that although animals have the same capacity to experience pain and suffering irrespective of their geographical location, they are not protected by the same legal standards

- People who commit crimes against animals face different penalties, depending on the state or territory in which the act of cruelty takes place. This means that although society might agree that it is morally wrong to harm an animal, there can be substantial variation in the punishment given to animal cruelty offenders

- As a general rule, state and territory police, designated government agencies, and animal welfare organizations such as the Royal Society for the Prevention of Cruelty to Animals (RSPCA) are the main bodies that respond to complaints about animal abuse. Despite this, the resources that government allocates to enforcement varies, with some state and territory inspectorates expected to monitor the treatment of large numbers of animals, or animals situated over vast distances on comparatively smaller budgets

- Some states and territories give third parties, such as private individuals or animal rights groups, the power to start proceedings under animal welfare laws, while others have passed laws designed to limit third party involvement. This means that it is easier to take action against someone who has harmed an animal in some parts of Australia than in others

- Some activities involving animals are banned in some states and territories but permitted in others. The Australian Capital Territory, for example, has prohibited rodeos and the use of certain wild animals in circuses, whereas these events continue to take place in other states and territories

Despite the variations in state and territory animal welfare laws, it is possible to identify a number of common themes. These themes are also found in the laws of many other industrialized countries. Some examples are set out below.

Animals as Property

First, as stated above, Australia's animal welfare laws have adopted the common law classification of animals as property. This is reflected in the wording of many acts and regulations, which define farm animals kept for economic gain as stock or livestock to be bought, traded, sold, and disposed of. Some laws and policies refer to animals kept for research or entertainment as specimens. This mirrors lawmakers' attitudes that animals are mere objects or resources, as opposed to living beings with complex needs and abilities. Until this way of viewing animals as property is changed, it seems likely that animals will never receive adequate legal protection.

Laws against Cruelty

Second, all of Australia's animal welfare laws prohibit acts of gratuitous or reckless cruelty, regardless of the commercial value of the animal. While much of today's society would consider this a basic ethical principle, these laws only came into effect in Australia in about the mid-19th century. Since that time, the definition of animal cruelty has been refined, and it varies between states and territories. Generally speaking, cruelty includes violent activities such as beating, mutilating, or torturing an animal. Some states and territories have introduced tougher punishments for people charged with aggravated cruelty, which are more serious acts resulting in the death, disabling, or serious deformity of an animal. The failure to provide food, water, shelter, and basic veterinary care is also generally seen as an act of cruelty.

Differential Treatment of Species and Implications for Farm Animals

Third, while Australia's animal welfare laws claim to apply to all animals, in practice many animals fall beyond the protective reach of the law. This happens because some animals, such as farm animals, are expressly excluded from legislative protection. In other words, certain acts which would constitute cruelty if performed on a dog or a cat are deemed acceptable if the victim is a farm animal. For example, both castrating a young piglet and dehorning a young calf without pain relief are considered defensible in the State of New South Wales.

Justifiable or Necessary Cruelty and the National Codes

As in other countries that have enacted animal welfare laws, animal suffering in Australia is considered lawful when it is judged to be necessary, reasonable, or justifiable. These words have no statutory definition, and are intended to be flexible in order to reflect changing community values. Since animals are classified as property, these words appear to have, by implication, sanctioned a range of practices that take an enormous physical and psychological toll on animals. This is particularly so when they are read in conjunction with the National Model Codes of Practice which underpin many state and territory laws.

The codes, which are primarily a joint initiative of Commonwealth, state and territory governments, set minimum standards for the treatment of animals from birth to slaughter in a range of industries. While they are characterized as representing industry best practice, in reality they help justify many practices that would otherwise constitute acts of cruelty to animals, since compliance with a code generally provides protection against prosecution for cruelty. Examples include:

- The permanent confinement or factory farming of millions of pigs used to satisfy Australia's appetite for ham, bacon, and pork. These sensitive and intelligent animals are kept indoors for the duration of their life, confined in sheds with thousands of others of their kind, and denied the opportunity to exercise many of their natural behaviors. Their mothers, female pigs or sows used for breeding, are generally treated as piglet producing machines. Under Australia's current animal welfare laws and policies, they may be confined in pens known as sow stalls, in which they can barely take a step forward or backward, for the majority of their reproductive cycle. Once they stop producing, they are considered of little utility and, with no meaningful legal rights to assert, they are sent straight to the slaughterhouse

- The factory farming of millions of chickens or battery hens, bred specifically to lay eggs. Under the current regulatory framework, battery hens may spend their entire lives standing on sloping wire bars in cages with between four to 20 cage mates. At 216 in (550cm) their allocated area is less than a letter-sized piece of paper. Australia's current animal welfare laws do not give them the opportunity to perform many of their natural behaviors

such as dust bathing, nesting, and foraging for food. They also do nothing to protect millions of male layer chicks who, since they cannot lay eggs, are generally considered waste products to be disposed of shortly after birth

- The carrying out of various procedures or mutilations on young animals, generally without pain relief. These procedures include the teeth clipping, castrating, and tail docking of piglets and the beak trimming (debeaking) or removal of one third of the beak of layer hens. These practices, which are carried out routinely in Australia's intensive or factory farming industries, would be considered acts of cruelty if they were carried out on companion animals, and would likely outrage a considerable proportion of the community.

Reform Efforts

In recent years, Australia's legal regime for the treatment of animals has come under increasing scrutiny from legal advocates for animals. This appears to be part of a broader international movement in animal protection law. The primary indicators of Australia's budding animal law movement include the increased availability of animal law as a course of study in universities, and the emergence of a community of legal academics and lawyers interested in debating and discussing present laws, and identifying potential areas for law reform.

Although law reform is not an overnight process, it seems likely that in the coming years legal advocates will challenge the many inconsistencies and inequities in Australia's regulatory framework for animals. Such action is essential if the widespread injustices perpetrated against animals are to be addressed.

Further Reading

Animals Australia. Major campaigns. Accessed from http://www.animalsaustralia.org/issues/ on October 14, 2008.

Australian Association for Humane Research, Inc. Statistics—Animal use in research and teaching, Australia. Accessed from http://www.aahr.org.au/index.html on October 14, 2008.

Australian Companion Animal Council, Inc. 2006. *Contribution of the pet care industry to the Australian economy,* 6th ed.

Australian Government/Australian Law Reform Commission, Reform. 2007/2008. Issue 91, *Animals,* Summer 2007–08.

Australian Government Department of Agriculture, Fisheries & Forestry. Animal welfare. Accessed from: http://www.daff.gov.au/animal-plant-health/welfare on October 14, 2008.

Australian Government Department of Agriculture, Fisheries & Forestry. 2008, June. *Australian animal welfare strategy* (Revised Edition). Accessed from: http://www.daff.gov.au/__data/assets/pdf_file/0008/749204/aaws-strategy-jun08.pdf on October 15, 2008.

Francione, G. L. 2000. *Introduction to animal rights: Your child or the dog?* Philadelphia: Temple University Press.

NSW Department of Primary Industries, Animal Ethics Infolink. Accessed from: http://www.animalethics.org.au/reader/arrp on October 14, 2008.

Sankoff, P., and White, S. eds. 2008. *Animal law in Australasia: A new dialogue.* Sydney: Federation Press.

Voiceless. Facts and stats. Accessed from: www.voiceless.org.au on October 14, 2008.

Wise, S. W. 2002. *Drawing the line: Science and the case for animal rights.* Cambridge, MA: Perseus Books.

Legislation

Animal Welfare Act 1992 (ACT)

Animal Welfare Act 1999 (NT)

Prevention of Cruelty to Animals Act 1979 (NSW)

Animal Care & Protection Act 2001 (QLD)

Animal Welfare Act 1985 (SA)

Animal Welfare Act 1993 (TAS)

Prevention of Cruelty to Animals Act 1986 (VIC)

Animal Welfare Act 2002 (WA)

Katrina Sharman

LAW AND ANIMALS: EUROPEAN UNION

Animals enjoy a sometimes high degree of esteem within European law. Historically speaking, legal animal protection began in Great Britain in 1822 with the passage of Martin's Act. Since then, laws on animal protection have spread their reach across Europe's breadth and width. A brief, and admittedly incomplete, survey confirms this and provides a springboard to consider legislation in certain European countries using a comparative approach.

Within the entire field of animal protection, animal welfare legislation plays an increasingly important role. In contrast to animal ethics, animal welfare law defines, with binding effect, how legal stakeholders should deal with animals with the help of the state. Animal protection law is characterized as legal means that protect animals from adverse effects upon their lives or wellbeing. This legislation can be divided into three categories. Animal welfare under private law regulates the proper classification of animals within

*Many thanks to Steven White, Sarah Kossew and Ondine Sherman for their review of this essay.

legal relationships between private people. Criminal law covers the punishment of animal torturers, and people who have committed other acts contravening animal welfare provisions. Administrative law regulates the appropriate interaction between people and animals with the help of enforcement measures.

Legal animal welfare is a fundamental part of protecting animals, and provides a decisive weapon in ensuring that this protection is ensured every day. Animal welfare law in Europe is a mixture of primarily administrative law with some criminal law provisions.

Actual Animal Welfare Law

Animal welfare laws directly regulate interaction with animals, and usually apply to domestic animals (such as agricultural food-producing animals, pets, and sport animals), those used in experiments, and wild animals. To a greater or lesser degree, they prescribe, among other things, how animals should be kept, cared for, fed, stimulated, transported or slaughtered, and specify when a public license is required, for example, for animal experiments or to keep particularly difficult animals.

Animal protection in Slovenia and some German-speaking countries is regulated by the national constitution. The majority of European nations have their own animal protection acts that apply to the country as a whole, or in the very least have animal protection laws on a regional level, or through relevant provisions on a national level. States like Germany, Switzerland, Austria, Sweden, Norway, and the Netherlands have a comparatively advanced level of animal welfare.

Here is a glimpse into some regulated areas:

Germany

Just as in Switzerland, Germany has also had an animal welfare act applicable throughout the country for several decades. This underwent a thorough revision in 1998, and was amended in 2001 to include the keeping of dangerous dogs. German animal welfare received a decisive boost when the principle of animal welfare was incorporated into the country's constitution in 2002. The new article, Article 20a, stipulates that:

Mindful also of its responsibility toward future generations, the state shall protect the natural foundations of life and animals by legislation and, in accordance with law and justice, by executive and judicial action, all within the framework of the constitutional order.

Thus, animal protection, just as in Switzerland since 1973, is something that the state is supposed to legislate on, and is considered to be an extremely important community asset. This is a significant decision on the value of animal protection as far as constitutional law is concerned, which the lawmaker while legislating, as well as the administrative authorities and courts during the interpretation and application of the law, has to take into account. This new clause within a state's objectives does not, however, lead to protecting animals without limits, but it does mean that henceforth this consideration must be weighed with other constitutional provisions and cannot be avoided when it comes to the unfettered exercise of art, religion, science, or teaching. Rather, animal protection represents, in principle, a legal equivalent to basic human rights and, therefore, must be respected in cases involving artists or researchers, for example, who invoke their own entitlement to basic rights when it comes to using animals.

The act protects the animal as a fellow being (*Mitgeschöpf*) as far as its wellbeing and life is concerned. The killing of a mammal without reasonable grounds is punishable; the law applies to non-mammals, too. Under various sections of the law, certain animal groups or species enjoy varying degrees of protection.

The act consists of 13 subclauses and 22 paragraphs, some very comprehensive. These regulate the keeping and killing of animals, interference, animal experiments, and husbandry. Also included are implementation rules, punishments, and fines, as well as transition periods. Compared with other international laws, certain provisions of this act are very advanced.

Austria

Austria's current revised animal protection act and related regulations were put in force at the beginning of 2005, replacing nine federal edicts. This act's purpose is to ensure that the life of an animal is protected, and prohibits killing both mammals and non-mammals without reasonable grounds. Pain-inducing breeding is expressly forbidden, as well as the import, export, and transfer of animals with marks of such breeding, and the display of dogs and cats in pet shops. Effective from 2009, keeping hens in battery cages, as well as housing cattle, horses, and goats permanently in tie stalls, is prohibited. At least 90 days in the open air is required.

France

France has no unified national legislation in this area; animals and their legal status are protected through various local decrees. Animals are divided into two groups: domestic and wild. However, the definition of domestic animals is broad and includes those that live, eat, and reproduce under human supervision and care.

Animal torture is punishable under the penal code (Articles 521–1, 521–2). These provisions, which were revised at the beginning of 2002, set out, among other penalties, a prison sentence of up to two years or a fine of a maximum of €30,000, permit a prohibition order on keeping animals and is against the abandonment of animals. Also punishable is the unjustified, deliberate or negligent killing or injuring of a pet (Articles R653–1, R654–1, R655–1).

Switzerland

Swiss legislation has been totally revised in the past few years, culminating in the new federal animal welfare act and ordinance on September 1, 2008. The main reasons for the overhaul were the significant gaps in the old law, in particular in implementation, as well as the need to align animal protection law with new scientific findings in the area of human-animal relations. The goal of the revision was also to improve practical implementation and to create the necessary related instruments.

In addition to the protection of an animal's dignity (see below), the responsibility held by animal-keepers stands at the heart of the new legislation. Whoever interacts with animals has to be aware of the latter's needs and know how to look after them properly. Awareness of a responsible and respectful interaction with animals is to be achieved through better training and information.

The act is 160 pages long and contains 46 articles on interaction with animals: keeping, breeding, and genetically modifying, trade, transporting, interference, animal experiments, slaughter, research, administrative measures, and complaints from authorities, as well as criminal and transitional provisions.

Animal Welfare Law in the European Union

The 27 member states of the European Union, to which Switzerland does not belong, have their own body of law. The EU is primarily an economic community. Animal protection is not of significance per se and is not listed in the catalogue of community activities. Despite this, the EU is concerned with technical matters relevant to animal welfare, as these are often closely linked to economic and trade-policy issues. A series of animal protection measures that are applicable to all EU members is to be found within the framework of the EU's Common Agricultural Policy (CAP). Specific guidelines, directives and regulations cover the protection of agricultural food-producing animals, layer hens, calves, and pigs, and the transport of slaughter animals. Others cover animal welfare during experiments and the keeping of zoo animals.

The various guidelines are extended by further legal instruments in some cases, particularly in regard to animal transport and experiments. The guidelines set out minimal requirements within animal protection, leaving it to member states to legislate more stringently on a national level. This, however, rarely happens. Sometimes member states are prevented from adopting more strict measures on a national level because of EU law.

General provisions on animal torture, using animals at sporting or cultural events, and keeping domestic or wild animals, fall under the competence of the national legislator.

Animal Welfare Law and the Council of Europe

The Council of Europe, not to be confused with the EU's European Council, has five important European conventions on the protection of agricultural food-producing animals, transporting animals, slaughter animals, those destined for experiments, and pets. They generally lay down minimal standards for animal protection that are less strict than animal protection legislation in many countries, particularly in Western and Northern Europe. However, seen within the context of Europe as a whole, the conventions are purveyors of considerable advancement due to their systematic declarations of intent.

Further International Treaties and Norms

Alongside European legal provisions there also exist other international regulations that are relevant to protecting animals and species. These include the Convention on International Trade in Endangered Species of Wild Fauna and Flora (CITES), the International Whaling Convention (IWC), and the global economic organization's OECD principles of Good Laboratory Practice (GLP), which aims for worldwide harmonization of testing methods in the field of chemical toxicology. The detailed Live Animals Regulations of the International Air Transport Association (IATA) apply to transporting animals by air. Also, the

World Organization for Animal Health (OIE) will in the future go beyond its current activities in battling animal diseases and into international animal welfare. Furthermore, discussions are taking place on whether import restrictions due to animal welfare considerations are permissible within the World Trade Organization's (WTO) framework. The areas affected include the international fur trade, a particularly controversial field from an ethical point of view, and the trade and transport of animals and animal products, the keeping of food-producing animals, ritual slaughter, and animal experiments.

Animal protection under private law Legal norms governing the classification of animals in legal relations between private persons, and the taking into account of animal interests using civil law, fall under private law.

Recent national regulations, such as German, Austrian, Swiss and French legislation, take into account the special legal standing of animals as being somewhere between objects and human beings as a norm for compensation. For example, in Switzerland, Germany, and Austria, adequate compensation and reasonable costs for veterinary care are to be paid in case of the injury or death of an animal. In France, a tenant may keep an animal on a rented property under certain conditions. In Swiss divorce cases since 2003, judges have been legally entitled to reassign pet ownership from one spouse to the other if the other person is a better pet keeper, even if he/she is not the owner. As far as lost and found pets are concerned in the UK, Austria, Italy, Spain, the Netherlands, Denmark, and Switzerland, there is a special delay until the finder of an animal becomes its owner. In Switzerland, a central office facilitates contact between the owner and the finder. In the UK, Germany, Austria, Switzerland, and Denmark, animals are protected from seizure if the owner is in debt. Although animals in Switzerland do not have legal rights, there are legal grounds for reinterpreting a last will in favor of an animal.

The new provisions are restricted mostly to the domestic arena and not to animals kept for business reasons, thus applying only to pets. This results in a change in their status as objects and is a step in the other direction.

Trends and the Future

Better Protecting the Needs of Animals The Swiss legislature attracted media attention in 2008 when it stated that animals belonging to social species should enjoy social contact with their own kind. This recognizes that an animal has a right to live its life well and in dignity, thus taking ethological studies into account.

Dignity of the Creature Traditionally speaking, animal welfare laws in Europe are aimed at protecting animals from unjustified pain, suffering, damage, and fear, and at preserving their lives. The basis for this concept is the capacity for suffering in animals. A fundamental development based on this took place in Switzerland in 1992, resulting in a worldwide first that resulted in animals' dignity being protected by the Swiss constitution. Swiss legislation on animal protection now states that the dignity of the creature must be respected when interacting with an animal. The animal's dignity is negatively affected if the animal is subjected in particular to pain, suffering, or injury, is made to feel fear or is subjugated, if its appearance or its abilities are strongly interfered with, or if it is disproportionately instrumentalized. Sexual relations with animals, or zoophilia, are also classified as disrespecting dignity. This is now punishable on the grounds of animal protection and not just because of morality.

The Animal Attorney Animal protection law now contains certain structures to ensure better animal protection in administrative and criminal law. The many concerns that crop up during the implementation of animal protection law have often been criticized. New implementation measures and the involvement of representatives from the world of animal protection should provide assistance. In Austria, for example, animals have had legal standing in administrative procedures through the animal protection ombudsmen since 2005, as provided in the animal welfare act. The animal welfare ombudsman can also challenge in court decisions taken by federal-state authorities once s/he has examined the case files.

The Swiss canton of Zurich created the post of animal attorney in criminal cases in 1992; the third holder of this post is the author of this essay. The position is anchored in the Animal Welfare Act of Zurich, which states:

In criminal procedures referring to violation of provisions in the national animal-welfare legislation, the administration of the Canton and a lawyer appointed by the cantonal government at the suggestion of the animal-welfare organizations safeguard the interests of the injured party.

This lawyer has unfettered access to all case files, investigations, and court proceedings. He must be informed in full of all decisions and can appeal against them. The animal attorney stands by animals as an independent representative in criminal procedures against animal torture, and provides dynamic support to the criminal investigatory authorities in their efforts to better protect animals.

Thanks to the creation of legal institutions and terms such as the dignity of the creature, as well as the presence of an animal attorney in criminal matters, the debate on animal rights and alternatives has new wind in its sails now.

Further Reading

Baranzke, H. 2002. *Würde der kreatur? Die idee der würde im horizont der bioethik.* Würzburg: Königshausen & Neumann.

Bolliger, G. 2000. *Europäisches tierschutzrecht—Tierschutzbestimmungen des Europarats und der Europäischen Union (mit einer ergänzenden Darstellung des schweizerischen Rechts).* Zürich/Bern: Schulthess & Stämpfli.

Goetschel, A. F. 1994. Der Zürcher Rechtsanwalt in Tierschutzstrafsachen. In *Schweizerische Zeitschrift für Strafrecht*, 64–85. Bern: Stämpfli.

Goetschel, A. F., & Bolliger, G. 2003. *Das Tier im Recht—99 Facetten der Mensch-Tier-Beziehung von A bis Z.* Zürich: Orell Füssli.

Goetschel, A. F., & Bolliger, G. 2007. The Animal in the Law—a Global Perspective.—update 2007. http://www.tierimrecht.org/en/PDF/IAHAIO_2007.pdf.

Herbrüggen, H. et al., eds. 2006. *Österreichisches Tierschutzrecht, Kommentar*, 2nd ed. Wien/Graz: Neuer Wissenschaftlicher Verlag.

Kluge, H.-G., ed. 2002. *Tierschutzgesetz—Kommentar.* Stuttgart: Kohlhammer.

Liechti, M., ed. 2002. *Die Würde des Tieres.* Erlangen: Harald Fischer.

Radford, M. 2001. *Animal welfare law in Britain—Regulation and responsibility.* Oxford: Oxford University Press.

Stohner, Nils. 2006. *Importrestriktionen aus Gründen des Tier- und Artenschutzes im Recht der WTO.* Bern: Stämpfli Verlag AG.

Teutsch, G. M. 1985. *Lexikon der Tierschutzethik.* Göttingen: Vandenhoeck & Ruprecht.

Wolf, J.-C. 2005. *Tierethik—Neue Perspektiven für Menschen und Tiere*, 2nd ed. Erlangen: Harald Fischer.

Antoine F. Goetschel

LAW AND ANIMALS: UNITED STATES

During the 1960s, vivid press coverage both of the kidnapping of family pets that were then sold for research, and also of the conditions under which dog dealers who sold animals to research facilities kept these animals, aroused the public's fear of having their pets kidnapped and sold for research. Congress reacted to these concerns by passing the Laboratory Animal Welfare Act of 1966, which mainly licensed and regulated animal suppliers but did little to assure the wellbeing of animals used in research. By the 1970s, however, more substantive concerns about animal research had surfaced in society. Growing public suspicions and misgivings about animal research were solidified in the early 1980s, when a number of serious examples of animal abuse in research facilities were revealed, including instances at the University of Pennsylvania Head Injury Laboratory and the laboratory of Edward Taub, both situations which involved abuse, improper care, and neglect of nonhuman primates. By the mid-1980s, public confidence in the research community's ability to regulate itself in the area of animal care and use was sufficiently eroded to demand federal legislation.

In 1976, a group of Colorado citizens consisting of two laboratory animal veterinarians, a humane advocate and attorney, and a philosopher began proposing legislation that would enforce self-regulation by local animal care and use committees. These committees would review research projects before they began, in order to make sure that everything possible was being done to assure that animal pain, distress, and suffering were minimized. The committees would also assure that facilities were adequate, and that systems of care assured proper animal husbandry.

In 1985, despite vigorous opposition from certain portions of the research community, the key concepts proposed by the Colorado group were passed by Congress as components of two pieces of legislation. The first piece of legislation was passed as an amendment to the Laboratory Animal Welfare Act and was entitled the Improved Standards for Laboratory Animals Act. The second piece of legislation, complementing the first, was the Health Research Extension Act. The major provisos of the Laboratory Animal Welfare Act amendment were as follows:

1. Establishment of an institutional animal care and use committee (IACUC) whose members must include a veterinarian and a person not affiliated with the research facility

2. A directive to the U.S. Department of Agriculture (USDA), which enforces the law, to establish standards for exercise for dogs

3. Establishment of standards for a physical environment for primates that enhances their psychological wellbeing

4. Establishment of standards of adequate veterinary care, including use of anesthetics, analgesics, and tranquilizers

5. Prohibition of the use of paralytic drugs without anesthetics for surgical procedures

6. Proof that the investigator has considered alternatives to painful procedures

7. Prohibition of multiple surgeries except for scientific necessity

8. The IACUC must inspect facilities at least semiannually, review protocols, and file an inspection report detailing violations and deficiencies

9. The USDA was mandated to establish an animal welfare information service at the National Agricultural Library to provide information aimed at eliminating duplicative animal research, reducing or replacing animal use, minimizing animal pain and suffering, and training animal users

10. Each research institution must train animal users in the items enumerated in (9), and in any other ways of minimizing animal suffering

11. The USDA should effect a working relationship with the National Institutes of Health (NIH)

The Health Research Extension Act turned NIH guidelines for proper care and use of animals into law. NIH had long promoted reasonable guidelines for animal care but had had no mechanism for enforcing them. Violations could result in seizure of all federal money granted to an institution. Between the two laws, virtually all vertebrate animals used in research in the United States, with the exception of farm animals used in agricultural research, and rats and mice used in private industry research, are now legally covered. Many IACUCs apply the same standards to agricultural researchers vis-à-vis pain and suffering as they do to animals used in biomedical research. Researchers are becoming increasingly sophisticated about animal pain, suffering, and distress, and how to control them in the face of federal law that assumes the existence of animal pain, thought, and feeling. Many researchers now admit that minimization of pain and distress results in better data. Researchers are also gradually becoming aware of the ethical issues in animal research. Consequently, researchers are increasingly looking into housing systems that better take into account animals' psychological and biological needs. The NIH *Guide to the Care and Use of Laboratory Animals,* currently being revised, is the bible for judging laboratory animal programs. The 1996 version urged environmental enrichment for all species used in research.

While the USDA initially looked only at pain control, once the research community had adapted to the use of analgesics, the USDA announced that it would begin auditing control of distress as well. Distress is a catchall phrase for a variety of noxious experiences that may be undergone by research animals in addition to pain—fear, anxiety, social isolation, boredom, etc. These concerns may be alleviated pharmacologically or by environmental modification.

Further Reading

Newcomer, Christian. 1990. Laws, regulations, and policies pertaining to the welfare of laboratory animals. In B. E. Rollin and M. L. Kesel, eds., *The experimental animal in bio-*

medical research, vol. 1. Boca Raton, FL: CRC Press.

Rollin, Bernard E. 1989. *The unheeded cry: Animal consciousness, animal pain, and science.* Oxford: Oxford University Press.

Rollin, Bernard E. 1995. Laws relevant to animal research in the United States. In A. A. Tuffery, ed., *Laboratory animals: An introduction for experimenters.* London: John Wiley.

Rollin, Bernard E. 2006. The Regulation of Animal Research and the Emergence of Animal Ethics: A Conceptual History. In *Theoretical Medicine and Bioethics* 27: 285–304.

Rollin, Bernard E. 2006. *Science and ethics.* New York: Cambridge University Press.

Rollin, Bernard E.. 2006. *Animal rights and human morality,* 3rd ed. Buffalo, NY: Prometheus Books.

Rollin, Bernard E. 2007. Animal research: A moral science. *Embo Reports* Vol. 8(6) 521–525.

Rollin, Bernard E. (forthcoming) The moral status of animals and their use as experimental subjects. In Peter Singer and Helge Kuhse. eds., *Companion to Bioethics,* 2nd ed. Oxford: Blackwell.

Russow, Lilly-Marlene. 1991. NIH guidelines and animal welfare. In James M. Humber and Robert F. Almeder. eds., *Biomedical Ethic Review: 1990,* 229–252. Clifton, NJ: Humana Press.

Bernard E. Rollin

M

MARGINAL CASES

The argument from marginal cases (AMC) has been one of the most powerful weapons in the contemporary debate about nonhuman animal rights. There are two basic versions of the AMC. The categorical version claims that so-called marginal humans, such as people with severe mental disabilities, have moral rights and concludes that nonhumans who are relevantly similar to these humans also have moral rights. The biconditional version maintains that the moral status of relevantly similar marginal humans and nonhumans is equivalent; the nonhumans have moral rights if and only if the humans have such rights. Several objections have been made to both versions of the AMC.

Some people are concerned that the argument is unfair to marginal humans. Many mentally disadvantaged humans are capable of speaking, going to school, learning trades, etc. These abilities are not possessed by any nonhuman animals, so far as we know. Defenders of the AMC can fully agree that many mentally disadvantaged humans are more capable than nonhuman animals. Nevertheless, quite a few severely damaged, sentient humans are far less capable than many nonhuman animals. Empirical evidence supports the contention that some humans and some nonhumans are roughly comparable in terms of their intellects, emotional capacities, and other capabilities. While some humans outstrip some nonhumans on this score, the reverse also appears to hold.

Another rather more serious charge of unfairness has been made against the AMC. Humans who become mentally incapacitated are unfortunate because they have been deprived of their personhood. Humans who are born with severe mental limitations are also unfortunate, one might argue, because they do not possess the potential to become normal members of their species. In contrast, the nonhumans used in laboratories and farms are likely to be normal members of their species. Thus there is a morally relevant difference between marginal humans and mentally and emotionally comparable nonhumans. Fairness dictates that we not add yet another huge burden to the unfortunate humans' lives. The normal nonhuman, then, rather than the marginal human, should be sacrificed to benefit persons. AMC supporters could respond as follows. The objection assumes that marginal humans are already morally significant. Only a morally significant being can be treated fairly or unfairly. But what makes them morally significant, in the context of the objection? It cannot be the misfortune itself, since this would make the objection circular. If it is the fact that they are capable of preferring pleasure to pain, this also holds for many nonhumans. Thus the latter would be morally significant also. In the case of two obviously morally significant beings, for example, two human persons who are alike apart from the fact that one of them is missing a leg and the other has two, we would not consider it justified to steal from the human with two legs rather than the human with one leg, because the latter is already more burdened than the former. A choice that would be fair to both individuals is the refusal to sacrifice either.

Another approach to criticizing the AMC is to deny moral status to both marginal humans and sentient nonhumans, but deny that unacceptable consequences would follow in practice. A. V. Townsend, for example, has argued that many humans, incapable of personhood in the strict sense, do not have rights, as is the case for similarly limited sentient nonhumans. Thus he rejects the categorical version of the argument, while accepting its biconditional form. But he does claim that persons must treat these humans as if they have rights. Otherwise, when distinctions among humans are blurred, genuine rights holders are threatened; this allegedly does not hold for the case of nonhumans. Peter Carruthers has made essentially the same argument. Animal rights supporters can counter that this is a textbook example of the slippery slope fallacy; without further evidence, it is assumed that treating marginal humans as we now treat nonhuman animals would lead to denial of persons' rights. Indeed, history and anthropology offer several examples of societies whose members had no difficulty in distinguishing between marginal and typical humans. After all, humans excel in their discriminatory powers, even when the characteristics chosen as the basis of that discrimination are morally irrelevant (e.g., race or gender). According to the final, very serious objection made by Alan Holland, the AMC is at best a useless addition to the case constructed for nonhuman animal rights, and at worst an unexploded bomb that could take out many humans as well as nonhumans. The biconditional version of the AMC claims the moral equivalence of marginal humans and sentient nonhumans. There is nothing in the argument to stop a person from rejecting the moral significance of both groups.

Although this last objection is strong, it cannot be concluded that the argument from marginal cases is rhetorically or psychologically superfluous. Both opponents and supporters of nonhuman animal rights should confront the following questions: If it were wrong to harvest the organs of a severely retarded human to save the life of a normal human adult, would it also be wrong to sacrifice a baboon or pig for the same purpose, assuming that transspecies transplants become medically feasible? In general, is it wrong to treat sentient nonpersons as resources for persons? Both versions of the AMC challenge all parties to the debate to do some very fundamental moral thinking.

See also Animal Rights; Sentience; Xenograft.

Further Reading

Carruthers, Peter. 1992. *The animals issue.* Cambridge: Cambridge University Press.

DeGrazia, David. 1996. *Taking animals seriously.* Cambridge: Cambridge University Press.

De Waal, Frans. 2006. *Primates and philosophers.* Princeton: Princeton University Press.

Dombrowski, Daniel A. 1997. *Babies and beasts.* Champaign: University of Illinois Press.

Frey, R. G. 1987. The significance of agency and marginal cases. *Philosophica* 39(1), 39–46.

Holland, Alan. 1984. On behalf of a moderate speciesism. *Journal of Applied Philosophy* 1(2), 281–291.

Narveson, Jan. 1987. On the case for animal rights. *Monist* 70(1), 31–49.

Nelson, James. 1986. Xenograft and partial affections. *Between the Species* 2(2), 70–80.

Pluhar, Evelyn. 1995. *Beyond prejudice: The moral significance of human and nonhuman animals.* Durham, NC: Duke University Press.

Pluhar, Evelyn. 2006. Experimentation on humans and nonhumans. *Theoretical Medicine and Bioethics* 27 (4), 333–355.

Regan, Tom. 1982. *An examination and defense of one argument concerning animal rights. All that dwell therein.* Berkeley: University of California Press.

Rollin, Bernard. 1982. *Animal rights and human morality,* rev. ed. Buffalo: Prometheus Books.

Savage-Rumbaugh, Sue, Lewin, Roger. 1994. *Kanzi: The ape at the brink of the human mind.* New York: John Wiley and Sons.

Singer, Peter. 1990. *Animal liberation.* New York: New York Review of Books.

Townsend, Peter. 1979. Radical vegetarians. *Australasian Journal of Philosophy* 57(1), 85–93.

Evelyn Pluhar

MEDICAL RESEARCH WITH ANIMALS

The topic of using animals in science in general and in medical research specifically is very controversial. Most people involved in the controversy focus on the ethics of using animals. Animal advocates, on the other hand, state that humans do not have the right to use sentient nonhumans for selfish purposes. Animal users claim that without using animals in medical research we would no longer see cures and treatments for diseases.

Differences

Animals are used for scientific purposes in essentially nine different ways (Table 1):

1. Animals as models for human disease
2. Animals as models for testing drugs destined for humans
3. Animals as spare parts
4. Animals as factories or bioreactors
5. Animal tissue to study basic physiological principles
6. Animals for dissection in education
7. Animals as a modality for ideas (heuristic)
8. To benefit other animals, such as in veterinary research
9. Knowledge for knowledge's sake

However, when people think about the use of animals in medical research, they usually think of number 1, using animals to model human diseases, and number 2, using animals in drug testing and development in order to predict human response. In these two areas, the differences between species and even between individuals become important.

Many drugs may be good for patients, provided they are given in the proper dose at the proper time. In the 16th century, this concept led Paracelsus to say: "The dose determines the poison." The same is true today. But today we should add a corollary: The genetic makeup also determines the poison.

For example, of ten medications withdrawn from the U.S. market between 1998 and 2001, eight were withdrawn secondary to side effects that occurred primarily in women (GAO, 2001). Among cigarette smokers, African Americans and Native Hawaiians are more susceptible to lung cancer than whites, Japanese Americans, and Latinos (Haiman et al., 2006). Identical, or monozygotic, twins do not always succumb to the same disease despite identical genetic makeup (Flintoft, 2005; Albert et al., 2005). Troglitazone, also known as Rezulin, was very effective for controlling diabetes in many patients, but in others it caused liver failure. Penicillin likewise has saved millions of lives, but causes life-threatening allergic reactions in some patients. Articles in *Nature* and *Science* have revealed that no two cancers are exactly alike (Associated Press, 2008; Kaiser, 2008).

Obviously humans are more similar to other humans than they are to mice, and yet one group of humans does not always respond the same as another, as these examples show. The genetic differences that result in different responses to drugs are multiplied when one considers different species.

Comparative genome research has revealed remarkable genetic similarities between humans and other animals like chimpanzees, dogs, and mice. But these and other studies have also revealed why the small differences are so important.

Single nucleotide polymorphisms (SNPs) are DNA sequence variations and are named such because they occur when a single nucleotide—A, T, C, or G—is changed. These changes can affect the activity of a gene by enhancing it, reducing it, or even inactivating it. When the protein the gene codes for is an enzyme involved in drug or toxin metabolism, the result can be a very large variation with respect to what that drug or toxin does to the body.

Copy number variations (CNVs) are an increase or decrease in the copies of a particular gene. CNVs can influence rates of drug or toxin metabolism so that a dose which is effective in one person may be ineffective in another (see Shanks and Pyles, 2007). In addition, CNVs also influence disease states and phenotypic variation. Variations with respect to SNPs and CNVs are very important even when studying humans.

Genes essentially come in two flavors: structural and regulatory. Regulatory genes tell the structural genes when to turn on and off and for how long. Humans and mice are virtually identical with respect to the genes regulating development, for example, the so-called *Hox* genes. Further, the mouse and human genomes do not appear to be qualitatively very different. They both contain about 30,000 genes, with mice having 300 genes that humans don't have and vice-versa. Humans and mice both have the genes that, in mice, result in a tail. In humans, the gene is turned off, while in mice it is turned on. Same gene, just regulated differently.

Consider pianos. All pianos have the same keys. But not all pianos play the same tune. The keys can be the same, but the music can be highly variable. The tune depends on the order and timing of the pressing of the keys, that is, how the keys are regulated by the person sitting at the keyboard. You might sit at the piano and play Ray Charles tunes, while someone else might play Chopin. Identical keyboards can give rise to very different musical phenotypes. Humans and mice develop from similar genetic keyboards, but the genetic analogs of the pianist's fingers are the regulatory genes.

The differences between the two species lies, in part, in the regulations of the same genes. Gene regulation also determines drug reaction and disease response. Because of differences in gene regulation, even identical twins may respond differently to diseases and medications.

SNPs, CNVs, and regulatory genes are not the only ways species and individuals differ.

Animals differ from humans, and humans differ from other humans, and hence manifest different responses to the same stimuli, due to the following:

- differences with respect to genes present
- differences with respect to mutations in the same gene
- differences with respect to proteins and protein activity
- differences with respect to gene regulation
- differences in gene expression
- differences in protein-protein interactions
- differences in genetic networks
- differences with respect to organismal organization (humans and rats may be intact systems, but may be differently intact)
- differences in environmental exposures
- differences with respect to evolutionary histories

These are some of the important reasons why members of the same species often respond differently to drugs and toxins, and experience different diseases. To many, these reasons also invalidate the use of animals as predictive models for human disease and drug testing.

Some courts are now recognizing that animal tests are not relevant to humans (see court cases listed in Further Resources).

Similarities

Despite the above differences, there is no doubt that animals can be useful in science. Numbers 3–9 in Table 1 are ex-amples of how animals can be used in a scientifically viable way.

Number 3 involves the use of animals for spare parts. For example, many people have had a damaged aortic valve replaced by the aortic valve from a pig. There is no doubt that pig valves function adequately in humans, and hence this is a scientifically viable use of animals.

Number 4 includes animals used as factories. For example, for decades insulin was harvested from cows and pigs at slaughter. Hepatitis B and C viruses and other viruses were grown in nonhuman primates and other animals so scientists could have a convenient reservoir of the virus for study. This was before the viruses could be grown in culture. More recently, mice have been used to produce monoclonal antibodies.

Number 5 relates to the fact that researchers frequently use tissues obtained from animals to study basic physiological processes. This is sometimes referred to as basic science research, as are numbers 7 and 9.

Number 6 is the use of animals for dissection, which most people have experienced in school. If a teacher wants a student to learn that nerves, arteries, and veins are found close to one another, dissecting animals will reveal this.

Number 7 is the issue of using animals as a heuristic, or exploratory, device or as a source from which to get new ideas. Of course, if a veterinarian or scientist wants to learn about diseases of cats, she can study cats as in number 8, where other nonhuman animals benefit from animal model use. And the final area, number 9, is knowledge for the sake of knowledge alone.

Numbers 3–9 are scientifically viable ways to use animals in science. The animal protection community objects to the use of animals in such a fashion, and may speak of alternatives for numbers 3–9. For example, aortic valve replacement can be performed with artificial or synthetic valves instead of obtaining them from pigs, and physicians rarely prescribe insulin from pigs and cows, since human insulin can be synthesized. Instead of using animal tissue to study basic physiological processes, human tissue is plentiful and better, if what the scientists actually want is more knowledge about humans. These are all alternatives to using animals. They are scientifically valid, and acceptable to the animal protection community.

The way veterinary students and researchers study animals can also be made less harmful to animals. The same criteria used in research involving humans can be applied to research designed to find cures for diseases in dogs and cats. Instead of creating models of cancer, researchers can study cancers that occur naturally in dogs. They can even break this down further and study the same cancer in different breeds, as there will be genetic differences between breeds just as there are differences between groups of humans.

A final consideration is the use of animals that lack either a central nervous system or a highly organized peripheral nervous system. These animals do not appear to be sentient, which is the basis of the animal rights and welfare arguments. Therefore their use should be acceptable to all. Some of the great breakthroughs in biological science are now coming from the study of invertebrates and organisms like yeast.

The level of examination has changed since the 18th and 19th centuries. As our examination of living systems has become increasingly fine-grained, we have found that, when it comes to explanations of biologic activity, subtle differences between organisms tend to outweigh gross similarities. Science could use animals to shed light on shared functions when we were struggling to understand living systems at the level of the organ; for example, the functions of the liver and heart are similar between species. But today we are studying drug response and disease at a level that defines not only a species, but in many cases the individual. Today, science studies human disease and drug response in light of complexity theory, evolutionary biology, gene expression, and gene regulation.

Arguments about ethics have been around for centuries. Ethics are somewhat subjective, but science should be less so. Scientific questions are eventually answered with more or less unanimous agreement. Such is the nature of studying the material world. In the final analysis, society will have to decide what role animals will play in scientific pursuits, indeed in all pursuits. But hopefully all sides will be able to agree on the scientific principles outlined above and use that knowledge to make intelligent decisions.

See also Alternatives to Animal Experiments: Reduction, Refinement, and Replacement

Further Reading

Americans for Medical Advancement at www.curedisease.com.

Associated Press. 2008. Gene map shows no two cancers alike. http://www.msnbc.msn.com/id/26546943/.

Bourne v. E.I.DuPont de Nemours and Company, 189F Supp. 2d 482 (S.D. W.Va. 2002).

Daubert v. Merrell Dow Pharms., 509 U.S. 579, 584 (U.S., 1993).

Flintoft, Louisa. 2005. Identical twins: Epigenetics makes the difference. *Nature Reviews Genetics* 6, 667; doi:10.1038/nrg 1693.

GAO. 2001. GAO-01–286R Drugs withdrawn from market. Washington, DC: US General Accounting Office.

Greek, J.S., and Greek, R.C. 2004. *What will we do if we don't experiment on animals? Medical research for the twenty-first century*. Victoria, BC, Canada: Trafford.

Haiman, C.A. et al. 2006. Ethnic and racial differences in the smoking-related risk of lung cancer. *N Engl J Med* 354:333–342.

Joiner v. General Elec. Co., 864 F. Supp. 1310, 1323 (N.D. Ga. 1994).

Kaiser, J. 2008. Cancer genetics: A detailed genetic portrait of the deadliest human Cancers. *science*, 321, 1280a-1281.

LaFollette, H., and Shanks, N. 1996. *Brute science*. New York: Routledge.

Shanks, Niall. 2002. *Animals in science*. ABC Clio.

Shanks, N., and Pyles, R. P. 2007. Evolution and medicine: The long reach of Dr. Darwin. *Philosophy, Ethics and Humanities in Medicine* 2:4, (April), http://www.peh-med.com/content/2/1/4.

Wong, Albert H. C., Gottesman, Irving I., and Petronis, Arturas. 2005. Phenotypic differences in genetically identical organisms: The epigenetic perspective. *Human Molecular Genetics*, Vol. 14, Review Issue 1 R11R18.

Ray Greek, MD

MICE

Contributing to stem cell research, the fast-moving world of mouse genetics has catapulted these small creatures into the forefront of science. The 2007 Nobel Prize in Physiology or Medicine was awarded for isolating and modifying embryonic stem cells and introducing specific genes into the mouse germline. Inserting viruses plus genes into the skin cells of an adult mouse transformed them into new stem cells, permitting the growth of new mouse skin and organ tissue. With the new knockout technology used in mice, termed targeted genomics, specific genes can be added or deleted to assess their effects on behavior and physiology. Mice are well-defined genetically, and procedures are available for manipulation of specific genes and control of reproductive outcomes. New mouse strains with specified genetic constructs are created to study disease processes. Cryopreservation (freezing) makes it possible to store embryos and blastocysts of mouse strains for later recovery, rather than needing to house colonies of valuable mouse strains. The new vocabulary for mice includes terms such as mouse engineering, chimera, targeted genomics for knockout and knock-in mice, molecular constructs, genetic analysis, and phenotyping. Expanding techniques for imaging can acquire significant information on the processes of disease from a small number of mice.

The most typical laboratory mammal, mice account for a large majority of all mammals used in research in the United States and Europe. Their genetic similarities to humans combine with a tiny body size and high reproductive rate to make them an economical, efficient option as models for studying the human body, the effects of diseases, and the feasibility of treatments for diseases. Specific genes can be added or deleted to examine the gene's effects. A further use of mice sometimes required by regulations is to evaluate the safety of new chemicals or products, drugs, and vaccines, as well as to measure the effects of limited or long-term exposure to a substance. Very few mice are used in education and teaching.

While it is difficult to know the exact numbers of mice used in scientific procedures, detailed records from the United Kingdom's Home Office show that mice account for 69 percent of the vertebrates

Millions of mice are used throughout the world in a wide variety of laboratory experiments that cause pain, suffering, and death. Although they display empathy for other mice in pain, mice are not protected from invasive experiments. Here, a Chinese scientist has grafted human cells on a white mouse to create an ear-shaped graft. (Associated Press)

used, and most of the remaining animals used are rats. These figures also show that the use of genetically modified animals, mainly mice, more than quadrupled between 1996 to 2007.

History of Breeding Strains of Mice

Long before their use in science, mice were specially bred for coat color and physical or behavioral traits for thousands of years. Before 1000 BCE, mice of special colors were bred in China. Mice have always been favored by some as pets, a form of petkeeping termed the mouse fancy. Historically, the mouse assumed religious importance in Egypt, Greece, China, and Japan, among others, despite also being a serious pest. In research, William Harvey, Joseph Priestly, and Antoine Lavoisier in the 17th and 18th centuries, were among the scientists who employed mice to make discoveries concerning anatomy and physiology.

In the 20th century, an expanding array of inbred strains of mice was bred starting with mice from the mouse farm of Abbie E. C. Lathrop in Granby, Maryland, which provided the ancestors of most of today's strains of mice.

Among the mouse pioneers, William E. Castle laid the groundwork for mammalian genetics. Clarence C. Little studied color inheritance in mice and became the first director of the Jackson Laboratory. Leonell Strong pursued cancer research using the mouse as his model. Ultimately, C57BL/6, BALB/c and C3H mice became the most common strains. Together with FVB and 129 mice for genomics research, they became termed the "big five" by Stephen Barthold.

Examples of Strains of Mice

Inbred mice are genetically identical due to inbreeding, and predisposed to getting a certain disease or genetic defect. *Transgenic* mice have been genetically engineered by injection of one or more genes, such as human breast cancer. *Immunodeficient* mice used in cancer and AIDS research have minimal immune function, and include nude mice and mice with severe combined immune deficiency (SCID). *Knockout* mice are engineered to lack a specific gene. *Pathogen-free* mice are free from all detectable viruses, bacteria and parasites.

Uses of Mice in Testing

Mice are used to evaluate the safety of new chemicals or products such as household cleaners and pesticides that may be potentially toxic to humans. Mice are also used to assess the safety of drugs and vaccines made for medical use. Toxicity tests are performed to measure the effects of limited or repeated long-term exposure of an animal to a particular substance. Other tests measure the extent to which the substance damages cells and causes cancer, mutations in DNA, and birth defects. Although mice are used to test the cancer-causing ability of substances, the number of whole animals used in carcinogenicity testing has diminished. Faster, short-term tests are now used to screen substances.

Uses of Mice in Research

Historically, the study of cancer and the production of vaccines and monoclonal antibodies are among the most widespread uses of mice.

Cancer research   Mice have been used in cancer research since 1894. Initially, mice were used for same-species tumor transplantations and drug treatment studies. In 1921, inbred strains that were predisposed to getting tumors were bred and disseminated among cancer researchers. More strains of mice originated from 1929 onward with the founding of the Jackson Laboratory in Bar Harbor, Maine, now the largest supplier of mice.

The inbreeding of mice predisposed to developing cancer led to specialized strains. In 1921, Leonell Strong established many inbred strains that frequently and spontaneously developed cancer. These inbred mice made it possible to study the growth and general characteristics of tumors.

The discovery in 1962 of the immunodeficient nude mouse led to human tumor transplantations without rejection, a valuable breakthrough for cancer research. Grafting human tumors onto these mice allows for the study of specific human cancers and the testing of new treatments in a whole animal system. A further breakthrough in the late 1980s led to transgenic mice, whose genes have been altered to produce a desired characteristic. Genes that cause cancer could then be studied in greater detail. In 1983, mice with severe combined immune deficiency (SCID) were discovered. SCID mice are even more immunodeficient than nude mice. Tumors from other species are easily transplanted into SCID mice and will grow without being rejected. SCID mice are used for the growth of *hybridomas in vivo* to produce a continuous supply of antibody. Sometimes referred to as reagents, antibodies are necessary for a wide range of diagnostic, clinical, and experimental procedures.

In the late 1980s, transgenic mice were engineered from genetically altered embryos, in which a gene or combination of genes is microinjected into developing oocytes. The genetic alteration can subsequently be transmitted to progeny. Through selective breeding, it is then possible to maintain a strain of mice consisting of individuals with particular traits of interest. A specific trait, such as a predisposition to develop a particular type of tumor, can be introduced into a mouse strain by injecting into the embryo a gene that causes cancer. Transgenic mice permit the study of cancer in specific tissues, including initial tumor development.

Vaccines Developing a new vaccine for a particular disease requires investigation of the efficacy and safety of the vaccine, both in the short-term acute phase, and also over the long term, to assure that birth defects or other delayed effects do not arise. Even after the vaccine is known to prevent infection with the disease and has been approved, batches of vaccine still need to be tested for safety.

Methods to develop new vaccines differ for each type of virus or bacteria. Animal experiments are usually required to select the initial materials in the formula, establish the stability and formulation of the vaccine, and determine the mode and frequency of administration. Experimental vaccines are tested for safety and efficacy on animals, chiefly mice, before clinical tests on humans begin. In the Netherlands in 1986, roughly two-thirds of the experimental animals used to make biological products were mice. In the actual vaccine production process, animal blood may be required for culturing media. Viruses are propagated in cells of animal or human origin. In the past, viruses were cultured *in vivo,* as in the production of smallpox virus on the brains of mice. Since 1949, primary cell cultures have largely been produced using *in vitro* methods.

Quality control is the most essential aspect of vaccine production. Since all vaccine batches are not the same, their content and effects must be tested regularly at selected stages of production to monitor safety, as required by federal regulations. Human lives have been lost when quality control has not been sufficient. The experimental animal is still a main indicator in the detection of the desirable and undesirable activities of newly-produced vaccine batches.

Safety testing assures that the vaccine product is inactivated and free from extraneous microorganisms or residual live virus. Other tests assess whether the vaccine causes development of tumors or is otherwise harmful. Also important are assays assuring that the vaccine is potent enough to induce protective immunity.

Monoclonal Antibodies In 1975, Kohler and Milstein first fused lymphocytes to produce a cell line which was both immortal and a producer of specific antibodies. The Nobel Prize for Medicine was granted in 1984 for the development of this *hybridoma,* used from about 1987 to produce monoclonal antibodies (MAbs) in rodents for diagnostics. These antibodies have exceptional purity and specificity, are components of the immune system, and can recognize and bind to a specific antigen. They are used diagnostically to measure protein and drug levels in serum, assess blood type, identify infectious agents, diagnose leukemias or tumor antigens, and assess hormones.

In vivo expansion of hybridomas in animals has become less acceptable due to humane and economic concerns. Several European countries have enacted legislation limiting antibody production in mice. MAbs are extensively produced *in vitro* in Switzerland and Germany, using cell culture systems. Although *in vivo* production is relatively inexpensive, ascites fluid extracted from mice may yield commercially unsuitable antibody. One popular alternative is bulk tissue culture in hollow-fiber bioreactor systems.

Preservation of Mouse Strains

Many thousands of distinct strains of mice now exist, some of which serve as models for specific human diseases. The mouse is the only mammal available in so many different genetic strains. Studying mice with specific genetic mutations can greatly advance studies of immune function, tumor growth, and various human genetic diseases. The Mutant Mouse Regional Resource Centers function as a repository system in the United States for the preservation and distribution of mice and embryonic stem cell lines, to make valuable genetically engineered mice available.

Imaging

Non-invasive imaging techniques are increasingly available for use with small animals. These permit monitoring the health or disease of an animal. Modalities include PET, SPECT, CT, MRI, ultrasound, autoradiography, and optical (fluorescence and bioluminescence).

Legislation

The United States Animal Welfare Act, as revised in 1985, includes most mammals, but excludes laboratory rats and mice. Research institutions can voluntarily seek accreditation by the Association for Assessment and Accreditation of Laboratory Animal Care International (AAALAC). Accreditation ensures that an institution is in conformance with the *Guide for the Care and Use of Laboratory Animals,* which applies to all laboratory animals, including rats and mice. This conformance with the Guide is a requirement for funding by many federal agencies, such as NIH. Hence, most academic institutions seek accreditation, and provide the same level of oversight for the care of mice as for other mammals.

For industries or testing facilities that do not seek funding and house only rats and mice, legislation and accreditation requirements do not apply. These institutions would only retain an institutional animal care and use committee as a proactive measure to assure optimal animal welfare, not as a regulatory requirement. One drawback of mice not being regulated is that no accurate figures are available concerning the numbers of mice used in the United States.

Patented Strains of Mice

Mice have been patented in the United States, Japan, Canada, and many European countries. The Harvard mouse, which carries a gene for breast cancer, was patented in 1988. The second patent was granted to Ohio University in 1992 for a mouse carrying a human gene that makes the animal resistant to viral infection, due to its continuous production of interferon, which attacks invading viruses.

Providing Welfare

The tiny body size, fast movements, and behavioral and sensory capacities of mice contrast with the traits of humans, and make it difficult for us to understand their behavioral needs. Mice are social animals and are most comfortable when surrounded by their own familiar odors. Research with mice requires infection control with special cages that are individually ventilated, which is not necessarily what the mice would prefer. Technicians are less likely to feel attached to mice in their care than to other species. They wear protective clothing, limiting tactile contact with the mice when cleaning cages, in order to protect the mice and also to reduce their own exposure to allergens from the mice.

Induced genetic defects and research procedures sometimes cause pain and suffering to laboratory mice, which may be somewhat alleviated by appropriate analgesia and anesthesia. Enhancing the quality of life for mice may partially offset some of their discomfort. For example, living in social groups would be a more normal situation for mice than solitary housing. Caregivers can make it more rewarding to work with mice and enrich their physical environments by enhancing their housing. Mice provided with hardwood shavings burrow and build nests. Placing hay or straw on racks above cages allows mice to pull material into the cage and arrange nests. Plastic tubes offer an artificial burrow space, perhaps shielding mice from illumination that may be too bright. Simple enrichments such as these can provide mice with some control over their environment.

One complication is that immunodeficient mice require sterile environments. All cage materials used for them, including bedding, food and water, must be autoclaved for sterilization before use. Thus, offering an improved quality of life requires more effort and cost when dealing with these mice that are especially valuable for their potential contribution to the scientific knowledge to improve human and animal health.

Further Reading

Barthold, Stephen W. 2002. Muromics: Genomics from the perspective of the laboratory mouse. *Comparative Medicine* 52: 206–223.

Critser, Greg. 2007. Of men and mice: How a twenty-gram rodent conquered the world of science. *Harper's Magazine* 315(December):65–76.

Fox, James G. 2007. *The mouse in biomedical research.* New York: Academic Press.

Golub, Mari S., Germann, Stacey L., and Lloyd, Kent C. 2004. Behavioural characteristics of a nervous system-specific erbB4 knockout mouse. *Behavioural Brain Research* 153:159–170.

Herzog, Hal A. 1989. The moral status of mice. *ILAR News* 31(1):4–7.

Morse, Herbert C. III. 1981. The laboratory mouse—A historical perspective. In H. L. Foster, J. D. Small, and J. G. Fox, eds., *The mouse in biomedical research*, pp. 1–16. New York: Academic Press, 1981.

Mouse Biology Program, University of California, Davis, accessed December 15, 2008, http://mbp.compmed.ucdavis.edu/.

Nobel Prize, The Nobel Prize in Physiology or Medicine 2007, accessed December 15, 2008, http://nobelprize.org/nobel_prizes/medicine/laureates/2007/press.html.

The Jackson Laboratory, accessed December 15, 2008, http://www.jax.org/.

UC Davis Center for Comparative Medicine, The Visible Mouse, accessed December 15, 2008; http://tvmouse.compmed.ucdavis.edu/.

Wood, Mary W., and Hart, Lynette A. Bibliographic Searching Tools on Disease Models: Locating Alternatives for Animals in Research, accessed December 15, 2008, http://www.lib.ucdavis.edu/dept/animalalter natives/diseasemodels.php.

Wood, Mary W., and Hart, Lynette A. The Mouse in Science, accessed December 15, 2008, http://www.vetmed.ucdavis.edu/Ani mal_Alternatives/mouse.htm.

Lynette A. Hart

MISOTHERY

The term misothery is derived from the Greek *misein,* to hate, and *therion,* beast or animal, and literally means hatred and contempt for animals. Since animals are so representative of nature in general, misothery can mean hatred and contempt for nature, especially its animal-like aspects. Tennyson, for example, has described nature as "red in tooth and claw," that is, bloodthirsty like a predatory animal. In another version of the same idea, we say, "It's a dog-eat-dog world." These are misotherous ideas, for they see animals and nature as vicious, cruel, and base.

The term misothery was constructed because of its similarity to the word misogyny, a fairly common word for an attitude of hatred and contempt toward women. The similarity of the two words reflects the similarity of the two bodies of attitudes and ideas. In both cases, the ideas reduce the power, status, and dignity of others. Misogyny reduces female power, status, and dignity, and thus aids and abets the supremacy of males under patriarchy. Misothery reduces the power, status, and dignity of animals and nature, and thus aids and abets the supremacy of

human beings under dominionism. Just as agrarian society invented beliefs to reduce women, it also invented beliefs or ideologies about animals that reduced them in the scheme of life. Among these are the idea that animals are too base and insensitive to feel physical pain or emotional suffering.

Further Reading

Fisher, Elizabeth. 1979. *Woman's creation.* Garden City, NY: Anchor Press/Doubleday.

Gray, Elizabeth Dodson. 1981. *Green paradise lost.* Wellesley, MA: Roundtable Press.

Nash, Roderick 1982. *Wilderness and the American mind,* 3rd ed. New Haven: Yale University Press.

Serpell, James. 1986. *In the company of animals.* London: Basil Blackwell.

Tuan, Yi-Fu. 1984. *Dominance and affection.* New Haven: Yale University Press.

Jim Mason

MORAL STANDING OF ANIMALS

Intelligence and adaptation in animals is often incomprehensible to us unless we attribute to them some form of understanding, intention, thought, imaginativeness, or form of communication. Many of their actions suggest adaptive and creative forms of judgment. To attribute these capacities to animals is to credit them with capacities analogous to human capacities, which suggests that animals merit at least some of the moral protections humans enjoy.

Historical Background in Darwin

Prior to Darwin, many biologists and philosophers argued that despite the anatomical similarities between humans and apes, humans are distinguished by the possession of reason, speech, and moral sensibility. Darwin thought, however, that animals often exhibit powers of deliberation and decision making, excellent memories, a strong suggestion of imagination in their movements and sounds while dreaming, and the like. He wrote about the intelligence, sympathy, pride, and love of animals. Darwin also criticized the hypothesis that only humans have significant cognitive powers.

The import of his theory is that complex biological structures and functions as well as cognitive abilities are shared in the evolutionary struggle. Darwin argued that despite differences in the degree of mental power between humans and apes, no fundamental difference exists in kind between humans and many forms of animal life. He thought that a greater gap existed between apes and marine life than between apes and humans. He judged that there are numberless gradations in mental power in the animal world, with apes and humans on the high end.

Moral Status

Questions of whether animals have higher-level cognitive capacities are closely connected to questions of moral and legal standing. Terms such as status and standing have been transposed into ethics from law, where standing is defined as "One's place in the community in the estimation of others; one's relative position in social, commercial, or moral relations; one's repute, grade, or rank" (*Black's Law Dictionary*). In a weak sense, standing refers to a status, grade, or rank of moral importance. In a strong sense, standing means to have rights, or the functional equivalent of rights.

To have moral status, then, is to deserve the protections afforded by the basic norms of morality. One popular view attributes a more significant standing to an animal by granting that it is relevantly similar to an intact adult human being. Its standing is still further enhanced by attributing personhood or autonomy. Defining it as an person or autonomous agent elevates the animal to a position approximating that occupied by those who have rights. A widely shared view today is that if animals have the capacity for understanding, intending, and suffering, these morally significant properties themselves confer some form of moral standing.

The Model of Cognitive Properties

Several philosophers have produced arguments along the following lines: One is a person if and only if one possesses certain cognitive properties. The possession of these properties gives an entity moral standing. As a corollary, anything lacking these properties lacks moral standing, and therefore does not possess rights.

Cognition here refers to processes of awareness and knowledge, such as perception, memory, thinking, and linguistic ability. The thesis is that individuals have moral status because they are able to reflect on their lives through their cognitive capacities and are self-determined by their beliefs in ways that nonhuman animals seem not to be. Properties found in various theories of this type include: self-consciousness (consciousness of oneself as existing over time, with a past and future); freedom to act and the capacity to engage in purposive sequences of actions; having reasons for action and the ability to appreciate those reasons for acting; capacity to communicate with other persons using a language, and rationality and higher-order volition. Many believe that more than one of these five conditions is required to be a person.

As long as high-level cognitive criteria are required, animals cannot qualify for significant moral standing. But if less demanding cognitive capacities are employed, animals might acquire a significant range of moral protections. For example, if a high-level qualifying condition such as speaking a human language is eliminated, and conditions such as intention and understanding are substituted, then it becomes plausible to find the cognitive capacities needed for moral standing in at least some animals.

Critics of theories based on human cognitive properties often argue that some creatures deserve moral status even if they do not possess a single cognitive capacity. They argue that a non-cognitive property may be sufficient to confer some measure of moral standing. The most frequently invoked properties are those of sensation, especially pain and suffering, but properties of emotion, especially those associated with fear and suffering, are also mentioned.

Animal Minds

At the root of many of these issues is a rich body of problems about animal minds. Little agreement exists about the levels and types of mental activity in many animal species or about the ethical significance of their mental activity. Humans understand relatively little about the inner lives of animals, or about how to connect many forms of observable behavior with other forms of behavior. Even the best scientists and the closest observers have difficulty understanding intention and emotion in animals. Neither evolutionary descent nor the physical and functional organization of an animal system, that is, the conditions responsible for its having a mental life, give us the depth of insight we would like to have to understand their mental states. The more we are in doubt about an animal's mental life, the more we may have doubts about its moral status and the issue of rights.

Further Reading

Cavalieri, Paola, and Singer, Peter. 1993. *The Great Ape Project.* New York: St. Martin's Press.

Frey, Raymond. 2003. Animals. in Hugh LaFollette, ed., *The Oxford handbook of practical ethics.* New York: Oxford University Press.

Griffin, Donald R. 1992. *Animal minds.* Chicago: University of Chicago Press.

Nussbaum, Martha. 2006. *Frontiers of justice: Disability, nationality, species membership.* Cambridge: Harvard University Press.

Orlans, F., Barbara, Beauchamp, Tom L., Dresser, Rebecca, Morton, David B., Gluck, John P. 2007. *The human use of animals: Case studies in ethical choice,* 2nd ed. New York: Oxford University Press.

Rachels, James. 1990. *Created from animals: The moral implications of Darwinism.* New York: Oxford University Press.

Regan, Tom, and Singer, Peter, eds. 1989. *Animal rights and human obligations,* 2nd ed. Englewood Cliffs, NJ: Prentice Hall.

Regan, Tom. 2003. *Empty cages: Facing the challenge of animal rights.* Lanham, MD: Rowman & Littlefield Publishers.

Singer, Peter. 1995. *Animal liberation,* 2nd ed. London: Pimlico.

Warren, Mary Anne. 1997. *Moral status.* Oxford: Oxford University Press.

Tom L. Beauchamp

MUSEUMS AND REPRESENTATION OF ANIMALS

All museums are human-centered, or anthropocentric. They are by their very nature monuments to human creations, concerns, collections, curiosities, explorations, memories, and attitudes. Even the most enlightened museum starts with the assumption of the primacy of the human animal. For our purposes here, the term museum includes art museums and galleries, natural history museums, history museums, historic houses, living history sites, children's museums, science centers, zoos, nature centers, and aquaria.

In art museums around the world, museum educators create tours called some variation of "Animals in Art" for school groups. Their docents troop these youngsters through their collections in a glorified scavenger hunt, as little ones gleefully point out the animals they spot. While sometimes these tours take the time to compare the elongated arms of the monkeys in a Chinese screen to the elongated necks of sculpted folk art birds for weathervanes or decoys, mostly they do not connect the dots beyond "Miss, I see one!" With the youngest groups, they avoid a bronze depicting animal savagery or Francis Bacon's terrifying dog, leaning far more heavily on richly painted depictions of Aesop's fables and versions of Edward Hicks' *Peaceable Kingdom,* more acceptable, non-nightmare-inducing material.

To some, this find and name process appears to be no more than youthful hunting, or a form of animal watching without any context. What attitude to the art, to the animals does it seek to engender in children? The differences between George Stubbs' horses, Alexander Calder's lions, and Northwestern Native raven masks are rarely discussed. While formalist concerns come to the fore with older students and adults, attitudes toward

animals as evidenced within the works of art remain largely ignored.

Even the most overtly political contemporary artists, such as Sue Coe or Walton Ford, are approached aesthetically, biographically, and contextually, within the framework of contemporary art. Meanwhile in parts of museum collections, like those of indigenous peoples, where adults could, if "Animals in Art" were revisited, look closely at the symbolic, totemic, and narrative, recurring imagery is usually thought too simplistic. Turtles, frogs, lions, dogs, and snakes go uncommented on, other than for their incorporation as design elements.

For many adults, museum memories relate to natural history museum visits. Of those, the memory may be of the American Museum of Natural History in New York, and specifically the center floor diorama of elephants that seem to dash from one end of the room to the other, majestic, terrifying, and, in the true sense of the word, awesome. Some may remember a sleepover under the whale. And though dinosaurs are ever popular, the art of the diorama, more than just the bones, is effective theater, and those charging elephants, not behind glass, but inhabiting the room, are startling enough to stick in the memory.

But museums in their texts and subtexts present more attitudes about animals than simply what a child, still determining alive from dead, real from fake, can perceive. For adults, museums present a broad spectrum of views about animals. While art museum depictions indicate attitudes that range from symbol and story to dominion and possession, there is also evidence of kinship, wonder, and catalog, as well as extensions of symbol to include totem, logo, and pure pattern and design.

A child gazes at a stuffed Tasmanian Tiger on display at the Australian Museum in Sydney. (AP Photo/Rick Rycroft)

While certain painters may be classified as Fauves or wild beasts, their newly and vividly colored foxes, wolves, and horses are a source of delight. Their dissonance, the struggle between what we know and what we see, makes them fascinating, all the more so because everything other than their color makes them completely understandable, and not particularly wild.

Perhaps because museums in America were born from the cabinet of curiosity and derive from a Victorian collector and cataloguer mentality, the sense in which many natural history museums seem like a Noah's Ark or exhaustive compendium is sometimes overwhelming. Many natural history museums, though, have sought to better acquaint their visitors with the animals of their region. On a museum tour in North Carolina, for example, visitors might see more snakes, living and dead, than at your average roadside snake farm. Near the end of this herpetological grand tour, a curator might note that the exhibit's purpose is not to show cobras and pythons for the sheer creepy excitement of it, but to show the snakes of North Carolina, all creatures their visitors might well encounter in the wild. Getting to know them, recognize them, and understand their dangers and benefits, is a survival tool, for both museum visitor and snake alike.

As natural history museums compete with zoos, botanical gardens, safari farms, and circuses, their serious mission, to educate about animals and their habitats, and the interrelatedness of all creatures and their habitats, has moved them into advocacy positions around endangered species, climate change, and global interdependence.

History museums and historic houses seem to catalogue or freeze old attitudes, sometimes reinterpreting them, often ignoring their implications, especially those whose impact is felt by animals. The historic homes of the early 20th century's very rich, many of them robber barons, for whom exploitation of people, animals, and all natural resources was a way of life, a source of their wealth. and motor of the nation's growth, sport animal head trophies from tours of the great American West and African safaris. Now many of these collect dust and act to mark their owners as belonging to a certain era, people who might well have caged or stuffed specimen humans had it not seemed somehow barbaric. Still, their dusty attitudes toward people and animals pervade the air.

Safari trophy heads at the George Eastman House in Rochester, New York, raise the interesting question of the difference between the shoot and the photo shoot, something the contemporary photographer James Balog takes aim at squarely in his work. While revisionism and political correctness have altered or appended the point of view of some history museums and historic homes, it is the rare historic house collection that comments with contemporary eyes on the morality of the day depicted. Such asides, in addition to being considered bad form, destroy whatever theatrical leap of the imagination the fully appointed house-as-time-warp might create. It is left to visitors to bring their own contemporary attitudes as they visit the past and to try to square the two.

Reconstructions, historic villages, farmsteads, and workplaces often stress domesticated animals and their care and feeding, as well as their uses in home, hearth, and community. For contemporary visitors, often totally unaware of the processes that go into the making of things they use all the time, exposure to those processes can only increase awareness, and sometimes even change attitudes. To watch shearing, carding, spinning, dying, weaving, and sewing is to understand clothing and, it is hoped, something about sheep, in a totally different light.

Since science centers and children's museums, though hands-on in their learning presentation, are not petting zoos, the experiences and attitudes they present about animals are constructed around dramatic play, sensory input, and some didactic presentations, these last an especially difficult and underutilized mode within an interactive context. They diverge largely because of their audiences. For science museums, biology, evolution, the structures of categorization, animal behavior, defense of animal testing for human uses, extinction, and mutual dependence and cycles, are all valid and often approached areas. With children's museums, in an attempt to garner empathy for animals, the focus is often on animal babies, experiencing the world through the animal's, mostly visual, senses, instructive help for young people and their pets, and initial steps at categorization.

Zoos, aquaria, and nature centers have the most insistent, if sometimes contradictory, stances toward the animal kingdom. While zoos and aquaria collect animals, many of them are actively involved in attempts to save species, breed them in captivity, and return them to their habitat. Zoos and aquaria often look to reproduce habitats so that humans can better understand the context from which a given animal or group of animals has been taken, as well as to aid in the animal's adjustment to captivity. The artificiality of the setting, captivity, and close contact with many other species and, of course, with human beings, thwart these admirable goals and often present visitors with animals in a psychologically damaged state, what critics characterize as slaves embracing their slavery.

Nature centers, giving visitors a chance to see animals in their own environs, expend immense efforts to make sure that their visitors, in their encounters, do not damage the habitat or the plant and animal life within it. In a real sense, encounters between people and animals in sites of this sort are the most primal, though groups of ten or more people with cameras, binoculars, and guidebooks are not harmless in or to a habitat.

The complexity of museums' attitudes towards animals, bound up as they are in each institution's mission and vision, sometimes also evidence attitudes often unspoken and unacknowledged. Sometimes these attitudes are the unwitting result of insensitivity—many museums are, in fact, no better in their treatment of indigenous peoples—and render them impossible to generalize. They are made all the more complex by new generations of political artists with advocacy positions and new understandings of the roles of zoos and aquaria in a world in which habitats and their creatures are rapidly disappearing.

While the elephants of the American Museum of Natural History, and contemporary Chinese-born artist Cai Guo-Qiang's *Inopportune: Stage Two*, a life-size, walk-through Chinese landscape with nine tigers pierced by hundreds of arrows, would seem worlds apart

in impact, if not in intent, they are similar. They destroy the boundary between viewer and viewed, remove the glass from the diorama, and in that simple act totally reorder our perceptions. People gasp when they walk into Cai's installation, until they realize that the tigers are fabricated. The gasp, the shock, is real. *Inopportune: Stage Two* has been shown at New York's Guggenheim Museum, the Massachusetts Museum of Contemporary Art, and other museums.

Photographer Jerry Uelsmann, in an untitled photograph from 1973, focuses on the diorama experience. The photograph presents a majestic landscape in black and white, with a backlit, theatrical jewel case of a diorama in a dark foreground, almost a television screen, but with more depth; this is 1973, no HDTV. Across the diorama walk deer, placid, unskittish, stopped dead in their tracks, a vignette within a larger story, a microcosm in a larger context and, really, a controlled way of seeing and experiencing, what is too big, too fast, too quiet, too other for many human beings to grasp otherwise.

See also Art, Animals, and Ethics

Further Reading
Balog, James. 1990. Survivors: *A new vision of endangered wildlife*. New York: Harry N. Abrams, Inc. Publishers.

Balog, James, & Pedersen, Martin B. 1999. *Animal*. New York: Harper-Collins.

Bell, Joseph. 1985. *Metropolitan zoo*. New York: Metropolitan Museum of Art and Harry N. Abrams, Inc.

Coe, Sue. 1996. *Dead meat*. New York: Four Walls Eight Windows.

Danto, Arthur Coleman. 1988. *Art/artifact: African art in anthropology collections*. New York: Center for African Art.

Fuller, Catherine Leuthold. 1968. *Beasts: An alphabet of fine prints*. Boston and Toronto: Little, Brown and Company.

Guo-Qiang, Cai. 2005. Essays by Laura Heon and Robert Pogue Harrison, interview with Jennifer Wen Ma. In *Inopportune*. Wilmington: MASS MoCA.

Karp, Ivan, & Lavine, Steven D. 1991. *Exhibiting cultures: The poetics and politics of museum display*. Washington and London: Smithsonian Institution Press.

Katz, Steve, & Kazanjian, Dodie. 2002. *Walton ford: Tigers of wrath, horses of instruction*. New York: Harry N. Abrams, Inc.

Ulrich, Laurel Thatcher. 2001. *The age of home-spun: Objects and stories in the creation of an American myth*. New York: Random House.

Weschler, Lawrence.1995. *Mr. Wilson's cabinet of wonder*. New York: Random House

Wilson, Fred. 1992. *Mining the museum*. Baltimore Historical Society museum exhibition.

James S. LaVilla-Havelin

N

NATIVE AMERICANS AND EARLY USES OF ANIMALS IN MEDICINE AND RESEARCH

Constitute reductionism is the idea that all living things in the organic world are essentially made of the same elements. Phylogenetic continuity is the concept, originally proposed by Charles Darwin, that differences between nonhuman animals and human animals are quantitative differences in degree, rather than qualitative differences in kind. This can be translated into the concept that physiological processes, including behavior and even cognition, share common properties across species. The bases of constitute reductionism and phylogenetic continuity, which are contained in the theory of evolution, provide the theoretical support for the use of animals in research to understand humans.

While these ideas may be rejected by some religions, Native American religions view the creation of humans as based on the transformation of an animal into a human form. Many Native American religions also stress the point that all items on the earth are related, and animals are not that different from humans. In Ojibwa culture, the creation of humans begins in the eastern great salt sea (the Atlantic ocean), where the back of the *megis*, a large ocean clam, is warmed by the sun and the Anishnabe (Ojibwa people) are brought to life. This legend shares some common ideals—life starting in the ocean, the relationship between human and animals—with the theory of evolution.

Many if not all pre-Columbian Native American nations used animals in medical treatments and education. The common view of Native medicine has been that it is shamanistic. Although ritual did, and still does, play an important role in Native American medicine, there was extensive use of practical therapy. The more practical therapies included the use of plants and animal parts to treat specific medical conditions.

Most Native American nations, with the notable exception of the Aztec, did not engage in internal surgical practices. Furthermore, in many Native American nations, postmortem examinations were not conducted on the dead for religious reasons. Most of the information Native Americans had about internal anatomy came from their dissection of animals during the butchering process. It has been documented that, from the analogy with animals, Native Americans knew the function of internal organs, and knew that the brain was the organ of thought.

In addition to anatomy lessons, animals were utilized in observational research. By noting particular animal behaviors, especially the interaction

Chief Arvol Looking Horse talks about buffalo in the Lakota language as Rosalie Little Thunder translates into English before a Spirit Releasing ceremony for the animal in Yellowstone National Park. Native Americans held a sacred ceremony in the park for those animals who were killed or removed as part of a livestock protection program. (AP Photo/Douglas C. Pizac)

between animals and plants, Native Americans gained information about the nutritional and medicinal properties of many plant substances. For example, the bear is a medicinal animal in Ojibwa culture, believed to be given the secrets of the *Mide* (medicine) by Kitshi Manido (Great Spirit). Because of this belief, the Ojibwa would carefully observe the bear in its environment. These examples demonstrate that, in addition to using animals for food and clothing, early Native Americans also used animals to gain information about themselves and their environment.

Although a number of Native American herbal remedies have been adapted by mainstream medical organizations, the use of animal products in medical treatment has not received the same attention. Animal products were used in a number of medical remedies in many Native American nations. Moose and bear fat were used by the Ojibwa to treat skin wounds, and to ensure healthy skin in extreme temperatures. Deer tendons were used as suture material by numerous tribes. The Yukon treated scurvy by ingestion of animal adrenal glands. Fish oil, because of its high iodine content, was used to treat goiters in Eskimo/Aleut nations. Some South American Nations treated epilepsy through shock treatment with electric eels. A type of injection device was used by some Native American nations well before the invention of the syringe in 1904. These were constructed from the bladder of a deer or duck connected to the reed or quill of the porcupine. These syringes were used to clean wounds or to inject herbal medicine into the wound.

The examples listed demonstrate that Native Americans' unique relationship with animals included their use in research and medicine. By documenting both the physiological and behavioral properties of animals, we as humans can learn more about animals, including ourselves.

Further Reading

Aronson, L.R. (1984). Levels of integration and organization: A revaluation of the evolutionary scale. In G. Greenberg & E. Tobach: *Behavioral evolution and integrative levels.* Hillsdale, NJ: Lawrence Erlbaum Associates.

Altman, J. (1966). *Organic foundations of animal behavior.* New York: Holt, Rinehart and Winston.

Hershman, M. J., & Campion, K. M. (1985). American Indian medicine. *Journal of the Royal Society of Medicine, 28,* 432–434.

Hoffman, W. J. (1885–86). *The Midewiwin or Grand Medicine Society of the Ojibwa.* Seventh Annual Report of the Bureau of American Ethnology. Washington, D.C. Government Printing Office, 149–300.

Major, R. C. (1938). Aboriginal American medicine north of Mexico. *Annals of Medical History,* 10(6), 534–4.

Vogel, V. J. (1970). *American Indian Medicine.* Norman: University of Oklahoma Press.

Lisa M. Savage

NATIVE AMERICANS' RELATIONSHIPS WITH ANIMALS: ALL OUR RELATIONS

The relationships between animals and Native Americans are as varied as the more than four hundred different tribal nations that existed in pre-Columbian North America. Native people were and in many cases still remain deeply tied to the particular ecosystems in their regions of the continent. Some based their lives on agriculture, some on the ocean and salmon fishing, others on the hunting of hoofed animals, some in all three. However, certain generalizations about the relationships between Native Americans and animals can be made. One of the most important generalizations is that animals are not seen by the American Indian as dumb beasts whose lives are ruled only by instinct, but as individuals—thinking, feeling beings with families, beings worthy of respect. They are the animal people.

In the truest sense of the word, animals are seen as relatives of human beings. Many Native traditions, such as those of the Cherokee or the Lakota, tell that certain animals were direct ancestors. The idea of clan often comes from a tradition of direct descendants from one animal or another—a frog, an eagle, a bear. If a person belongs to the Bear Clan, it may be that the clan's origin is in the form of a bear who married a human woman and produced offspring. The border between the worlds of the animal people and human beings is easily crossed. A human being may go and live among the animals and become a bear or a deer as easily as an animal may take on human shape and live among human beings. Sometimes these animal people have great power and are to be feared. Through the Midwest and West, tales are still common of the Deer Woman who comes to gatherings to lure off young men and harm them. Beneath her long dress she has hooves, not feet. Such beliefs are extremely widespread and are reinforced by stories and ceremonies.

Animals often appear as teachers in traditional stories. Humans can learn many things from the animal people.

Traditional stories tell us how flute songs came from the birds, how medicine plants were shown to the humans by the bears, and how humans were taught how to work together and to care for their children by watching the behavior of wolves as they hunted and cared for their cubs.

Native American people have found it necessary to hunt animals to ensure their own survival. However, even hunting is seen as cooperation with the animals. Although the animal's body is killed, its spirit survives, and it may punish a disrespectful or greedy hunter. It is only through animals' consent that they allow themselves to be hunted. Further, the hunting of animals that are pregnant, or caring for young ones whose survival depends upon the mother, is usually forbidden. Many current game laws, closed seasons, and limited harvesting of game animals have their roots in Native American traditions that have existed for thousands of years.

Animals are frequently kept as pets or companions. In the Northeast, among the Iroquois, orphaned beavers were often suckled by Native women and adopted into the family. Across the continent, dogs were kept as pets and used for hunting. According to the traditional stories of the Abenaki, the dog was not domesticated, but chose to live with the human beings because it liked them. To this day, the dog in a Native American household is often viewed not as a possession but as a family member. The fact that in some Native American cultures dogs were sometimes eaten or sacrificed, as in the Seneca White Dog Sacrifice, so that the dog's spirit could take a message to the Great Spirit, did not diminish the respect for the dog or its place in the household.

In the traditions of the many different Native peoples of North America, animals are almost universally seen as equal to humans in the circle of life. The word circle is especially appropriate, for all living things, animals and humans alike, are viewed as part of a great circle. No part of that circle is more important than another, but all parts of that circle are affected when one part is broken. In the eyes of the Native American, animals are all our relations.

Further Reading

Brown, J. E. 1992. *Animals of the soul: Sacred animals of the Oglala Sioux*. Rockport, MA: Element.

Caduto, M., and Bruchac, J. 1991. *Keepers of the animals*. Golden, CO: Fulcrum Publishing.

Cornell, G. 1982. Native American contributions to the formation of the modern conservation ethic. Ph.D. dissertation, Michigan State University.

Hughes, J. D. 1983. *American Indian ecology*. El Paso: Texas Western Press.

Vecsey, C., and Venables, R. W. 1980. *American Indian environments: Ecological issues in Native American history*. Syracuse, NY: Syracuse University Press.

Joseph Bruchac

O

OBJECTIFICATION OF ANIMALS

In 1995, the Summit for Animals, a loose confederation of national and grassroots animal protection organizations, passed a resolution stating, in part, "We resolve to use language that enhances the social and moral status of animals from objects or things to individuals with needs and interests of their own." Collectively called the linguistic turn, a current view in several academic fields holds that language plays an important formative role in the way we see, think about and, ultimately, treat entities in both the cultural and natural world.

Numerous areas that need change have been identified. The most important and perhaps the most difficult to bring about is the use of the term animal, which has come to mean "as distinguished from human." In this use, the conflicting terms human and animal deny that human beings are part of the animal kingdom. More critically, this usage reinforces the invidious comparison of animal as inferior to human. Although a number of suggestions have been made to correct this, for example, retaining the term animal to refer to all animals including humans, and anymal to refer to animals other than humans, none has gained common usage.

Other linguistic habits support the lower status of animals. In many settings, such as the farm and the research laboratory, individual animals are not named. Further, they are referred to as it rather than he or she and which or that rather than who. These uses decrease the value of animals by depriving an animal of individuality, including their identities as members of a particular gender. This practice is also seen in language used by hunters and wildlife managers when they refer to deer as a species rather than a group of individuals.

In farm and laboratory settings, language operates to deprive animals other than humans of even this identity as members of a particular species. Rather than the rat or the monkey, investigators typically refer to animals in the lab as the animal. A final decrease in value occurs when they are referred to as less than even this already-weakened notion of animal. On the farm, the individual animal is referred to as beef or meat on the hoof, while in the laboratory the individual rat is an organism, a generic living being, or a preparation, a living physiological or behavioral process.

In the scientific laboratory setting, additional practices support the devaluing of animals. Many scientists use the term anthropomorphism as a criticism of both scientific and popular accounts that use psychological terms to describe animals other than humans. For example, terms like intended, anticipated, felt, and attributions like play, grief, and deceit to animals other than humans are avoided, because their use commits the error of anthropomorphism. This prohibition against terms implying consciousness in animals other than humans is a legacy of the philosopher Descartes, in whose view animals were mechanical beings, without psychology, without minds. Consistent with this view, the pain, suffering, and death attendant to either the conditions of an experiment or the conditions under which animals in the laboratory are kept is typically not described as such. For example, an animal is said to be food deprived rather than hungry, or subjected to aversive stimulation rather than experiencing pain. The death of an animal is obscured by various terms such as collected, harvested, or sacrificed, or anaesthetized and then exsanguinated.

Further Reading

Birke, L., and Smith, J. (1995). Animals in experimental reports: The rhetoric of science. *Society and Animals, 3,* 23–42.

Dunayer, J. (2001). *Animal equality: Language and liberation.* Derwood, MD: Ryce.

Jacobs, G., and Stibbe, A. (Eds.) (2006). *Language matters, society and animals, 14,* whole issue.

Shapiro, K. (1989). The death of an animal: Ontological vulnerability and harm. *Between the Species, 5,* 4, 183–195.

Kenneth J. Shapiro

P

PAIN, INVERTEBRATES

While most people assume that vertebrates (animals with backbones) perceive pain, this is not as clear for most invertebrates (animals without backbones). However, the common octopus, with its large central nervous system and complex behaviors, has been given the benefit of the doubt in Great Britain and is now protected under the Animals (Scientific Procedures) Act of 1986.

Some argue that insects do not perceive pain and others that it is difficult to be certain. One may be uncertain about insect pain but still believe they should be given the benefit of the doubt. The conclusion that insects do not perceive pain is based on several lines of reasoning.

First, although insects have complex nervous systems, they lack the well-developed central processing mechanisms found in mammals and other vertebrates as well as the octopus, which appear to be necessary to feel pain. Second, it is not apparent that insects have a nerve fiber system equivalent to the nociceptive fibers found in mammals. However, this does not mean that they do not have some nerve fibers that carry aversive signals. Third, the behavior of insects when faced with noxious or harmful stimuli can usually be explained as a startle or protective reflex. In some cases, for example, locusts being eaten by fellow locusts, insects display no signs that the tissue damage that is occurring is aversive.

The conclusion that insects do not perceive pain appears to contradict the claim that pain confers important survival advantages. However, simple neural reflex loops producing an aversive startle reflex that involves no pain perception could confer sufficient evolutionary advantage in short-lived animals like insects that rely on a survival strategy involving the production of very large numbers of individuals.

If insects and most other invertebrates do not perceive pain, this would be relevant for ethical systems that rely on sentience as an important criterion of moral consideration. However, it would not necessarily indicate that insects should be accorded no moral consideration. Moral arguments that rely on reverence for life considerations—for example, the Jain or Schweitzerian systems—or ecosystem values would still regard insects as deserving some moral consideration.

Further Reading

DeGrazia, D., and Rowan, A. 1991. Pain, suffering, and anxiety in animals and humans. *Theoretical Medicine* 12: 193–211

Eisemann, C. H., Jorgensen, W. K., Merrit, D. J., Rice, M. J., Cribb, B. W., Webb, P. D. et al. (1984). Do insects feel pain?—a biological view. *Experientia* 40: 164–167.

Fiorito, G. (1986). Is there pain in invertebrates? *Behavioral Processes* 12: 383–386.

Rome and other Italian cities have adopted some of the world's strictest animal rights laws, including banning the boiling of live lobsters. (morgueFile)

Mather, J. A. (2001). Animal suffering: An invertebrate perspective. *J. Appl. Anim. Welfare Sci.* 4: 151–156.

Sherwin, C. M. (2001). Can invertebrates suffer? Or, how robust is argument-by-analogy? *Anim. Welfare* 10: 103–118.

Somme, L. S. (2005). *Sentience and Pain in Invertebrates.* Norwegian Scientific Committee for Food Safety at http://jillium.nfshost.com/library/pain.htm.

Wells, M. J. (1978). *Octopus.* London: Chapman and Hall

Wigglesworth, V. B. (1980). Do insects feel pain? *Antenna* 4: 8–9.

Andrew N. Rowan

PAIN, SUFFERING, AND BEHAVIOR

Suffering can be defined as: A negative emotional state, which may derive from various adverse physical or physiological or psychological circumstances, and which is determined by the cognitive capacity of the species and the individual being, as well as its life's experience.

This proposed definition addresses the mental distress that may be caused in some animals through their perception of the external environment, particularly through senses such as smell, sight, and sound, as well as their internalized individual predicament through feelings such as pain, or an instinct to carry out certain behaviors, for example migration in a wild bird when it has been caged. This mental distress will also be affected by an animal's experiences in life and the ability to recall them and recognize contextual similarities.

What Animals?

Only animals that have the neurological development and capability to experience adverse states are the subjects of concern here. More primitive forms such as amoebae, simple multi-cellular organisms that lack a complex organized nervous system are unlikely to feel, although they may well react and even possess simple programming mechanisms. However, they are unable to interpret novel circumstances. When animals possess a level of consciousness that allows them to assimilate new information and to apply general learned principles to novel circumstances, they are more likely to anticipate the future and therefore possibly suffer more than an animal that does not have these faculties. As far as we are know, this level of awareness is generally found in vertebrates, but not in invertebrates. However, new evidence and a reinterpretation of existing data for invertebrates suggest that this is not entirely true. For example, the octopus seems to have an ability to recall adverse experiences and use avoidance behaviors. Moreover, the ability to feel pain and other adverse states varies between different phyla and, if two key questions regarding sentience are phrased differently, a different view emerges. Specifically, does the animal possess similar or homologous neural pathways, neuropeptides, and hormones that might indicate sentience? And does it behave as it if is sentient in response to what would be a noxious stimulus to vertebrates?

Not surprisingly, sentience progressively develops throughout gestation or incubation. In humans this seems to be somewhere between 18 and 26 weeks, probably later rather than earlier. Furthermore, the development of the nervous system in some mammalian species has been shown not to stop at birth. For example, the descending inhibitory pathways that control the passage of nociceptive impulses up to the brain, which is known as gating, because it serves as an obstacle to the continuing passage of impulses up to the brain, continue to develop for several weeks after birth. This has led to the speculation that neonatal and young animals feel more pain than they will later in life when their nervous systems have fully matured and the gating mechanism is fully developed. Finally, the development of self-awareness and therefore the ability to reflect on one's own circumstances, could add another dimension to any experience of suffering, and appears to develop at around two years of age in humans, but there is little data in animals other than that it may be present or absent in adults. This entry discusses those beings that are sentient, that is, capable of feelings such as pain, fear, frustration, boredom, and possibly other feelings such as happiness, pleasure, grief, and guilt.

"A Negative Emotional State . . ."

Animals that are sentient can feel positive and negative emotions, and suffering may occur when these feelings are overwhelmingly negative. In some situations there may be a mixture of positive and negative states. Obtaining food at the price of an electric shock may still be an overall positive experience, and an animal may return to such a situation to maintain its homeostasis, that is, to satisfy its inner feeling of hunger. It is obvious that animals can experience a range of emotions, from those indicating pleasure and happiness in some way (dogs wagging tails, cows eager to get out to grass even though they have ample food before them, cats purring) to the other end of the spectrum, where animals may deliberately avoid situations that they have found unpleasant (a puppy returning to the veterinary clinic, sheep avoiding a shed where they have undergone electro-immobilization in the past, a surgical procedure such as foot trimming in dairy cows). Such negative experiences can be recalled by an animal from an earlier event, and animals may take avoiding action when given a chance. However, not all negative experiences require prior exposure, for example, thirst, or the desire to migrate, mate, or play.

". . . Which May Derive from Various Adverse Physical or Physiological or Psychological Circumstances . . ."

Examples of adverse physical states include environments that induce abnormal behaviors, or where abnormal lengths of time are spent carrying out normal behaviors. These are closely linked with the mental or psychological health of animals. Barren environments like cages or pens where animals have little opportunity to carry out instinctive behaviors such as dust-bathing in chickens, digging in rabbits and gerbils, nest-building for mice and sows seem to set up an internal conflict for the animal akin to a feeling of frustration when they are unable to satisfy their instincts, and lead them to subconsciously carry out stereotypic behaviors. Wild animals also often show repeated escape behaviors when caged, and which seem to be exacerbated when an animal has known freedom as opposed to being born and bred in captivity. Examples might in-

clude polar bears and wolves pacing in their concrete pens in zoos, horses weaving at the door of their stable, rabbits pawing at the back of their cages, wild birds looking for an escape route from their cages. These poor environments lead to psychological responses that are internally driven, but the physiological changes are less obvious. We can start to examine aspects of the environment that may be better for animals by observing what environments they choose to be in and how hard they will work to get there. For example, rodents work hard in choice tests to reach a particular type of environment. They prefer solid floors to grid floors, and certain types of bedding substrates are preferred to others. A word of caution, however. Animals may not always choose what is good for them in the long term, nor does it really tell you what they want, as humans may not offer that particular choice. Animals that carry out stereotypic behaviors due to an impoverished environment, to the point of causing tissue damage to themselves, will obviously suffer additionally.

Adverse physiological circumstances would include poor health, for example, due to an infection that, in humans, leads to feelings such as discomfort or malaise. Similar signs are also seen in animals when their behavior changes to inactivity, poor appetite, and possibly a change in disposition from docility to aggression. Animals in pain, for example horses with colic, animals with fractured bones, slipped discs, or arthritis, cats with an aortic thrombus, any animal with unrelieved post-surgical pain, all lead to various changes in behavior and physiology. Animals can also suffer with nonpainful diseases such as cancer, diabetes, or epilepsy.

" . . . and Which Is Determined by the Cognitive Capacity of the Species and the Individual Being"

The development of the central nervous system is manifestly different between species; consider the development of key areas such as the cerebral cortex. The evolutionary older part of the brain dealing with emotions, the limbic system, is present in all sentient species. Perhaps it is the interaction between the cerebral cortex and other areas of the brain, for example, the hippocampus, which determines the level of cognitive ability and hence the ability of an animal species to suffer. However, it is also apparent that individuals within a species will have had different life experiences, and this too will have an effect. At one end of the spectrum, a human being in a permanent vegetative state will be unable to suffer, as their cerebral cortex has been irreversibly damaged; for others, such as anencephalics, it may not have developed, or development has been restricted or retarded, for example, mentally impaired through hypoxia at birth. At the other end might be a highly sensitive and imaginative person who will suffer more mental anxiety than most. Animal species represent a range of development of the central nervous system, but those that are self-aware, that is, having the ability to become the object of their own attention, and self-conscious, that is, having the ability to be aware of one's own existence especially in relation to others and over time, may suffer more through an anticipation of the future based on a mix of past experience, natural instincts, and intuition. Thus a captive animal that has known what it is like to live in the wild may have internalized that experience and so suffer more when kept confined, compared with an animal that has not had that experience. The basic instincts to carry out certain behaviors are still there, but now there is the extra dimension of physiological integrity, prior experience, and memory.

Animal Well-being

So how does this affect human assessment of animal suffering, that is, on concern for animal welfare or an individual animal's well-being? The following description can be used to help decide what might be good and poor welfare: "Welfare is dependent on and determined by an animal's physiological and psychological wellbeing in relation to its cognitive capacity and life's experience."

At one level, an animal's well-being is reflected in its subconscious attempts to cope with an aversive environment, that is, the homeostatic reflex, through activation of the autonomic nervous system, the hypothalamic-pituitary axis, and the adrenal glands. But this is not the stuff of suffering that is being described here; suffering is more an animal's conscious attempts to deal with its specific predicament. When an animal feels threatened in some way, it usually tries to take avoidance action. This self-preservation is universal in all vertebrates as far as we can tell, and has been conserved through evolution. Many laws and research guidance notes state something to the effect of "It should be assumed that persistent pain or distress in animals leads to suffering of animals in the absence of evidence to the contrary" (OECD, 2001). Many believe that the same should be considered for all areas where humans use or exploit animals for their own ends.

Further Reading

OECD (2001) Environmental Health and Safety Publications Series on Testing and Assessment No. 19 Guidance Document on the Recognition, Assessment, and Use of Clinical Signs as Humane Endpoints for Experimental Animals Used in Safety Evaluation Environment Directorate. http://www.oecd.org/ehs/or contact: OECD Environment Directorate, Environmental Health and Safety Division 2 rue André-Pascal, 75775 Paris Cedex 16 France E-mail: ehs cont@oecd.org.

Stamp-Dawkins, M. (1992). *Animal suffering: The science of animal welfare*, 2nd ed. London: Chapman & Hall.

Stamp-Dawkins, M. (1993). *Through our eyes only. The search for animal consciousness.* Oxford: W. H. Freeman Spektrum.

David B. Morton

PAINISM

Painism is a term coined by Richard Ryder in 1990 to describe the theory that moral value is based upon the individual's experience of pain (defined broadly to cover all types of suffering whether cognitive, emotional, or sensory), that pain is the only evil, and that the main moral objective is to reduce the pain of others, particularly that of the most affected victim, the maximum sufferer. Painism opposes the prejudice of speciesism.

The unit that experiences pain is the individual organism, the whale, the human, or the mouse. It is not the gene nor the herd nor the species itself that feels pain. This is an important point for animal protection and is crucial for ethics generally. Yet it is routinely overlooked.

Pain is the main subject of animal protection and, ultimately, of all ethics. Whatever is deemed to be bad—injustice, loss of liberty or infringement of other's rights—is bad because it causes pain to individual organisms. Pain and suffering are the great impediment to happiness, and they underlie all rational concepts of evil.

The two main systems of modern secular ethics, utilitarianism and rights theory, have both helped to improve the treatment and status of nonhuman animals over many years, but both systems have major faults: rights theory principally because of its difficulty in resolving inevitable conflicts of rights, and utilitarianism because it allows the infliction of agony on one or a few individuals if that action causes mild pleasure to such a large number of others that the total of their pleasures outweighs the pain of the victim(s). Thus a gang rape may be considered a good thing by utilitarians if all the pleasures of the rapists add up to more than the severe suffering of the victim. The same sort of argument can be used to excuse severely painful vivisection or torture.

Painism aims to avoid such pitfalls. It does so by denying the validity of adding up the pain or pleasure of separate individuals. Ryder says that pain or pleasure has to be experienced to truly be pain or pleasure, but that no one individual actually experiences these added-up totals. They are theoretical and not real. One does not add up the feelings of surprise of separate people and say the total of surprises is meaningful, so why claim that the totaled feelings of pain or pleasure felt by several separate individuals has meaning? Painism measures the badness of a situation neither by the total of pain, nor by the number of sufferers, but by the quantity of suffering experienced by the most painfully affected sufferer.

It is important to emphasize that painism still permits the tradeoff of pain and pleasure between individuals, so it may be permissible to force mild discomfort, let's say taxation, on A, if this action releases B from agony, for example, through free medical treatment. But it is not permissible, according to painism, to trade off the pain of A against the added up pleasures of A plus B plus C, etc. By rating rights according to their efficacy as pain-reducers, painism can also help to deal with conflicts of rights, giving preference to whichever right reduces the most pain or produces the most pleasure.

Ryder, as the inventor of the concept of speciesism (1970), is concerned with pain regardless of whether it is felt by animals, humans, or others. So X amount of pain in a pig matters as much as X amount of pain in Socrates! Ryder has described the effects of painism in animal protection, politics, and generally, claiming that it may resolve the main problems of other ethical theories and even of democracy itself. He sees democracy as being based upon a cobbled-together mixture of utilitarianism (the dictatorship by the majority) and rights theory, both imperfect theories, and he proposes painism as a more consistent approach. Painism is counterintuitive in challenging the usual everyday assumption that large numbers of sufferers matter more, morally speaking, than lesser numbers of sufferers. We are accustomed to thinking that the maiming and murder of ten people is morally worse than the maiming and murder of one. Painism questions this. For painism, the agony of one counts for more than the mere discomfort of many. The question is not how many were harmed by an action, but how much was suffered by the maximum sufferer. So painism focuses on the intensity of suffering of each individual and not on the number of sufferers.

See also Speciesism

Further Reading

1999. Painism: Some moral rules for the civilised experimenter. *Cambridge Quarterly of Healthcare Ethics* Vol 8: 1.

Chadwick, Ruth, ed.. *The encyclopedia of applied ethics.* New York: Elsevier.

Painism versus Utilitarianism. 2009. In *Think,* 21, Vol 8. Cambridge: Cambridge University Press.

Ryder, Richard D. 1989 and 2000. *Animal Revolution: Changing Attitudes Towards Speciesism* revised ed. Oxford, UK: Basil Blackwell Ltd., 1989, and Basingstoke, UK: Berg, 2000.

Ryder, Richard D. 1998. *The political animal: The conquest of speciesism.* Jefferson, NC: McFarland.

Ryder, Richard D. 2001. *Painism: A modern morality.* London, UK: Opengate Press.

Ryder, Richard D. 2006. *Putting morality back into politics.* Exeter, UK: Imprint Academic.

Richard D. Ryder

PEOPLE FOR THE ETHICAL TREATMENT OF ANIMALS (PETA)

People for the Ethical Treatment of Animals (PETA) is an international non-profit charitable organization based in Norfolk, Virginia, with affiliates in the United Kingdom, Germany, the Netherlands, India, and the Asia-Pacific region. Dedicated to establishing and defending the rights of all animals, PETA operates under the principle that "animals are not ours to eat, wear, experiment on, or use for entertainment" ("PETA Guide to Animals and the Dissection Industry," 2008).

Origins

PETA President Ingrid E. Newkirk previously served as a deputy sheriff in Montgomery County, Maryland; as a Maryland state law enforcement officer with a high success rate in convicting animal abusers; director of cruelty investigations for the second oldest humane society in the United States; and as Chief of Animal Disease Control for the Commission on Public Health in the District of Columbia.

While Newkirk was working at a Washington, D.C. animal shelter, she read the book *Animal Liberation,* written by Australian philosopher Peter Singer. Inspired by the concepts set forth in Dr. Singer's book, Newkirk founded PETA in 1980 with a small group of friends. PETA has grown into the largest and, many consider, the most influential animal rights organization in the world, with more than two million members and supporters (PETA, www.peta.org/about/).

Investigations

In 1981, PETA embarked on its first undercover operation, when an investigator took a job in a laboratory in Silver Spring, Maryland, where a group of monkeys was kept, all but one of them having been captured as infants from their native habitat in the Philippines. The nerves in the monkeys' spines had been cut, affecting their ability to control their arms. The animals' limbs were also injured and fingers torn off from getting caught in the rusted and broken cage wires. The researcher had converted a small refrigerator into a shock box inside which the monkeys were punished if they failed to pick up objects with their damaged limbs.

The investigation found grossly unsanitary conditions, with cages cleaned so rarely that fecal matter rose to a height of some inches in places and fungus was

growing on it. The monkeys had not been given food bowls, so food thrown into the cage fell through the wire, requiring the monkeys to pick the food out of the waste collection trays in order to eat. The evidence provided by PETA's investigation resulted in the first search and seizure warrant served on a laboratory for cruelty to animals, the first arrest and criminal conviction of an animal experimenter in the United States on charges of cruelty to animals, the first confiscation of animals from a laboratory, the first cancellation of a government animal research grant, and the first U.S. Supreme Court victory for animals in laboratories (PETA, stopanimaltests.com/investigations/asp).

Subsequent PETA investigations have led to further protections for animals, including these examples:

- An undercover investigation ended scabies experiments on dogs and rabbits at Ohio's Wright State University and led to charges by the USDA of 18 violations of the Animal Welfare Act
- PETA released photographs and videotapes showing ducks being force-fed on a foie gras farm in New York, resulting in the first-ever police raid on a U.S. factory farm, as well as ending the sale of foie gras at many restaurants
- Investigations at pig-breeding factory farms in North Carolina and Oklahoma revealed substandard conditions and regular abuse of pigs, including one pig who was skinned alive, leading to the first-ever felony indictments of farm workers
- A California furrier was charged with cruelty to animals after a PETA

investigator filmed him electrocuting chinchillas by clipping wires to the animals' genitals

- PETA's undercover investigation of a Florida exotic-animal training school revealed that big cats were being beaten with ax handles, which resulted in the USDA's developing new regulations governing animal training methods
- PETA campaigned successfully to have car companies replace all use of animals in crash tests with mannequins

Campaigns

PETA's investigation of a contract testing laboratory in Philadelphia to many leading companies, such as Avon, Revlon, and Estée Lauder, permanently banning animal tests (PETA, www.stopanimaltests.com). PETA now lists hundreds of personal and household-care companies that do not test products on animals.

PETA's work to promote vegan diets and reduce the living and dying conditions of animals on industrialized farms is the largest of the group's campaigns. According to its Web site, PETA distributes hundreds of thousands of free vegetarian starter kits each year (PETA, http://www.peta.org/about/). As a result of PETA's campaign efforts, industry-leading companies, including Burger King, Safeway, and McDonald's have reduced the suffering of animals used and killed by their suppliers (PETA, http://www.goveg.com/corpcampaigns.asp).

Through its "Fur Is Dead" campaign, PETA has brought attention to the ways in which animals are trapped, raised, and killed in the fur industry and con-

vinced retailers, including J. Crew, Wet Seal, Forever 21, and Ann Taylor to stop selling fur in their stores. Top designers such as Ralph Lauren, Marc Bouwer, and Stella McCartney have banned the use of fur in their designs. PETA also convinced dozens of companies, including Adidas-Salomon, Gap Inc., Eddie Bauer, Nike, and Reebok, to refuse to use Indian and Chinese leather in their products, after their investigation of the overseas leather industry revealed severe abuses during the transport of cattle to slaughter (PETA, http://www.peta.org/mc/factsheet_display.asp?ID=107).

A PETA campaign against the circus industry led some corporations, including General Mills, Liz Claiborne, MasterCard, Ford Motor Company, and Sears, Roebuck and Co., to stop sponsoring the Ringling Bros. and Universoul circuses (PETA, http://blog.peta.org/archives/2008/01/dennys_victory.php).

CDW, Puma, Honda, and Subaru are among the companies that have withdrawn ad campaigns using great apes as a result of PETA's "No More Monkey Business" campaign (PETA, http://www.nomoremonkeybusiness.com/campaignUpdates.asp). Other efforts on behalf of animals in entertainment include campaigns against zoos, and the use of animals for rodeos and blood sports such as bullfighting, dog fighting, and cockfighting.

Although PETA frequently works directly with companies and governmental bodies, the organization's campaigns primarily focus on making individuals aware of issues affecting animals and encouraging them to take action. PETA's International Grassroots Campaigns department works with local activists to organize demonstrations and to send letters urging companies and individuals to make changes for animals.

Companion Animals

PETA works to address issues affecting dogs, cats, horses, birds, and other companion animals. PETA's cruelty caseworkers investigate cruelty to animals, and alert district attorneys to the link between cruelty to animals and violent acts against humans, urging them to prosecute abuse cases.

Since its inception in 2001, PETA's mobile clinic, SNIP (Spay and Neuter Immediately, Please), has sterilized tens of thousands of dogs and cats at a reduced cost, preventing the births of unwanted animals. PETA also builds and delivers free doghouses and gives away bales of straw in order to provide better shelter to dogs forced to live, often chained, outdoors.

Tactics

PETA has been a pioneer in using provocative tactics to attract attention to its messages. Their "I'd rather go naked than wear fur" ads, featuring eye-catching images of largely unclothed models, were an early success for the organization. PETA ads and demonstrations also often feature colorful costumes or the involvement of sympathetic celebrities. PETA maintains that attention-getting stunts are necessary to attract the notice of the media in order to reach the public, even if they alienate some people.

PETA's media-friendly tactics have also led to a growing involvement by younger animal advocates. A 2006 youth marketing survey by Label Networks found that PETA was "the #1 overall non-profit organization that 13-24-year-olds in North America would volunteer for" by a nearly two-to-one margin over the second-place finisher, and that, "The younger the demographic, the higher the percentages who

Members of the animal rights activist group, People for the Ethical Treatment of Animals (PETA), hold up signs in front of pedestrians in protest against the Ringling Bros. and Barnum & Bailey Circus at the Los Angeles Sports Arena in Los Angeles on Wednesday, July 21, 2004. PETA is part of the growing animal rights movement that is concerned with the safety of animals around the world. (AP/Wide World Photos)

would volunteer for PETA, peaking among 13–14-year-olds at 29.1 percent of this age group" (Generation Vegan, 2006).

Further Reading

Generation Vegan. 2006. PETA—Kids' Favorite Nonprofit Organization. http://www.generationv.org/index.php?m=200607.

Hawthorne, Mark. 2008. *Striking at the roots: A practical guide to animal activism.* Berkeley, CA: O Books.

Mathews, Dan. 2007. *Committed: A rabble-rouser's memoir.* New York: Atria.

Newkirk, Ingrid. 2005. *Making kind choices.* New York: St. Martin's Griffin.

Newkirk, Ingrid. 2008. *One can make a difference* Cincinnati, OH: Adams Media.

PETA. 2001. *The PETA guide to animals and the dissection industry.* www.petakids.com/pdf/lanimaldisindust.pdf.

PETA. http://blog.peta.org/archives/2008/01/dennys_victory.php; http://www.peta.org/about/;

http://www.peta.org/mc/factsheet_display.
asp?ID=107; http://www.nomoremonkeybusi
ness.com/campaignUpdates.asp; http://www.
stopanimaltests.com/investigations.asp;
http://www.goveg.com/corpcampaigns.asp.

Singer, Peter. 1990. *Animal liberation.* New
York: Random House.

Ingrid Newkirk

PET RENTING

In 2007, pet renting companies received considerable publicity. Arguing that they met a societal need by providing dogs for people who could not commit to owning one in the long term, they advertised pools of dogs online. Some offered a membership scheme that allowed members to book dogs in advance and have them for any period.

Often claiming that their dogs are checked regularly by a veterinarian for physical health and mental well-being, although it is not stated how the latter can be determined by a veterinarian conducting a routine examination, some rental companies assert that they screen members to ensure that the dogs will be properly looked after. Furthermore, they argue that because their members must undergo a brief training session and meet the dogs in the presence of a certified dog trainer, the dog's welfare while under hire is somehow assured.

It is unclear whether these businesses are commercially viable, but their emergence raises some important questions about the nature of guardianship of animals. The hiring of animals is not a new concept. For example, for centuries, horses have been hired for riding, with attendant concerns about the quality of equitation and the unwelcome cumulative effects of relentless and poorly timed pressure from the legs and hands of in-competent riders. More recently, horses have been hired for draft work, for example, for weddings and horse-drawn caravan holidays, with and without hired equestrian expertise, respectively. In the UK, retired racehorses make guest appearances at supermarket openings. Exotic, and sometimes potentially dangerous, animals can be hired for staged appearances to glamorize an event, though these animals are not generally left unattended with those hiring. In contrast, the pet renting business leaves the dog with its hiring humans for as long as they are willing to pay for it. Responsible animal shelters screen potential owners and may subsequently inspect the premises in which potential adoptees are to be housed. The same level of scrutiny is not afforded to hired dogs, and there may be financial disincentives for dog rental companies to find anything amiss. Unless inspections of the dogs' temporary destinations are undertaken and the credentials of the renters are fully established, the owners of rented dogs appear to be failing in their duty of care.

What is particularly novel about this venture is that the providers argue that they are leasing dogs not as accoutrements but as companions. This implies that humans and dogs can develop a transient bond in a very brief period, and that the dissolution of the bond at the end of the period of hire has no deleterious effects on the dogs. The formation of transient bonds seems possible because dogs are opportunists and, given the right set of enticements and rewards, some will accompany unfamiliar humans and readily forsake their familiar territory and guardians. The notion of bonds being broken without any costs, on the other hand, seems less plausible. Admittedly, the next human in a hired dog's life could come loaded with even better enticements and therefore win the dog over. However, the chances are that the dog will establish a valued routine and profound bonds with its relatively permanent caretakers between leasings. It would be interesting to see whether dogs that are serially hired become sensitized or habituated to this social flux. Physiological evidence from shelter dogs indicates that habituation to novel environments takes at least four days. So, unless this period is taken as the minimum, dogs are unlikely to adjust fully to each context.

The motivation of people who rent dogs is worth consideration. This maybe reflected in the breeds in demand for this purpose. Among several engaging breeds, some pet-renting sites offer Afghan hounds for hire. This breed is often described as relatively high maintenance and difficult to train; the Afghan hound breed standard describes them as having an aloof temperament. In other words, they are not the most personable of breeds. Afghan hounds are generally released from shelters to new homes only after exhaustive questioning of the adopting family's knowledge of and expertise in dog care. These observations suggest that, when acquired on a temporary basis, Afghans have an appeal chiefly as status symbols rather than as companions. A dog's role as a status symbol can be more demanding than that of a companion, since it may require parading in ethologically challenging contexts. While a companion dog may accompany its erstwhile caregiver on a recreational trip to a dog park, a status dog may be taken shopping.

Dogs have a behavioral need for stable social groupings. Reflecting this undeniable reality, leasing operators often claim that their dogs live with them when not on visits. Although this approach seems humane, it is not necessarily as worthy as it appears, because the dogs are repeatedly withdrawn from the security of their owner's home. One could argue that the dogs somehow know that they will return to their base. This knowledge can only come with repeated experience of separation and reunion. In the absence of any data to show that dogs can reliably predict the future return of their preferred companions, we should err on the side of caution and assume that repeated disruptions in a dog's social network are likely to compromise its welfare.

Dog hiring companies claim that they source their dogs from pounds, and so assert that they have saved numerous canine lives. Yet many veterinary behaviorists believe that dogs that have spent time in a shelter need a stable rather than a transitory base. Noting that a history of having been acquired from a shelter is a risk factor for separation-related distress, they propose detailed protocols to reduce the recently adopted dog's chances of becoming distressed when left alone. Repeated fragmentation of established bonds with humans may compromise a dog's ability to cope during periods of separation from its primary attachment figure. This means that rescued dogs are likely to be among the least suitable candidates for the social flux typified by pet renting.

Unsurprisingly, dog hiring companies also offer their dogs for sale. This seems to acknowledge that there is a ready supply of replacement dogs, that little training is required to prepare them for their home-hopping existence, and that some humans will reliably feel pity for dogs with no permanent home. People contemplating this offer should consider bypassing the operators and visiting a shelter directly.

There it will become clear that many other dogs can be walked on a voluntary basis, as a form of environmental enrichment, and that foster homes are always needed. Thus dogs' needs can be met without the need to pay for the privilege.

Paul McGreevy

PETS

See Companion Animals

PIGS

Domestic pigs are canny and sensitive animals, with strong urges to forage, explore, and interact socially. These characteristics were inherited from their ancestor, the Euro-Asian wild boar (*Sus scrofa* L.). Historically, pigs were either herded in woods, housed in pens, or roamed scavenging around human dwellings. In Euro-American civilization, they were often regarded with some scorn, which was sometimes connected with rough treatment. Their of way life has been altered during the last 60 years by intensive husbandry and selective breeding. Through selection for large litter size, fast growth, and high-yielding carcass characteristics, pigs became heavier and more muscular, whereas the relative weight of bones and heart decreased. Pigs are prone to overheating and even heart failure in stressful situations, as well as to leg problems, especially if they have little exercise and/or when they are housed on slippery or rough slatted floors. Breeding for fast growth also boosts pigs' appetites. While growing pigs and lactating sows can be fed to satiation, gestating sows cannot, since they will get fat. Hence, they must be kept in a permanent, even if only "subjective," state of hunger.

Nevertheless, domestication also brought about changes such that pigs do not miss the challenges of the wild life. Although no behavioral pattern is known to have disappeared from the pig repertoire, quantitative changes have occurred during domestication that make domestic pigs inherently less active, less excitable, and less aggressive than their wild ancestors. Therefore, if the keeper provides quality food with adequate doses of fiber, an environment structured and spacious enough to meet exploratory, foraging, resting, and thermoregulatory needs, and arranges for stable social company, then pigs can live a contented, and perhaps even happy life in human care. Most current systems are far from this ideal, because mass consumption of and therefore massive demand for cheap pork, combined with tough low-price marketing competition among big retailers, pushes farmers into very slim economical margins with little space for any improvements that will make their meat production even slightly more expensive. Nevertheless, partial legislation-based changes in the EU and market-based shifts in North America have improved the welfare of pigs slightly in these two world regions in the last ten years. Much depends on whether these regions will be able to maintain the pace or even get other regions, like Latin America, on board, under the pressure of imports from countries with little or no welfare legislation or self-regulation.

Most pigs today are housed in barren environments which conflict with their behavioral make-up. The most pressing problems are:

Absence of bedding and inadequate flooring: Straw, which in older housing systems provided dry floor comfort, an outlet for exploratory and foraging activities, and a source of dietary fiber, has disappeared from most piggeries. Lack of bulky or high-fiber food for restrictedly fed sows is not only associated with frustration, but also increases the incidence of painful stomach ulcers. However, starting in 2013, sows and gilts in EU will have access to manipulable material, and intensive research is going on to establish which properties of the material are most important for pigs. Slatted or partly slatted floors that are in common use for all categories of pigs bring increased levels of claw, leg, and limb injuries as well as shoulder lesions in lactating sows, and carpal skin lesions in suckling piglets. Full-time housing on deep straw litter, on the other hand, leads to overgrown claws and the risk of lameness, unless the claws are properly trimmed. Thus, optimal flooring combines soft dry bed for resting and hard solid surface in activity areas for abrasiveness.

Restriction of movement: The majority of pregnant sows in North America and many in Europe are confined in small crates. This, combined with hunger and the absence of bedding, leads to continual chewing on bars or other repetitive stereotypic behaviors, and causes constant stress, as revealed by elevated levels of corticosteroid levels. Oral stereotypies could be reduced by a high-fiber diet. There has been positive development both in EU, where gestation stalls will be phased out by 2013, and in the United States, where some states have banned crates, and big pork suppliers, under the demand of large fast food chains, are shifting towards group housing of pregnant sows. In small piglets, spatial limitation and lack of contact with other litters suppresses social play, which may hamper normal development of their social skills.

Thermoregulation: For adult pigs, temperatures above 25°C (77°) pose a

A pig standing in his pen on an Iowa farm. (AP Photo/Charlie Neibergall)

challenge, as pigs cannot sweat. In nature, they cool themselves by rolling in mud or wallowing. For pigs kept indoors, maintaining the temperature below 25°C is important, while pigs kept outdoors should be provided with shadow and wallowing opportunities.

Body cleanliness: If space allows, pigs defecate and urinate in one location and never lie in a fouled place. They are forced to do so, however, when kept in groups of high spatial density, or confined in crates, or if they are exposed to high temperatures and cannot use other wallowing material.

Social behavior: When unfamiliar pigs meet, they perceive each other as intruders, and intense fighting invariably begins. Numerous, although superficial, injuries are inflicted by biting. As confined spaces prevent the losing individuals from fleeing, attacks may last for several hours or even days, with the losers becoming extremely distressed. The composition of the pigs in a group should be changed as little as possible. Pigs have a strong need to eat synchronously, and if access to food is disturbed, low-ranking pigs can suffer from bullying by pen mates.

Farrowing and nursing: Hormonal changes preceding parturition prompt the sow to seek a half-hidden place and build a nest. Most parturient and lactating sows are housed in unbedded farrowing crates. This prevents locomotion and nest-building and results in agitation, futile nest-building movements, and elevated levels of the stress hormone cortisol. Individual bedded pens give more freedom, but may result in a higher number of piglets crushed by the mothers (see below). Some farmers use straw-bedded indoor group housing systems where lactating sows can freely enter small individual farrowing pens, but this system demands highly skilled management. Outdoor huts, which are a good option under some climatic and soil conditions, provide better welfare for farrowing and suckling sows, given that protection from extreme weather and disease is guaranteed.

Piglet mortality. About 15 percent of live-born piglets die within the first three days of life, even on well managed farms, due to low birth weight, insufficient milk intake, and crushing by the sow. This is a direct consequence of very large litter size (median around 12, maximum up to 16 piglets). Selection for further increase in litter size is therefore clearly undesirable from the pig welfare point of view.

Surgery on small piglets: The majority of piglets are subjected to tooth-trimming and tail-trimming, and the males are castrated. No anesthesia is given. Tooth-trimming is performed to prevent damage to sows' teats and to littermates, and tail-trimming is performed to prevent mutual tail-biting. As these are husbandry-related risks, the objective should be to treat the causes and thus avoid the need for these practices.

Weaning: While the natural age of weaning is four months, piglets on factory farms are most often weaned at 3–5 weeks. The method of weaning at 8–16 days, based on strict hygiene and mandatory antibiotics in food, is becoming more common. However, weaning before three weeks of age causes intense distress reactions and disturbed behavior among the piglets, such as suckling-related belly-nosing and nibbling of age mates.

Human-swine interactions: Rough treatment, such as hitting, kicking, and using pain-inflicting devices, makes pigs fearful of humans. They are then difficult to handle, get easily excited, and produce less well in terms of growth and reproduction. Working with such animals is an unsatisfying job, produces negative attitude towards them among the personnel, and a vicious circle occurs. Improvement both in welfare and in performance can be reliably achieved by educating the personnel about the principles of pig behavior and positive ways to handle them.

Transport: Transportation is stressful to pigs. The strain may be severe or even fatal if pigs also experience exposure to extreme temperatures, long durations without water, food, and rest, mixing with alien pigs, overcrowding, and slippery floors. Regulation concerning animal transport are being gradually imposed, but unacceptable practices are still common, often in defiance of existing regulations.

Slaughter: Most industrialized countries require instantaneous stunning of pigs before slaughtering. It is the pre-slaughter handling and housing of pigs rather than the slaughter itself that causes considerable suffering because of its large scale, total anonymity, and the tendency among the personnel to depreciate the suffering. Education and setting firm standards for procedures and equipment can eliminate unnecessary suffering at slaughterhouses.

Further Reading

Appleby, M. C., and Hughes, B. O. 1997. *Animal welfare*. Wallingford: CAB International.

European Food Safety Authority. 2007. Animal health and welfare aspects of different housing and husbandry systems for adult breeding boars, pregnant, farrowing sows and unweaned piglets[1]—Scientific Opinion of the Panel on Animal Health and Welfare. *The EFSA Journal* 572, 1–13.

Faucitano, L., and Schaeffer, A. L. 2008. *Welfare of pigs. From birth to slaughter.* Wageningen: Wageningen Academic Publishers.

Marchant-Forde, J. M. 2009. *The welfare of pigs.* New York: Springer.

Velarde, A., and Geers, R. 2007. *On farm monitoring of pig welfare*. Wageningen: Wageningen Academic Publishers.

Writing of this contribution was supported by Grant #MZE0002701402.

Marek Špinka

PLEASURE AND ANIMAL WELFARE

The desire to feel good dictates much of what we do in our lives. The foods we eat, our choice of companions, our career choices, and our hobbies all can bring feelings ranging from satisfaction to joy. While it may be more important to avoid pain and suffering where we can, it is the pursuit of pleasures that fills more of our waking time.

Despite its central place in our lives and, as I shall argue, in the lives of other animals, the study of pleasure has suffered neglect. Today, the titles of at least nineteen scholarly English-language journals contain the word pain, yet there are no journals dedicated to the study of pleasure.

The neglect of pleasure has been even more profound in the study of animals. One reason for this is that pleasures are so-called private experiences, and therefore difficult to demonstrate clearly, especially in another being who doesn't use our sort of language. Also, for much of the last century it was considered bad science to even suggest that nonhuman animals were conscious or had feelings. Fortunately, in recent decades, the momentum has shifted, and time is now ripe for the pursuit of pleasure to be understood as an important element of animals' day-to-day lives, as it is for ours.

The Basis for Animal Pleasure

Pleasure is a product of evolution. There are good reasons that sentience, the awareness of pain and pleasure, evolved. Because animals, unlike plants, are able to move freely, the capacity to feel allows them to be pleasure-seekers and pain-avoiders. The individual is rewarded for performing behaviors that promote survival and procreation. Pain's unpleasantness teaches the animal to avoid bad behaviors that risk the disaster of death. Similarly, pleasure encourages animals to behave in good ways, such as choosing high quality foods, seeking good mates, and finding a comfortable shelter.

In appreciating the pleasure that other animals feel, we also have the advantage of relating their experience to our own. Because we know the enhanced taste of food when we're hungry, the scent of a flower, the thrilling touch of an intimate partner, or the experience of slipping into a warm bed on a chilly night, perhaps with a cat or dog snuggled against us, it is easier to recognize that other animals can have comparable sensations. Human languages contain rich vocabularies for good feelings, which attest to the diversity of both the physical and emotional pleasures we can feel. It follows that some animals, having evolved to dwell in diverse environments, from flying to burrowing to living submerged in the oceans, might also be able to experience realms of pleasure unfamiliar to humans. For example, the echolocation abilities of bats and whales, electric communication in fish, birds' tuning into the earth's geomagnetic field to help navigate, may not explicitly involve pleasure, but they illustrate the potential for sensory pleasures unknown to us. Several leading scientists have recently suggested that other animals may experience some feelings more intensely than we do.

Examples of Animal Pleasure

Animal pleasure can be studied through both observation and carefully designed experiments. To illustrate, let's look at some examples relating to the realms of play, food, sex, and touch.

Play Play behavior is widespread in mammals, and has also been so far described in about half of all families of birds. Behavior suggestive of play has also been observed in reptiles, fish, and at least one invertebrate, the octopus. Because play tends to occur spontaneously and unpredictably, it is difficult to measure, and most published studies of animal play are anecdotal. But more systematic studies are possible. For example, a three-year study of aerial drop-catching behavior by herring gulls in Virginia concluded that it was play. These birds will drop clams onto hard surfaces to smash them and access the soft parts; but they will also swoop to catch clams and other objects they have dropped before they hit the ground. The latter behavior appears to be playful, because drop-catches were performed more by younger birds, drop-catches were not necessarily made over a hard substrate, sometimes non-food objects were dropped/caught, and it occurred more often during warm, windy weather.

There are good adaptive reasons for the existence of play. Playing games of chase is no doubt beneficial for young animals who need to be prepared to flee from a lurking predator, as it is for young predators who will need to catch their food. Yet, animals including humans do not consciously play for ultimate reasons; they play because it is fun to do so. Enhanced survival can be seen as a positive reward in the evolutionary sense, but not in the sense of experiencing a pleasurable sensation. Several species are known to calibrate the boisterousness of their play, apparently to sustain the activity, which also suggests that they are enjoying it.

Food There are innumerable clues that animals favor the flavor in their food. Individual food preferences are well documented in both domesticated and wild animals, as are the anticipation of food and individual tastes that change over time. Language-trained apes and parrots can actually tell us their enthusiastic reactions to food. It has also been shown that animals produce pleasurable compounds known as opioids during both the search for food (the pleasure of anticipation) and its consumption.

Facial responses to tastes are similar in rodents, and humans and other primates, suggesting shared evolutionary origins. Enjoyably sweet flavors elicit characteristic licking responses, while bitter tastes cause gaping and head shaking. These responses are accompanied by activity in shared hedonic hotspots of the brain. This linking of brain activity with positive patterns of behavior points to the conscious experience of pleasure.

A study of juvenile green iguanas showed that these animals would trade-off the palatability of a bait (lettuce) with the disadvantage of having to venture into a very cold area to retrieve it. As the temperature near the bait was lowered, the lizards visited the bait less often and for shorter periods, choosing instead to stay under the heat-lamp where nutritionally complete reptile chow was freely available. Moreover, time interval between sessions with the lettuce bait, ranging from one to eight days, had no effect on the duration of stay on the bait, suggesting that the lettuce was more of a luxury rather than an indispensable nutritional food source. Rats respond conversely, shunning convenient but dull laboratory chow and running into a cold environment to consume highly palatable foods.

Sex It is hard to overestimate the importance of reproduction to an organism. Without it, species would cease to exist. Because reproduction is so important, natural selection should strongly favor behaviors that promote mate-seeking, mating and, where necessary, the raising of young. Unfortunately, sexual activity in animals is usually portrayed as all business and no pleasure. From journal articles to textbooks to television documentaries, the idea that animals may be enjoying themselves is not explicitly rejected; it is just generally avoided.

One piece of evidence for the sensual aspect of animal sex is that a good deal of animal sexual behavior is not procreative. Many animals routinely copulate or engage in other sexual activities outside of the breeding season, including during pregnancy, menstruation, and egg incubation. Such non-procreative activity constitutes a large proportion of the sexual behavior expressed by such animals as common murres, proboscis monkeys, addax antelopes, rhesus macaques, wildebeest, golden lion tamarins, and mountain goats, to name a few. Another variation on the theme of wasteful sex is group sexual activity wherein few if any participants are passing along genes. Spinner dolphins, gray and bowhead whales, swallows, and herons are known for their orgies. Other common examples in normal wild animals include various forms of non-copulatory mounting; stimulation with hands, paws, flippers or mouth; same-sex mating, interspecies sexual couplings, and self-stimulation.

Touch Touch sensitivity, while indispensable neither to survival nor to reproductive success, is very useful.

It allows animals to react adaptively to their environments. Being able to detect water movements helps fish orient themselves in murky streams or on migration routes, and to detect the movements of other nearby animals, including potential predators. But for group-living species especially, the pleasure of touch acts as a social lubricant, strengthening friendships and defusing tensions. For chimpanzees, macaques, and other primates, grooming occupies up to a fifth of their waking time. The release of pain-relieving endorphins has been shown in grooming primates.

Few investigators have addressed the pleasure of touch. In some cases, animals' liking of tactile contact may reveal itself by accident. For example, in a study in which dolphins could request rewards by pressing plastic symbols on a keyboard with the tips of their beaks, some animals favored getting a rub to getting a fish. When human researchers experimentally groomed Camargue horses, the animals' heart rates slowed significantly more, an indication of pleasurable relaxation, when the touch was directed at areas of the neck that horses prefer when grooming each other.

Young rats show a mirthful response to the touch of a trusted human. Trained to expect a friendly tickle when the human hand is introduced to their space, these rats pursue the hand. No food reward is provided; touch is the reward. Tickled rats run to the hand about four times as quickly as do control rats trained to expect a gentle stroke on the neck. Tickled rats also make about seven times more high-pitched chirps during play and other presumably fun activities. Brain imaging reveals that a tickled rat shows similar brain activity patterns to those of a human who is enjoying a good laugh.

The Well-being of a Pleasure-Seeker

Animal pleasure has weighty moral implications. Being a pleasure-seeker adds considerably more to one's interests than if one were merely a pain-avoider. Being able to feel good means being able to enjoy life. There is more at stake, more to be gained, and lost.

Philosophers for centuries have recognized the significance of pleasure to ethics. Utilitarianism, originating in the 18th century, favors actions that optimize pleasurable outcomes while minimizing negative ones. Its founder, Jeremy Bentham, regarded animals as serious objects of moral concern, based on their capacity for both pain and pleasure. Peter Singer argues that sentient animals have interests, and that those interests involve not just avoiding physical pain and/or psychological suffering but also the experience of pleasure. Tom Regan emphasizes the intrinsic value of sentient organisms. An individual who can experience good feelings has a life that is of value to that individual, independent of any value it could have to another, such as a source of entertainment or revenue. American veterinarian Franklin McMillan adds that such an individual has a quality of life.

Regarding the human-animal relationship, it is the denial of pleasure, not its bestowal, that has moral weight. One has no obligation to provide pleasures to another, be they another animal or a fellow human. Bringing flowers to a friend is an act of kindness, but it is not an injustice if I decide to keep them for myself. If, however, my friend has flowers and I take them away, then I am violating my friend's interests, albeit rather trivial ones in this example. The pleasures we deny animals are more serious. When we keep animals in factory farms, laboratory cages, fur farms, and other settings, we not only inflict pain and suffering, we deny them the opportunity to express natural behaviors. Animals confined for generations in laboratories and in factory farms retain high levels of motivation to engage in activities natural to their species; thwarting them leads to frustration, physical stunting, and psychological illness.

A more profound way in which we may deny animals pleasure is in killing them. An untimely death denies the victim the opportunity to experience the rewards that life would otherwise offer them. It may be claimed that a dead animal misses nothing. But the main reason that our criminal system treats murder so seriously is not that the victim may suffer, though that certainly compounds the crime. Murder is wrong because life, specifically that portion of life yet to be experienced, has value. Thus, killing is the greatest harm that can be done to conscious, autonomous beings, and pleasure is firmly rooted in the harm committed.

By and large, the harms humans cause animals are not necessary. Most animals killed by humans are killed to be eaten, and with rare exceptions humans can choose plant-based diets, including highly palatable faux meats. Similarly, we use animals in laboratory experiments and tests because we can, not because we must. The same goes for other consumptive uses of animals by man: hunting and fishing, blood sports, fur and leather fashions, classroom dissection, etc. As a society we can change our laws and policies toward animals, and such changes are beginning to happen. As individuals, we can effect immediate change by making lifestyle choices that don't aid and abet the industries that harm animals and deprive them of pleasure. If animals felt only pain, that would be a worthwhile aim. That they also feel pleasure makes it more so.

Further Reading

Bagemihl, B. 1999. *Biological exuberance: Animal homosexuality and natural diversity.* London: Profile Books, Ltd.

Balcombe, J. P. 2006. *Pleasurable kingdom: Animals and the nature of feeling good.* London: Macmillan.

Bekoff, M. 2007. *The emotional lives of animals.* Novato, CA: New World Library.

Burgdorf, J., Panksepp, J. 2001. Tickling induces reward in adolescent rats. *Physiology and Behavior* 72:167–173.

Burghardt, G. M. 2005. *The genesis of animal play: Testing the limits.* Cambridge: MIT Press.

McMillan, F. D. 2005. The concept of quality of life in animals. In McMillan F. D. (ed.), *Mental health and well-being in animals* 183–200. Ames, IA: Blackwell Publishing.

Regan T. 1983. *The case for animal rights.* Berkeley, CA: The University of California Press.

Singer, P. 1975. *Animal liberation: A new ethics for our treatment of animals.* New York: Random House.

Jonathan Balcombe

POETRY AND REPRESENTATION OF ANIMALS

In poetry, the distanced relationship between modern and contemporary poets and the animal kingdom is a clear example of the slow movement of civilization that separates people from the natural world. In the poetry of indigenous peoples around the world, in the crystalline evocations of a moment, the presentness of the *haiku* poets (Issa, Basho, Shiki), and in the incantatory works of visionary poets (William Blake), one discovers a charged closeness to animals.

In the poetry of childhood, nursery rhymes and instructive books of childhood verse, one encounters poems which wean children away from a world view in sympathy with animals to the controlled and distanced relationship the adult world maintains.

In his preface to *Technicians of the Sacred,* Jerome Rothenberg notes that, of the primitive poetries from around the world which he has collected, ". . . above all there's a sense-of-unity that surrounds the poem, a reality concept that acts as a cement, a unification of perspective" (Rothenberg, 1968, p. xxii).

In their poetry, Native Americans spoke in the voice of the deer spirit, as well as their own, the hunter's. John Bierhorst's *In the Trail of the Wind* includes a whole section of poems "The Deer" from the Papago, Pima, and Chippewa. The poems construct a sort of conversation between animal and man grounded in respect, belief, and connection (Bierhorst, 1971, pp. 51–57). It is in their understanding of the connectedness of all living things, even the hunter and the prey, that their relationship to the animal kingdom is expressed.

Japanese *haiku* poets Basho, Issa, and Shiki, from the 17th, 18th, and 19th centuries respectively, address animals (frogs, crickets, cicadas, and others). In some poems they speak in the voices of animals, and throughout their work they show a consistent, tangible awareness of animal presence. In his introduction to *The Penguin Book of Japanese Verse,* Geoffrey Bownas notes that among Basho's construct of rules for *haiku,* the poet "should so express the nature of the particular as to define, through it, the nature of the world." (Bownas, 1964, p. lxvi). While Issa's poems directly address a cricket, a lanky frog, and insects, with a question or a warning, Basho's in

many instances use a creature's sound or presence on which to hang a mood, a comment, an instant. In Shiki's poems, the slightness, to American sensibilities, of the *haiku* form seems to turn toward the poetic equivalent of a snapshot. In all of these poets, the presence of and connectedness to the animal kingdom is unmistakable.

Incantatory and magnificent, William Blake's "Tiger, tiger burning bright" ("The Tyger") is poetry's most startling creature, feverishly real, devastatingly powerful, and alive, addressed, in fact questioned, throughout the poem as to what "could frame thy fearful symmetry?" (Blake, 1958, pp. 49–50). American poet Robinson Jeffers shows the same reverence and respect for the animal kingdom, for hawks, skunks, deer, stallions, and "the bird with the dark plumes in my blood" (Jeffers, 1959, p. 196). The question of the extent to which visionary poets employ animals as symbols or metaphoric constructs lies outside this consideration. For our purposes, the creatures are as vibrant and staggeringly real as they are meant to be.

The poetry of childhood is instructional, memorable and, at its best, able to capture the world from a child's perspective. Start with Mother Goose and her animal tortures. Four and twenty black birds are baked in a pie. Everywhere that Mary goes, so does the lamb. The mouse runs up and down the clock, and finally three blind ones have their tails cut off with a carving knife by the farmer's wife. Luckily, thankfully, the poet James Stephens asks in his poem "Little Things" that the "Little creatures everywhere" forgive us all our trespasses (Bogan & Smith, 1965, p. 19).

Edward Lear's *Nonsense Books* are filled with animals, some of them little more than personifications or anthropomorphisms, and quite Victorian at that. His Owl and Pussycat are a prime example. Similarly, Lewis Carroll's creatures, the Snark and the Jabberwocky among them, are imaginary, vivid, magical, and requiring no empathy. And Robert Louis Stevenson's *A Child's Garden of Verses is* devoid of animal life.

It takes Christopher Smart and William Butler Yeats, writing each about their house pet cats, to refocus children's caring and connection. Smart's gloriously celebratory "My Cat Jeoffry" in its penultimate line reminds us " For he is an instrument for the children to learn benevolence/ upon." (Grigson, 1959, pp. 120–121). Yeats' Minnaloushe in "The Cat and The Moon," is "Alone, important and wise" (Bogan & Smith, 1965, pp. 245–246). But perhaps Thomas Hardy, hardly a poet of childhood, best sets the example when in "Snow in the Suburbs" he writes of a black cat, stray in the snow which "comes, wide-eyed and thin; And we take him in" (Bogan & Smith,, 1965, p. 254).

Is it too easy to posit that life in the modern world—largely urban, separated in most cases from food gathering processes, enclosed in concrete, steel, and glass—is the culprit in modern and contemporary poetry's seeming lack of connection to animals? There is no poetic equivalent to the visual tirade against the mistreatment of animals in the meat industry by the artist Sue Coe. And though there are poets who show some awareness of the animal kingdom, appreciation for animal beauty, a fistful of poems about birds and butterflies, and some Zen-inflected poets like Robert Bly, Gary Snyder, and W.S. Merwin, who show quick recognition of animal being, for the most part, with a few notable exceptions, the work, especially in English, of

the last hundred years, speaks more than anything to the break in the connection we once had with animals.

In her rich and instructive anthology *We Animals: Poems of Our World,* Nadya Aisenberg collected poems from around the world, and developed a classification system for the relationships they display, which are reverence, dominion, fraternity, communion, and fantasy (Aisenberg, 1989, p. 3). Each of these relationships is reflective and reflexive, indicative of a loss from the connectedness described earlier. When D. H. Lawrence in "Snake" throws a "clumsy log" at a snake he is watching, aware of his ambivalence, fear, and fascination, in the last stanza he bemoans his loss, having "missed his chance with one of the lords/Of life" and indicates his need to make amends for his small-mindedness (Lawrence, 1965, pp. 95–98; Aisenberg, 1989, pp. 22–24). How much more clearly can the lost connection be expressed or mourned?

Contemporary Native American poets, caught between a traditional world and modern life, often write of mending that broken connection. For many of them, poems move slowly back to an older knowledge. Maurice Kenny's "Late Summer in the Adirondacks" from his *In The Time of the Present,* celebrates, with echoes of traditional poets in its repetitions of the phrase "they have come," the arrival of blue jays to the land. Peter Blue Cloud's *Winter with Crows* lets the crows of his home take their rightful places in his poems, considered, described, and honored.

But what of American poets, modern and contemporary, in a world disconnected? Can they capture presentness or connection with the animal kingdom, a belief, made apparent in their words, in the rights and welfare of animals? Wallace Stevens' "Thirteen Ways of Looking at a Blackbird" works with an almost mathematical sharpness, seemingly a construction, an exercise, but one which deepens and opens up. Stevens' meditation not only consciously asks questions about the relationship between us and the blackbirds, but acts as a measure, not of us against the blackbirds, but of their presence in the world, albeit and however reluctantly, our world. The blackbirds live beyond symbol, memory, myth, and number in the poem.

Denise Levertov gracefully and gratefully acknowledges all animals and their role in the world in "Come into Animal Presence." She speaks with hope, the hope that as a consciousness of animal rights and welfare grows, so it will grow in our poems, returning to something we have lost, or as Levertov puts it, "An old joy returns in holy presence" (Aisenberg, 1989, p. 43).

Further Reading

Aisenberg, Nadya, ed. 1989. *We animals: Poems of our world.* San Francisco: Sierra Club Books.

Bierhorst, John, ed. 1971. *In the trail of the wind: American Indian poems and ritual orations.* New York: Farrar, Straus and Giroux.

Blake, William, & Bronowski, J. ed.1958. *William Blake—The penguin poets.* Baltimore, MD: Penguin Books.

Blue Cloud, Peter. 2008. *Winter with crows.* Potsdam, NY: Potsdam College Press.

Bly, Robert. 1971. *The sea and the honeycomb: A Book of Tiny Poems.* Boston: Beacon Press.

Bogan, Louise, & Smith, William Jay, comps. 1965. *The golden journey: Poems for young people.* Chicago: Reilly & Lee Company.

Bownas, Geoffrey & Thwaite, Anthony, trans. 1964. *The Penguin book of Japanese verse.* Baltimore: Penguin Books.

Grigson, Geoffrey, ed. 1959. *The cherry tree.* New York: The Vanguard Press.

Jeffers, Robinson. 1959. *The selected poetry of Robinson Jeffers.* New York: Random House.

Jeffers, Robinson. 1963. *Robinson Jeffers: Selected poems.* New York: Vintage, Random House.

Kenny, Maurice. 2000. *In the time of the present.* East Lansing: Michigan State University Press.

Lawrence, D. H. 1965. *Selected poems.* New York: The Viking Press.

Lear, Edward. 1967. *Nonsense books.* New York: Grosset & Dunlap Publishers.

Opie, Iona, & Opie, Peter. 1955. *The Oxford nursery rhyme book.* London: Oxford University Press.

Rothenberg, Jerome, ed. 1968. *Technicians of the sacred.* Garden City: Doubleday & Company, Inc.

Stevens, Wallace, & Stevens, Holly, ed. 1972. *The palm at the end of the mind: Selected poems and a play.* New York: Random House.

James S. LaVilla-Havelin

THE POLITICAL SUBJECTIVITY OF ANIMALS

Although nonhuman animals are the objects of legislation governing their welfare, they seem *prima facie* to lack political subjectivity, which is to say that they do not seem to be agents who can represent themselves politically. Thus, it would seem that humans must speak on behalf of nonhuman animals, representing them in the exclusively human political domain.

This exclusion of nonhuman animals from the political sphere was of course classically signaled by the ancient Greek philosopher Aristotle when he defined man as the *zoon politikon,* the political animal, therefore implying that other animals are not political, which is to say, they cannot play a part in the life of the *polis,* the city, the basic unit of Greek civil life. Now, it was not only nonhuman animals who were in principle excluded by

Aristotle from political life, but also the mass of humans who were non-Greeks, and even the vast majority of Greek humans who were female, slaves, and/or children. It may thus be argued that the exclusion of nonhuman animals from political participation might be ended, just as women and the common people have ended their exclusion by acquiring political suffrage in modern democratic societies.

Certainly something like this claim seems to be true for animal rights. At first, we had the rights of man, then human rights, extended not only to men but also to women, children, and what were once regarded as inferior races, and now animal rights, which have actually been enshrined in law. The rights of animals indeed give them a form of political subjectivity under the law.

Types of Political Rights for Animals

Although he is not interested in animal rights so much as animal liberation, Peter Singer is one among the many who have argued that there is an historical progression at work here. The key concept of this view of the political status of animals is Richard D. Ryder's concept of speciesism, adopted most famously by Singer. This concept condemns the exclusion of animals from political consideration because of their species, just as racism has excluded some humans through the subcategorization of some humans.

The anti-speciesist animal liberationists do not, however, argue for the extension of full political rights to nonhuman animals. It seems that there is still a level that nonhuman animals cannot attain, namely, participation in political decisions. Certainly they do not

have the rights of suffrage, the right to vote or participate otherwise in political processes, even though these processes claim the right to create legislate about animals. In this they are in a similar position to human children. Both groups are held to lack sufficient rationality to determine their own futures, as was once held to be the case also for slaves and women, hence they are barred from playing a formal role in the political process.

However, this lack of a de jure role in political processes does not mean that nonhuman animals and infant humans are not de facto political agents. It is clear that human children in fact have a nontrivial political influence, both through influencing their parents and other enfranchised humans and by influencing things more directly, carrying out small acts of resistance, organizing politically within schools, and so on.

Some of the more explicitly political actions of human children, such as joining political youth groups or participating in school governance, are of course not undertaken by nonhuman animals. However, animals are able to undertake actions that have political import.

Peter Singer's views have tended to contradict the prevalent view of nonhuman animals as passive political objects, by claiming that animals have desires or rather preferences which they manifest and which can readily be discerned. Thus, Singer argues that when an animal tries to escape captivity, it is expressing its preference not to be captive in a readily discernible way. In this, Singer accords more importance to the agency of animals than does Ryder, whose painism emphasizes the capacity for suffering of animals as the source of our ethical obligations to them.

Although animals resist our control enough to show that they do not want to be controlled, their lack of political subjectivity in the full sense possessed by human animals seems to be confirmed by their inability to resist effectively on the human level. Animals' resistance is such that they are readily contained by now-perfected measures. By changing the animals themselves through selective breeding, and building environments, fences, cattle prods, cages, and so on, we now control domesticated animals to the extent that their resistance, although still commonplace and obvious, is apparently neutralized. Unlike humans, nonhuman animals in such situations seem incapable of, for example, secretly organizing to stage an uprising against their captivity, though such a scenario is the premise of several works of fiction, most prominently George Orwell's *Animal Farm.*

This inability is in fact a major cause of a certain contemptuousness on the part of the traditional political left against the placement of animal liberation on the same plane as the liberation of humanity, because they see political struggle as being an exclusively human affair. They are therefore out of sympathy with talk about how animals are exploited in much the same way as human workers, despite the fact that animals are often exploited in the same facilities as humans and by the same people, because animals are not seen as possible allies in the organized human struggle.

However, the fact that animals are incapable of political organization in a narrow sense does not necessarily mean that their resistance has been entirely negated. Just as African elephants are actually farming grass on the African savannah by their habit of uprooting trees, animals have political agency via the actual political effects of their actions. Specifically, through their expressions of anguish during human maltreatment, animals can influence humans to act to protect them. Such animal actions may certainly be seen as an essential cause of the discourse of animal rights and animal liberation itself.

Certainly, animals manipulate humans the way children manipulate adults, which is to say that it is naïve and lacking in cynicism. Indeed, part of the reason why humans are moved to help animals can be their very innocence.

Everyday Power Relations

While we can say that, although on a macro-political level it does seem that animals lack political subjectivity because they cannot participate in government, on a micro-political level, as most prominently put forward by French philosopher Michel Foucault, this is by no means obvious. In a household, for example, animals seem quite capable of exercising power, defined by Foucault as the ability to act on the actions of others. For example, a cat is quite capable of behaving in such a way as to purposefully motivate its owners to give it food, in much the same way a human can with another human. Pets and other animals are able to enter into power relations with humans in which they entice, seduce, or threaten humans or are in turn cajoled or seduced by humans. It would even be possible to link this micro-political subjectivity to a macro-political influence; pets owe their very survival to an ability to bond with and command the loyalty of owners, which can in turn lead to owners taking relevant political action for the needs of pets.

Such recent poststructuralist thinking about political agency tends to turn the tables on traditional thinking about political subjectivity. Thinkers such as Singer merely see an existing progressive trend broadening in the future to include respect for animals in addition to the rights of man, without challenging our notions of subjectivity themselves. Poststructuralist approaches to subjectivity, on the other hand, are suspicious of such a progressive view of history, and instead attempt to undermine our way of thinking by exploring our notions of subjectivity themselves and showing how in fact we can, in this case, understand animals as politically engaged subjects.

Of course, it still appears to be the case that animals' political capacities are inherently and permanently limited to a level lower than that of most adult humans. As Singer has frequently pointed out, however, there are adult human adults who are handicapped and therefore have similarly limited political capacities.

See also Ethics and Animal Protection— Political Action Committees (PACs) for Animal Issues

Further Reading

Aristotle. (1995). *Politics.* Oxford: Oxford University Press.

Carruthers, P. (1998). Animal subjectivity. *PSYCHE, 4*(3). Available online at http://psyche.cs.monash.edu.au/v4/psyche-4–03-carruthers.html

Garner, R. (2004). *Animals, politics and morality* (2nd ed.). Manchester: Manchester University Press.

Kelly, M. (2008). *The Political Philosophy of Michel Foucault.* New York: Routledge.

Singer, P. (1975). *Animal liberation: A new ethics for our treatment of animals.* New York: Random House.

Williams, A. (2004). Disciplining animals: Sentience, production, and critique. *International Journal of Sociology and Social Policy, 24*(9), 45–57.

Mark G. E. Kelly

POLYISM

Polyism is the phenomenon whereby a given standard of care is lower because of the numbers of animals involved, and also partly because of the size of the animal. It is particularly noticeably in intensive farming systems for pigs and poultry, where literally tens or hundreds of thousands of animals are kept in a single shed or similar confinement, compared with, say, dairy cattle where the herd size is measured in hundreds or less. It is partly due to the impossibility of observing each animal individually, the financial value of each animal as a unit of production, and the numbers of animals showing adverse effects, such as lameness in chickens towards the end of the growing period. Farm personnel would likely pay more attention to a lame dairy cow than to a lame chicken or pig.

David B. Morton

PRACTICAL ETHICS AND HUMAN-ANIMAL RELATIONSHIPS

Our relationship with other animals is complex, and our treatment of them is as controversial as our treatment of other human beings. Questions about the ethics of human-animal relations are thus an ongoing concern, and have implications for humanity's interaction with wild, companion, farm, and research animals. The topic of animal ethics is therefore at the core of the emerging field of animal studies, the discourse of environmental studies, and in many subfields such as animal geography. One manner of addressing such issues is through practical ethics, a mode of moral understanding that is well suited to grappling with our responsibilities in a more-than-human world.

Ethics

Definitions of ethics can differ vastly. Most of these differences are rooted in attempts to explain ethics in terms of something else. For example, various academics have tried to associate ethical concerns with personal preferences, emotional responses, religious beliefs, social expectations, and genetic determinism. Personality, empathy, spirituality, social custom, and science may all enrich ethics at various points and times. Yet we should be careful not to let this obscure the meaning and importance of ethics itself.

To discover the meaning of ethics, we can look to Socrates, a Greek philosopher whose definition of ethics has been at the core of ethical thought for several thousand years. Socrates saw himself as a gadfly and midwife. As a gadfly he pushed people to think harder. As a midwife he helped them develop their thoughts to a higher level of rigor. For him and his followers, ethics was about how we ought to live (Plato *Republic*, Book 2, p. 312d). What this brief statement means is this: ethics is about the moral values that inform or should inform our lives. When we engage in ethics, we are not only exploring our ideas about what is good, right, just, and valuable, we are also articulating principles of conduct based on these ideas.

Note that ethics is not only for human beings. People may be the only creatures on Earth who have abstract systems of thought labeled ethics. In this sense, ethics is an artifact of human culture. This does not mean our ethical considerations must exclude other creatures. The moral community is a mixed one, populated by humans and other animals, all of whom share an intrinsic value and moral standing alongside the rest of nature. In addition, individuals and groups, ecosystems and societies, represent different foci and scales of ethical reason. People, animals, and nature all have characteristics that ethics helps us appreciate and protect.

Theoretically Rich and Empirically Situated

The world's moral complexity and the kind of ethical reasoning necessary to grapple with it was no secret to Socrates. He practiced a form of moral reasoning that was fully engaged with the empirical world, and differs markedly from the standard ways in which ethics is often practiced today.

In the standard model of ethics, moral truth is determined through abstract argument prior to one's engagement with concrete moral problems. This is sometimes called theoretical ethics. These claims are then applied to concrete cases in a top-down, linear, and deductive manner. This is what is meant by applied ethics.

Practical ethics proceeds differently. Rejecting an easy division between theoretical and applied ethics, it does not determine moral truth ahead of time. Rather, it seeks out a situated truth by integrating what we learn from a concrete case, in conjunction with the conceptual insights that help us best understand and resolve that case. Practical ethics looks to diverse moral principles, rooted in the empirical reality of cases, to triangulate on the reasons for and resolutions to our moral concerns. Put another way, practical ethics is a situated moral understanding, an ethics that is simultaneously conceptually rich and situated in real life.

Several features about practical ethics should be emphasized here.

Pluralism. For the practical ethicist, moral concepts are plural and complementary. The more concepts we have, the deeper our reservoir of potential insights. Thus the practical ethicist is not precommitted to a single concept that she uses over and over in all situations. She is free to choose from a constellation of concepts. Ideally, her choice reflects those concepts that are most useful in resolving a moral problem. Moral concepts that are commonly used in practical ethics are recognized by such terms as good, right, fair, just, and valuable.

Triangulation. Ethical concepts cannot be applied by rote, like some grid of latitude and longitude from which we can read the correct moral position. Rather, moral understanding is akin to triangulating on the best ethical position. When triangulating over land or sea, one needs several reference points to properly plot one's position. These reference points may be stars or landmarks. The same applies to practical ethics, where the reference points are well developed moral concepts.

Principles and Maxims. Moral concepts can be used as either principles or maxims. A principle is a moral concept used to clarify our thinking. It provides guidance to our reasoning about how we ought to live. A maxim is a moral concept used to clarify our actions. Maxims provide more focused guidance than principles. The intrinsic value of people and animals is an example of a principle. The golden rule, treat others as you want to be treated, is an example of a maxim. Overall, principles justify the use of certain maxims that guide our conduct, while maxims align our actions with principles.

Rule of Thumb. Moral concepts are not rigid or absolute laws. They are a rule of thumb that helps us distinguish better from worse ways of thinking and acting. Both principles and maxims actively and dynamically reveal the ethical issues at stake, and provide guidance on what we ought to do about them. They do not, however, make moral decisions for us. Rather, they are the tools through which we exercise moral judgment.

Praxis. The term praxis refers to putting theory into action. Praxis is not a one-way relation where one deductively reasons from theory to action. It is a two-way relation where theory and action are reciprocally informing. In practical ethics, the principles and maxims we use to reveal ethical issues and guide our subsequent actions are selected in light of the case at hand. It is a form of practical reasoning where theory and reality are not disengaged from each other.

Context. Concrete moral problems are situated in space, time, nature, and culture. All ethical issues therefore have a geographical, historical, environmental, and cultural context. The stock of moral concepts in use and the actions that a moral agent can take are enabled and constrained by the context in which one operates. These are the sites and situations in which moral problems, the controversies that swirl around them, and their possible resolutions exist.

Judgment. The proper matching of principles, maxims and cases takes experience and skill, a feature that practical ethicists refer to as judgment. Having good judgment means one can correctly match the most appropriate moral concepts to the case at hand. This is best done when we integrate the facts on the ground with our best ethical understanding. From this point we can make moral decisions and chart a course of action from there.

Truth. From the standpoint of practical ethics, there is rarely a single, indisputable judgment that is right or wrong. Reasonable people will differ on the best principles and/or maxims for understanding a particular case. They may also differ on what a reasonable course of action might entail. Recognizing that absolute truth (veracity) is rarely possible, practical ethics seeks the best account of truth that is possible (verisimilitude).

Situating. The recognition that absolute moral truth is very difficult to come by is not a reason to endorse ethical relativism. With its emphasis on praxis and context, practical ethics is not only situated in the world, but takes the creative middle ground and situates itself between absolutist and relativist interpretations of ethics. It does so in the belief that we can distinguish better from worse moral reasoning or courses of action. We do so in light of the evidence at hand, and the rigor of our thinking.

Examples of Principles and Maxims

Below are a few examples of how a practical ethics regarding nonhuman animals would work.

Principles (Guidelines for Thought)
Geocentrism—We should acknowledge the moral value and standing of people, animals, and nature. This means that we value animals and their habitats, while encouraging recognition of humanity's membership in a wider moral community. Geocentrism incorporates the insights of anthropocentrism (the moral value of people and their communities), biocentrism (the moral value of individual people and animals), and ecocentrism (the moral value of biodiversity and ecosystems).

Equal Consideration—We should give equal consideration to the well-being of people, animals, and nature. This principle helps us actualize geocentrism by identifying and balancing our responsibilities to people, animals, and their mutual habitats. Note that equal consideration does not imply equal treatment. When creatures differ in their capacities and modes of life (for example, people, foxes, voles), then equal consideration requires appropriate differences of treatment.

Hard Cases—Our universal need for geographic space—habitat, resources, etc.—makes win/lose conflicts a fact of life. When faced with a situation pitting humans against animals, we must first solve the underlying problem, then look for alternatives and, as a last resort, choose a geographic compromise that protects the entire community's well-being. This principle helps us think through the complications raised when we give equal consideration to the well-being of humans and nonhumans.

Moral Carrying Capacity—People should live within an overall carrying capacity that protects the well-being of nonhumans, biodiversity, and landscapes. This principle is crucial, as it helps us avoid the hard cases mentioned above. While technology and social organization may mitigate the upper limit on the earth's carrying capacity for humans, there is a definite and negative impact of societal growth and consumption on the nonhuman world. Humans must take responsibility for limiting their use of the earth's carrying capacity.

Precaution—The idea behind precaution is similar to the medical principle: First, do no harm. This is a principle for dealing with the uncertainty that pervades questions of both ethics and science. Precaution states that a lack of certainty is not an excuse for actions that are irreversible or may create harm. In the face of uncertainty, precautions should be taken to minimize the risks to people, animals, and the rest of nature. One has no inviolable right to engage in activities with risk of harm (for example, polluting a water source) simply because the range and extent of that harm is not yet well documented.

Maxims (Guidelines for Action)
Integrity—We should endeavor to respect the psychological, physical, and social integrity of wild and domestic animals by minimizing stress, using noninvasive and nonlethal techniques in cases of conflict, and avoiding the disruption of their social organization and ecological relationships.

Graduated Response—In cases of human-animal conflict, there is a continuum of responses, from the nondestructive and nonlethal through the destructive and lethal. We should seek to resolve a problem with nondestructive and nonlethal responses first. Where one starts on this continuum depends on the severity of the problem.

Harm-Benefit Ratios—During the design phase of research, policy, or management strategies regarding nonhuman animals, we should calculate harm-benefit ratios for each action. Such ratios help us explore whether the probable benefits to science, society, or nature can outweigh the foreseeable harms to animals as individuals, groups, populations, or species.

Mutual Benefits—Whenever possible, we should adopt those actions that provide mutual benefits for people, animals, and nature. Vague assertions about human benefits or risks to public health are rarely sufficient reasons to sacrifice the well-being of animals. This is a more positive and proactive principle than the harm-benefit ratios mentioned above.

Reduction, Refinement, Replacement (the 3Rs)—When using invasive or harmful procedures in the laboratory or the field, we should practice the 3Rs—a reduction in the number of procedures, refinements in their technique, and replacement with noninvasive and non-harmful procedures.

End-Points—Invasive or harmful actions should specify humane endpoints, so that if an action proves harmful, we know when to stop. When an action based on a policy or management strategy is proving harmful, it should have a predefined endpoint. After the action is brought to a halt, the situation should be reassessed to produce a better course of action.

In modern times, variations on the practical approach to ethics have been advocated by philosophers such as Hans-Georg Gadamer, Karin Lauria, Arne Naess, Stephen Toulmin, and Anthony Weston.

It was Mary Midgley, however, who set the tone early on in animal ethics as well as animal studies, a discipline which might also be dated to her publications. In her book *Animals and Why They Matter* (1984), Midgley carefully explores the dominant theories of animal ethics. She does so as part of an appreciative critique, seeking out conceptual insights, while at the same time noting shortcomings when a theory or concept is misapplied. She does not ask her readers to choose one theory per se, but to appreciate and carefully use the full range of concepts that are made available through a diversity of theories. In other words, she asks that we generate a situated moral understanding, one that takes both moral concepts and the facts on the ground as equally important and mutually informing. Midgley's practical approach to ethics models the use of principles and maxims to triangulate on better versus worse accounts of how we ought to live. This is the task of practical ethics.

Further Reading

Gadamer, Hans-Georg. 1993. *Truth and method.* 2nd ed. New York: Continuum.

Lauria, K. 2009. Christian theologies of animals: Review and implications for a new theology of animals. *Journal for the Study of Religion, Nature and Culture,* forthcoming.

Lynn, W. S. 2006. Between science and ethics: What science and the scientific method can and cannot contribute to conservation and sustainability. In D. Lavigne, ed., *Gaining ground: In pursuit of ecological sustainability,* 191–205. Limerick: University of Limerick Press.

Lynn, W. S. 2007. Practical ethics and human-animal relations. In M. Bekoff, ed., *Encyclopedia of human-animal relationships,* 790–797. Westport, CT: Greenwood Press.

Midgley, M. 1998. *Animals and why they matter.* Athens: University of Georgia Press.

Naess, A. 1989. *Ecology, community and lifestyle: Outline of an ecosophy.* Cambridge: Cambridge University Press.

Toulmin, S., and Jonsen, A. R. 1988. *The abuse of casuistry: A history of moral reasoning.* Berkeley: University of California Press.

Weston, A. 2006. *A practical companion to ethics.* New York: Oxford University Press.

William S. Lynn

PREDATOR CONTROL AND ETHICS

In the United States, more than 120,000 native carnivores are killed each year by the federal government as part of the U.S. Department of Agriculture's Wildlife Services predator control program. Taxpayer dollars subsidize this carnage to the tune of tens of millions of dollars, even though the killings are intended primarily to benefit private livestock operators. Killing native carnivores has been a common practice since European colonists arrived in North America nearly four centuries ago. The colonists viewed native carnivores as a threat to livestock and as competition for game species. So prevalent was this view that a bounty on wolves was enacted shortly after the founding of the Plymouth Bay Colony in Massachusetts in 1630.

As settlers pushed west into the Great Plains in the 1800s, they slaughtered native carnivores to open the land to livestock and farming. Ranchers, bounty hunters, and professional trappers killed millions of coyotes, wolves, bears, and mountain lions. Large-scale cattle grazing resulted in the widespread depletion of vegetation and the wildlife that consumed it, thereby reducing the numbers of prey available for native carnivores. With less natural prey, the remaining coyotes, wolves, bears, and cougars turned to livestock, which only bolstered predator eradication campaigns. The federal government became officially involved in predator control in 1915 when Congress allocated $125,000 to create the Branch of Predator and Rodent Control within the Department of Agriculture's Bureau of Biological Survey. Their mission was to carry out official strychnine poisoning and trapping campaigns targeting wolves, mountain lions, coyotes, foxes, bears, and eagles on the public domain lands of the West. Later, during the Hoover Administration, livestock operators and hunters pressured Congress to pass the Animal Damage Control Act in 1931. This Act, still in effect today and largely unchanged, authorized the "suppression, eradication, and control" of wild animals that caused injury to agriculture, horticulture, forestry, and animal husbandry.

The methods employed by USDA Wildlife Services, formerly the Animal Damage Control Program, include poisons, steel-jaw leghold traps, strangulation neck snares, denning (the killing of coyote pups in their dens), hounding, shooting, and aerial gunning. Critics of the program argue that it perpetuates an endless cycle of conflict and killing with an emphasis on nonselective methods, that it lacks accountability to the public, needlessly kills millions of animals for the benefit of a relatively small number of livestock producers, and fosters a dependence on taxpayer-funded assistance instead of promoting effective long-term solutions to conflicts (O'Toole, 1994; Treves and Karanth, 2003; Mitchell et al., 2004; Fox and Papouchis, 2005; Robinson, 2005; Feldman, 2007) (see Table 1).

TABLE 1 Coyotes killed by U.S. Department of Agriculture Wildlife Services in 2007, by state

Texas	19,123
Wyoming	10,915
Montana	9,251
California	7,759
Nevada	7,447
Oregon	6,492
Oklahoma	5,544
Utah	4,888
Idaho	4,783
New Mexico	4,568
Colorado	2,738
North Dakota	1,899
Nebraska	1,858
Arizona	1,218
Washington	608
West Virginia	400
Virginia	368
Wisconsin	74
Louisiana	69
Florida	66
Georgia	61
Michigan	56
Ohio	54
Minnesota	44
Kentucky	30
Illinois	17
Tennessee	17
Missouri	12
Massachusetts	12
Mississippi	10
North Carolina	6
Pennsylvania	5
Indiana	5
New York	4
South Carolina	3
New Jersey	3
Iowa	3
Alabama	3
Maine	2
Kansas	1

Wildlife Services reported that no coyotes were killed during FY 2007 in Alaska, Connecticut, Delaware, Washington, D.C., Maryland, New Hampshire, South Dakota, Vermont, Rhode Island, or Arkansas. Coyotes are not found on Hawaii. During FY 2007, a total of 90,416 coyotes were killed by Wildlife Services.

Project Coyote. www.ProjectCoyote.org. More information about Marin County's Strategic Plan for Protection of Livestock and Wildlife and predator control can be found on the Web site.

Impacts of Lethal Predator Control

Scientists are only just beginning to fully comprehend the ecological impacts and potential long-term consequences of broad-scale removal of large carnivores from the landscape. By studying the effects of their removal on ecosystems, biologists have found that large carnivores can function as keystone species, playing a pivotal role in maintaining ecological integrity and preserving species diversity. The disappearance of a keystone species can trigger the loss of other resident species, and the intricate connections among the remaining residents begin to unravel, dramatically changing the habitat. In a domino effect, species losses cascade through the ecosystem, as the disappearance of one species prompts the loss of still others. As argued by conservation biologists, "Our current knowledge about the natural processes that maintain biodiversity suggests a crucial and irreplaceable role of top predators. The absence of top predators appears to lead inexorably to ecosystem simplification accompanied by a rush of extinctions" Terborgh et al., 1999).

Remarkably, USDA WS has never attempted to calculate the overall environmental costs of its predator control programs or its impact on ecosystems and the biota therein. Indeed, we may never be able to accurately and fully assess the extent of its impact. Soulé et al. postulate that the failure of wildlife management agencies to incorporate a doctrine of "best conservation practices based on the best science," and to consider the ecological value of maintaining large carnivores on the landscape, is due to these agencies still functioning under anachronistic laws and policies that are based on old and simplistic scientific concepts, for example, that predators are bad and need to be eradicated (Soulé, Estes, Miller, and Honnold, 2005).

The ecological impact and ethical implications of broad-scale lethal predator control would be a serious cause for concern even if such programs were effective in their apparent aim of reducing livestock losses. However, when ecological systems are damaged by ineffective programs, this compounds the tragedy. More than a century of killing predators has done little to diminish overall livestock losses. This is largely because lethal control does not address the underlying cause of livestock predation, which is the presence of an attractive prey, for example, domestic sheep, in the habitat of opportunistic carnivores. The large size of livestock and the absence of defense against predators provide a sizable meal for relatively little effort, especially in terms of domestic sheep unaccompanied on open range far from human activity, as occurs on public lands throughout the West. Further, livestock consume and trample the vegetation needed to survive by most of the predators' natural prey (Crabtree, and Sheldon, 1999). When these species are depleted, predators may turn to livestock, leading to increased lethal control efforts and an endless and ultimately futile killing cycle.

Evidence of the futility comes from a recent study by biological economist Kim Murray Berger, who examined predator control in the United States in relation to sheep production. Berger suggests that the decline of the sheep industry is more closely associated with unfavorable market conditions than predation, and raises serious questions about the effectiveness of traditional lethal predator control programs (Berger, 2006). Berger also found that despite Wildlife Services's killing of

While coyotes have adapted to lived side by side with humans in urban landscapes, people are often less tolerant of America's native wild dog. (John Harrison)

five million predators at a cost of $1.6 billion from 1939 to 1998, the effort had little effect on sheep industry trends. Even though the agency has been killing predators for nearly a century, she points out, 85 percent of U.S. sheep producers have gone bankrupt.

Attempts to reduce coyote populations, the main emphasis of Wildlife Services's predator control program (more than 90,000 coyotes were killed by the agency in 2007; see Table 1), have failed because coyote populations exhibit strong compensatory responses to lethal control. While lethal control may result in short-term reductions in the number of coyotes in a specific area, the vacuum is soon filled by coyotes emigrating from surrounding areas and by increased litter size and pup survival in remaining populations. Lethal control disrupts the social

hierarchy of coyote packs, causing pack members to disperse and allowing more females to breed. Females in exploited populations tend to have larger litters because competition for food is reduced and more unoccupied habitat is available. Lethal control also often selects for coyotes that are more successful, wary, nocturnal, and resilient, what some biologists call a super coyote (Fox, and Papouchis, 2005).

Hence, lethal control of coyotes may actually exacerbate livestock conflicts by stimulating improved reproductive success and pup survival in the remaining coyote population (Connolly, and Longhurst, 1975). Despite research conducted over 30 years ago showing that suppression of a coyote population over the long term requires removing more than 75 percent of the population annually because of the reproductive rate of the species (Connolly, and Longhurst, 1975), USDA WS continues to emphasize lethal coyote control in its national livestock protection program. While millions of coyotes have been systematically killed through subsidized predator control programs over the last century, their range has expanded three-fold since 1850 (Crabtree, and Sheldon, 1999). Even in the most extreme cases, when measures are taken to eliminate entire coyote populations, the loss in species diversity that results from killing predators to protect livestock can lead to increased problems for ranchers. Researchers at Texas Tech University reported in 1999 that removing nearly all of the coyotes in a 5,000-hectare area caused a severe decline in the diversity of rodent species and a significant increase in the numbers of jackrabbits, badgers, gray foxes, and bobcats (Henke, and Bryant, 1999). They concluded that removing coyotes to protect livestock could actually be counterproductive:

> Increased jackrabbit density caused by a lack of predation could cause increased competition for forage between jackrabbits and livestock . . . consequently, a reduced stocking rate [of livestock] may be required to offset competition, which may financially negate the number of livestock saved from predation. (Henke, and Bryant, 1999)

Tools of the Trade

Many of the lethal methods used to kill native carnivores are inhumane, indiscriminate, and a threat to public safety. The primary killing tools employed by Wildlife Services include leghold traps, strangulation neck snares, poisons, denning (the killing of coyote and fox pups in their dens), and aerial gunning. Increased public, scientific, and Congressional scrutiny has led to greater awareness and widespread condemnation of lethal predator control methods which are often used prophylactically prior to lambing season. In 1995, largely as a result of public outcry, Congress directed the General Accounting Office to investigate Wildlife Services' predator control activities in the field. The GAO found that: "ADC [Wildlife Services] personnel in western states use lethal methods to control livestock predators despite written USDA policies and procedures giving preference to the use of nonlethal control methods where practical and effective" (GAO, 1995). Then in 1999, the American Society of Mammalogists passed a resolution stating that the

> common methods of predator control are often indiscriminate, preemptive, lethal measures, particularly in relation to state- and federally funded livestock protection programs . . . and often result in the needless killing of animals that are not contributing to the problem, as well as many non-target species. (ASM, 1999)

They called on the federal government to "cease indiscriminate, preemptive, lethal control programs . . . and to focus on the implementation of non-lethal control strategies, compensatory measures, and sound animal husbandry techniques" (GAO, 1995). As the ASM and other scientists have pointed out, not all predators kill livestock (Treves, and Naughton-Treves, 2005). In fact, many of the animals killed through predator control programs, up to 81.3 percent according to one study that looked at lethal carnivore management programs across the globe, are non-offending animals (Treves, and Naughton-Treves, 2005). However, the dominant practice in the United States is based on the theory that by killing a large number of predators the offending animal will be among the casualties (Wagner, 1988). Wagner suggests that the federal government's approach is "something of a sledge-hammer one: If enough coyotes are shot, trapped, and exposed to M-44s . . . their numbers can be reduced and the chances are that the offending animal(s) will be among those taken and the losses reduced" (Wagner, 1988).

In addition to the U.S. federal government, several states carry state-sponsored predator control programs that range from bounty and contest hunts to aerial hunting and carnivore snaring programs. For example, in Alaska, Governor Sarah Palin announced in 2007 that the state would pay wolf hunters $150 when they bring in the left forelegs of wolves taken from any of several designated control areas. When wildlife advocates challenged the program, the state insisted it wasn't a bounty; however, the judge presiding in the case ruled that the program was indeed a bounty and ordered it to stop. Then in June 2008, the state issued a press release announcing that it had successfully used a helicopter to kill 28 wolves on state lands near Izembeck National Wildlife Refuge to boost caribou numbers for hunters. Newspaper reports on July 19 revealed that 14 of the 28 wolves killed by the Alaska Department of Fish and Game were actually pups. Conservation groups are challenging the action and maintain that the pups were illegally killed because denning is an illegal practice.

More than 90,000 coyotes were killed by federal agents in 2007 as part of a government-subsidized lethal predator control program. (John Harrison)

Paradigm Shift Despite clear scientific evidence demonstrating the futility and counterproductiveness of indiscriminate lethal predator control, many state and federal wildlife managers continue to promote prophylactic killing as the best method to address conflicts. An increasing number of scientists, however, have begun to speak out publicly against lethal control. As discussed above, their studies show that coyotes and other large carnivores play a vital ecological role and that their removal can have a devastating impact on species diversity and on the health and integrity of native ecosystems.

But scientific evidence is not enough. Many scientists believe that a new paradigm is needed for the way humans treat native carnivores, indeed all wildlife, one that recognizes the ecological importance of these species as well as their intrinsic value as individuals. If the money and efforts used to kill predators were redirected toward cost-effective, nonlethal methods such as public education, better landscape development, improved fencing, and guard animals, conflicts could be significantly reduced without the need to kill indiscriminately (Fox, 2006). Ultimately, wildlife managers will be forced to make this ethical shift as communities across North America demand humane solutions to wildlife conflicts that consider the importance of individual animals as members of a larger integrative community that includes both humans and nonhumans alike.

An Alternative to Lethal Predator Control: The Marin County Model

In 1996, in the bucolic northern California county of Marin, community-wide controversy arose when wildlife advocates learned that Marin was to be one of three counties where the deadly poison Compound 1080 would be pilot tested to kill coyotes. The proposed plan led to a rancorous debate about management of native carnivores in a community known for its environmental consciousness and strong support of agriculture (Fox, 2001). On one side were animal advocates and conservation groups who questioned the ethics of using taxpayer dollars to employ a federal trapper to kill native wildlife with predator poisons, denning, and body-gripping traps. On the other side were sheep ranchers who argued that federal assistance with predator management was necessary and that loss of such assistance would put them over the edge in a market that was already being undermined with cheap imports from overseas.

After a series of roundtable discussions organized by the Marin County Agricultural Commissioner that included ranchers, animal advocates, conservationists, and local public officials, the Marin County Board of Supervisors attempted to reach a compromise with Wildlife Services. The Supervisors said they would renew the contract with the federal agency, but stipulated that neck snares and other lethal methods could only be used a last resort after nonlethal methods had been tried and proven unsuccessful (Fox, 2001). When Wildlife Services refused to operate under the county's guidelines, the Marin County Board of Supervisors decided it was in the county's best interest to cease contracting with the agency. The decision, however, did not prevent ranchers from removing predators on their own land to protect their livestock.

In place of the traditional WS program, the Supervisors approved a program put forth by a coalition of animal advocacy and conservation organizations and later more fully developed by the Marin County Agricultural Commissioner's office with input from the ranching community. The plan, called the Strategic Plan for Protection of Livestock and Wildlife, redirected the county's $30,000 annual cost for WS, to assist qualified ranchers in implementing nonlethal techniques including livestock guard dogs, llamas, improved fencing, lambing sheds, and night corrals.

To date more than 80 percent of all Marin sheep ranchers participate in the program, and initial data indicates that livestock losses have declined since implementation of the program (Fox, 2008). More important, the program provides a model that has successfully addressed and embraced ethical concerns as well as differing values expressed by both the animal protection and ranching communities.

See also Wildlife Abuse; Wildlife Services; Trapping, Behavior, and Welfare

Further Reading

American Society of Mammoligists (ASM). 1999. Mammalian predator control in the United States. Resolution passed at the University of Washington, Seattle, Washington, June 20–24, 1999.

Berger, K. M. 2006. Carnivore-livestock conflicts: Affects of subsidized predator control and economic correlates on the sheep industry. *Conservation Biology* 20:751–761.

Connolly, G. E. 1978. Predator control and coyote populations: A Review of Simulation Models. In M. Bekoff (ed.), *Coyotes: Biology, behavior, and management,* 327–345. New York: Academic Press.

Connolly, G. E., and Longhurst, W. M. 1975. The effects of control on coyote populations: A simulation model. *Div. of Agricultural Sciences, University of California, Bulletin* 1872:1–37.

Crabtree, R. L. and Sheldon, J. W. 1999. Coyotes and canid coexistence. In T. W. Clark et al., (eds.) *Carnivores in ecosystems:*

The Yellowstone experience, 127–163. New Haven: Yale University Press.

Feldman, J. W. 2007. Public opinion, the Leopold Report, and the reform of federal predator control policy. *Human-Wildlife Conflicts* 1:112–124.

Fox, C. H. 2006. Coyotes and humans: Can we coexist? *Proceedings of the Vertebrate Pest Conference* 22: 287–293.

Fox, C. H. 2008. Analysis of the Marin County strategic plan for protection of livestock & wildlife: An alternative to traditional predator control. Master's thesis, Prescott College, Prescott, Arizona.

Fox, C. H. and C. M. Papouchis. 2005. *Coyotes in our Midst: Coexisting with an adaptable and resilient carnivore.* Animal Protection Institute, Sacramento, California. 64 pp Fox, C. H. 2001. "Taxpayers say no to killing predators." *Animal Issues* 31:26–27.

Fox, C. H., and C. M. Papouchis. 2005. *Coyotes in our midst: Coexisting with an adaptable and resilient carnivore.* Sacramento, CA: Animal Protection Institute.

General Accounting Office (GAO). 1995. Animal damage control program—Efforts to protect livestock from predators. *GAO Report* B-261796, October 1995.

Henke, S. E., and Bryant, F. C. 1999. Effects of coyote removal on the faunal community in Western Texas. *Journal of Wildlife Management* 63 (1999):1066–1081.

Mitchell, B. R. Jaeger, M. M., and Barrett, R. H. 2004. Coyote depredation management: Current methods and research needs. *Wildlife Society Bulletin* 32:1209–1218.

O'Toole, R. 1994. Audit of the USDA Animal Damage Control Program. Cascade Holistic Economic Consultants Research Paper Number 31.

Parker, G. R. 1995. *Eastern coyote: The story of its success.* Halifax, NS: Nimbus Publishing.

Project Coyote. www.ProjectCoyote.org.

Robinson, M. 2005. *Predator bureaucracy: The extermination of wolves and the transformation of the West.* Boulder: University of Colorado Press.

Soulé, M. E., J. A. Estes, B. Miller and D. L. Honnold. 2005. Strongly interacting species: Conservation policy, management, and ethics. *BioScience* 55:168–176.

Terborgh J. et al. 1999. The role of top carnivores in regulating terrestrial ecosystems.

Chapter 3, in M.E. Soulé, and J. Terborgh (eds.), *Continental Conservation: Scientific Foundations of Regional Reserve Networks,* Washington, DC: Island Press.

Treves, A. R., and Karanth, K. U. 2003. Human-carnivore conflict and perspectives on carnivore management worldwide. *Conservation Biology* 17:1491–1499.

Treves, A. R., and Naughton-Treves, L. 2005. Evaluating lethal control in the management of human-wildlife conflict. In R. Woodroffe, S. Thirgood, and A. Rabinowitz (eds.), *People and wildlife: Conflict or coexistence?*, 86–106. Cambridge, UK: Cambridge University Press.

Wagner, F. H. 1988. *Predator control and the sheep industry.* Claremont, CA: Regina Books.

Camilla H. Fox

PUPPY MILLS

Puppy mills are one of America's biggest secrets. Buyers don't know it, but often the adorable lumps of fur frolicking in the pet store window or posing on the Web are produced by the millions, like cash crops, in dark cages and sheds. Victims of inbreeding and poor care, puppy mill dogs often emerge frightened of ordinary noises, grass, even the touch of a human hand. They may suffer seizures, autoimmune disorders, and other illnesses. It's not uncommon for a puppy mill dog to die within weeks after a family has taken it home.

The mother dogs have it worse. Confined to cages for years at a stretch, they may be forced to bear litters of puppies every six months before they are rendered useless and put to death or, if they're lucky, turned over to rescue groups.

Puppy mills have proliferated over the last half century or so, from a handful of operators to an industry of more than 5,000 licensed commercial breeders, concentrated in the Midwest. The Humane Society of the United States estimates that when unlicensed kennels are included, the true number of puppy mills is closer to 10,000. Large-volume breeders insist that they produce dogs under optimum conditions; that to do otherwise would be bad business. Critics maintain that many puppy mill operators cut corners on humane treatment and churn out as many dogs as they can.

Puppy mill dogs typically live in cramped cages in dirty conditions, often exposed to the elements. The may be given inadequate food and water and almost no medical care. They're called purebreds, but often suffer a range of debilitating illnesses and conditions that are far below the standards most would associate with purebred dogs. Puppy millers often dispense with veterinary care and perform necessary procedures like cesarean sections themselves, sometimes without anesthesia. Reckless breeders have crudely amputated the legs of dogs trapped in wire cages. Dogs confined to wire cages for years at a stretch can and do go mad. One 13-year-old Sheltie kept in a kennel piled with feces for nine years limped in circles even after he was rescued. It was a habit he'd developed to stave off boredom.

The irony is that puppy mills thrive in a country where, according to the American Pet Products Manufacturers Association, 44 million households embrace dogs as members of the family. Dog owners have become so devoted to their pets that they buy plaid berets for their bulldogs, send their malamutes to doggy daycare, and spend hundreds of dollars on pet portraits. Beneath all this affection lies the shadowy world of puppy mills, but most dog lovers have no clue that they exist.

A bright yellow sign reading "puppies" still beckons the visitor to a now-shuttered puppy breeding facility Tuesday, March 24, 2009, in Seneca, Missouri. In February rescuers found more than 200 dogs living in their own excrement, crammed into weather-exposed single cages and hutches, many of them contaminated and hairless at the facility. Missouri is the "puppy mill" capital of America, home to more than 4,000 shoddy and inhumane dog-breeding businesses, by one estimate. But now the state is trying to shed its reputation, with the chief of the Agriculture Department pledging to do more to crack down on bad breeders. (AP Photo/Jeff Roberson)

The federal government has done little to address the issue. A 1966 *Life* magazine spread exposing puppy mills outraged readers and helped galvanize support for a law governing commercial dog-breeding. Yet 40 years later, the U.S. Department of Agriculture has turned a blind eye to puppy mill abuses and has actually encouraged large-volume dog breeding as a way for retired chicken and pig farmers to earn new income.

Most puppy breeders are paid $100 to $200 for each puppy. Brokers turn around and sell the dogs to pet stores for $200 to $300 each. Pet shops, and breeders selling directly via the Internet, are able to charge buyers up to $2,000 for a dog.

Even if the USDA wanted to crack down on the problem, it has just 96 in-spectors overseeing 4,700 licensed facilities. The head of the department's Animal and Plant Health Inspection Service conceded several years ago that barely half of all licensed commercial dog breeders met even minimal standards set by the federal government. Thousands more puppy mills operate on the sly, absent a license, which enables them to escape any scrutiny. And the federal law has an enormous loophole: it doesn't apply to dogs sold directly to the public.

Consequently, thousands of dogs are now available through such Web sites. By one estimate, at least 10,000 of them a year are flown into the United States from countries as far away as Hungary, Poland, Russia, and Lithuania. New designer breeds like labradoodles, a

Labrador-poodle mix, and puggles, a cross between a beagle and a pug, also fuel the demand.

The result is a consumer's nightmare. Regardless of their origin, puppy mill pups are often weaned too early and shipped in large groups, exacerbating the likelihood they will contract a transmittable disease such as parvovirus, parasites, or distemper. Customers who have purchased puppy mill dogs report dismaying experiences in which the dogs arrive sick or injured, often fatally, with no way to get their money back. Buyers in some states have some recourse if they purchased a dog at a pet store, but they still may have paid vet bills and experienced heartache. By the time most owners realize their dog is sick, they've bonded with the animal. They don't want to trade it in for a new puppy: they want to do everything possible to help their new pet recover.

When he testified before the U.S. Congress in 2006, the Humane Society of the U.S.'s Chief Executive Officer Wayne Pacelle cited three typical examples of puppy mills:

- In Berry, Kentucky, officials found 108 dogs covered in feces, with frozen water bowls. One dog had frozen to death
- In Macomb, Missouri, an Internet dealer had 147 live dogs and four dead ones, all with severely matted fur, suffering eye ailments, hair loss, deafness, blindness, and tumors
- At the home of a breeder in Vero Beach, Florida, authorities discovered 151 dachshunds and springer spaniels so emaciated they were skin and bone

Those puppy mills pale in comparison to more recent discoveries: Nearly 700 malnourished dogs in Lyles, Tennessee (http://www.pet-abuse.com/cases/13942/TN/US/), 750 Chihuahuas and other small breeds at a puppy mill near Tucson, Arizona (http://www.azstarnet.com/metro/229469), and 1080 dogs at a puppy mill in Hillsville, Virginia (http://www.pet-abuse.com/cases/12626/VA/US/).

States are beginning to crack down on puppy mills. Louisiana and Virginia now limit the number of dogs that can be kept in kennels, and Pennsylvania has passed one of the toughest laws in the country; it mandates larger cages for dogs in large-volume kennels, requires breeders to provide exercise for the dogs, and requires that dogs in large-volume kennels undergo twice yearly veterinary exams. Breeders have protested in vain that the new rules would be costly and time-consuming and would drive responsible breeders out of business. Dogs were livestock, they argued, and with livestock it was only natural to expect the occasional deadstock.

In the meantime, there are ways to avoid buying puppy mill dogs. Customers should never buy a dog from a pet store or online. When dealing with breeders, they should insist on visiting the kennel and meeting the parents of the puppy they're looking to buy, and be prepared to walk away from any dealer who refuses to let them. Better yet, they should check out the prospects at their local animal shelter or on www.petfinder.com, which profiles thousands of abandoned dogs in need of a new home.

Further Reading

Humane Society of the United States. Inside a puppy mill. Available at http://stoppuppy mills.org/inside_a_puppy_mill.html

Kohl, Jana. 2008. *A rare breed of love, the true story of Baby and the mission she* inspired to help dogs everywhere. New York: Fireside.

McGowan, Katherine. 2006. How much is that doggie in the window suffering? A lot—but you can help. *Animal Sheltering*, September/October 2006, 32–45.

Williams, Libby. You may know better than to go to a pet shop for a new puppy. But should you go to a breeder instead? Available at http://www.njcapsa.org/index.php?option=com_content&task=view&id=19&Itemid=32

The following Web sites also have information about puppy mills:

www.petshoppuppies.org
www.unitedagainstpuppymills.org
www.pet-abuse.com

Carol Bradley

Q

QUALITY OF LIFE FOR ANIMALS

Content, Richness, and Value

By content, philosophers and others refer to the subjective experiences of nonhuman animals, especially the higher animals. That the higher animals have experiential lives with unfolding sets of experiences is widely accepted today. The nature of these experiences and of the lives that contain them have come to be important for three reasons.

Moral Standing

Some accounts of moral standing or moral consideration turn upon cognitive abilities in human and nonhuman animals alike, and if decisions about how to treat creatures in part turn upon their moral standing, then the cognitive abilities of animals matter. It is sometimes claimed, for example, that in order to have moral standing a creatures must be (1) autonomous, or (2) able to make choices about how to live its life, or (3) able to plan out its life over time, or (4) able to act for reasons, or (5) capable of agency. Depending upon how these notions are unpacked, some creatures will be incapable of these intellectual feats. Thus, this way of conferring moral standing runs into the argument from marginal cases, that is, unfortunate humans, since some human beings are also incapable of these feats. Accordingly, if we nevertheless extend moral standing to these humans, then what reasons do we have for not extending it to at least the higher animals? If we do not extend moral standing to humans with radically impaired lives, then how ever many of them gain entry into the moral domain through the interests of other humans, they count for nothing, morally, in their own right, and so arguably can be treated in the way that other creatures who are not members of the moral community are treated at the present time.

Value of Life

Increasingly today, on all sides, it is recognized that quality of life, not life itself, is what matters essentially. The value of a life is determined by the quality of the life being lived. There is debate over how to determine quality of life, not least over whether the issue is primarily a subjective or an objective one. One of the central difficulties with objective accounts is that, while by objective criteria a life could be going well, by subjective criteria it might be going badly. A person might have all the calories needed to function well, yet still not think there lives are going well. The subjective element is about how the life looks from the point of view of the creature living it, and the subjective element seems to require some account of the subjective experiences of creatures in order to be properly understood. What we want to know in essence is how rich a life is from this point of view, and by richness we refer to such things as the variety, depth, and extensiveness of kinds of experiences.

To hold that we have absolutely no access to the interior lives of animals seems false, at least if we take scientific work by ethologists, cognitive scientists, biologists, and others seriously. Again, to hold that we cannot know exactly what these interior lives are like does not mean that we cannot know a good deal about them, and so can make some very provisional or, indeed, even more permanent, judgments about them. Playing fetch with a dog enriches its life is a case in point.

Of course, in discussing the richness of animal lives, we must not apply criteria appropriate to human lives as if they applied straightforwardly, without further defense, to animals. This would be a second-order form of speciesism. Yet something here does set a kind of presumption of where both empirical science and argument must occur, for it does seem clear that richness of content in our lives is tied in large measure to our capacities for enrichment. Where these capacities are impaired or missing, as with the loss of a sense, a life appears less rich than an ordinary adult life which contains the kinds of experiences that that capacity makes possible. This does not mean that another capacity for richness cannot compensate for this loss, but it does mean that we must be convinced of this.

Thus, at the end of life, when we look back and say of a human that they lived

In this photo provided by the animal rights group Mercy for Animals, chickens in a cage at a California egg farm are shown during a news conference in 2008. The group released a video showing chickens at a major California egg farm being mistreated by workers and housed in cages so small they could not spread their wings. (AP Photo/Handout-Mercy for Animals)

a rich and full life, we refer to an array of kinds of experiences that characterize the lives of normal adult humans. At this level, we take ourselves to mean something far beyond what we would mean were we to say this of the life of a dog, for we take ourselves to have capacities for enrichment that far outstrip anything the dog has. Nothing is settled, of course, by this presumption of argument; it simply means that something must be said in the dog's case, by way of compensation, to make us think that the richness of its life approaches that of the normal adult human. Again, nothing is prejudged. Perhaps one can point to features of one of the dog's capacities that transform its life. That is, is there any single dimension of a dog's life analogous to our lives all our various capacities? If one thinks only of the role of culture, or marriage, or accomplishment of chosen ends in our lives, however, those who wish to contend that the dog's life is as rich as the lives of normal adult humans have a case to make.

Comparative Value of Human and Animal Lives   Everything here is cast in terms of normal adult humans, for the obvious reason that it is false that all humans live lives of equal richness. Some human lives are so wanting in richness and scope of enrichment that we strive mightily to avoid them for ourselves and our families. We do not, despite most religions and devotees of the argument from marginal cases, appear to hold that all human lives are equally valuable. Rather, a quality of life view commits us to another view: if human lives are not approximately equally rich, they are not of equal quality, and if they are not of approximately equal quality, they are not of equal value. In fact, what such a view suggests is that some animals' lives can be of a richness and quality higher than some human lives, such as the brain-dead and anencephalic infants, and so can be of greater value. There is nothing speciesist about this conclusion.

Empirical work on the subjective lives of animals can be seen as necessary for these reasons. It must fit in with a philosophy of mind that makes intelligible to us ways of understanding and appreciating animal experience, with a moral philosophy that enables us to fit animal experience into our account of the value of a life.

R. G. Frey

R

RABBITS

Wild rabbits (*Oryctolagus cuniculus*) have been hunted for fur and for meat in Europe and Asia for thousands of years, but it wasn't until the Middle Ages that rabbits were domesticated. These rabbits were kept in large pens for food and fur, and for hundreds of years bred on their own; later, their keepers selectively bred them for size, temperament, color, and other characteristics. By the early 20th century, following the popularity of Gregor Mendel's work on the inheritance of traits, dozens of breeds of rabbits were created, primarily for the meat, fur, and newly developing show markets. While some people kept rabbits as pets, probably going back hundreds of years, true pet-keeping and the pet industry did not develop until the Victorian era. Today, rabbits are purpose-bred to fulfill the needs of four primary industries: pets, meat, fur, and vivisection.

Rabbits have been kept for meat longer than for any other purpose. In the United States, rabbit breeding remains a cottage industry, and typically takes place in backyard farms, unlike the massive factory farms that produce this nation's chickens or pigs for the table. Also unlike other meat producers, rabbit farmers are relatively unregulated; USDA inspectors only inspect rabbit production facilities when requested to do so by the operator. According to the USDA, these federally-inspected facilities sold two million rabbits for meat in 2001, out of a total of at least 8.5 million rabbits overall in the United States. Worldwide the total is about 800 million per year, primarily centered in France, Italy, and China, where rabbitries are much larger than in the United States (USDA, 2002).

Rabbits bred on meat farms live short, brutal lives. Weaned at four weeks so that their mothers can be bred again, sometimes as soon as 24 hours after they give birth, baby rabbits live together in very small cages until they are slaughtered at 12 weeks. Breeding adults live their entire lives (about two years, as opposed to a pet rabbit's life expectancy of ten or more years), on the other hand, in solitary cages, which is difficult, since they are a social species like primates or dogs, preferring to spend their time in the company of others. In addition to behavioral deprivation, living in small cages for one's entire life leads to disease, broken bones, damaged paws, and other problems. Once the rabbits are ready for slaughter, they are shipped to processing plants in small crates loaded onto trucks, and many die from stress or injury along the way.

Rabbits, like chickens, are not considered to be livestock under the USDA definition of the term. This means they are exempt from the USDA's Humane Methods of Slaughter Act, which is meant to ensure that meat animals are rendered insensitive to pain before they are slaughtered. Rabbits, then, can be killed when fully conscious, by breaking their necks, by hitting them with a blunt object, by decapitation, or by any other means.

The rabbit fur industry in the United States is the smallest of the industries that uses rabbits for profit. Rabbit fur is not considered a luxury fur, and has never had the cachet of mink or sable. On the other hand, because it is cheap to produce, it is often considered a fun fur and is used on everything from cat toys to the trim on cheap clothing aimed at young people with limited disposable income. Today, most rabbit fur and, indeed, a large percentage of fur in general, used in American clothing and products is imported, primarily from China, although the number of rabbits killed for fur annually is not known. Wherever it is produced, rabbit fur is not a byproduct of the meat industry. Instead, fur is taken from rabbits that are slaughtered at six months, while rabbits killed for meat are killed much earlier, at three months.

The newest industry that uses rabbits is the vivisection industry. It wasn't until the mid-20th century that rabbits and other animals began to be purpose-bred specifically for laboratory use. Today, rabbits used in medical experimentation and product testing come from a handful of large laboratory animal suppliers which supply labs with millions of animals per year. Of the animals that must be reported to the USDA (rodents, birds, amphibians, and reptiles are excluded from reporting requirements), rabbits are the most popular laboratory animal in the United States, with 239,720 used in the United States in 2006 (USDA, 2006). The number of rabbits and other animals used in labs every year is dropping, as these animals are being replaced by non-animal substitutes as well as by genetically-modified rodents.

Rabbits are popular for scientific use because they are cheap, as little as $30 apiece, small and relatively docile, and they have short reproductive cycles. Rabbits are used for fertility studies, for product testing, especially toxicity tests on skin and in eyes, and for their antibody production.

While living in the lab, whether at a university, private testing facility, or government-run laboratory, rabbits typically lead lives of isolation. Because most are not surgically sterilized, they are kept in small, single cages to prevent fighting and unwanted reproduction, and typically have nothing to play with and nothing to do. Rabbits, like other laboratory animals, are often observed engaging in stereotypic behavior associated with emotional and psychological deprivation, such as bar licking or paw chewing, and sitting in a hunched position for hours at a time.

On the other hand, some laboratories provide environmental enrichment for their laboratory animals in order to try to meet the animals' psychological and physical needs as well as the requirements of the Animal Welfare Act, which only mandates enrichment for primates and dogs, but recommends it for other animals.

The USDA Animal Welfare Information Center provides resources for the voluntary enrichment of all animals. Suggestions for rabbits include social housing, the ability to forage and dig, and opportunities to run and play. Dozens of studies have been published in the past fifteen years on the benefits of enrichment for rabbits. It's impossible to know, however, how many laboratories have actually implemented any of these suggestions.

The pet rabbit industry is certainly the most seemingly benign of all of the industries that use rabbits for profit, in that it produces rabbits to be purchased as companions in families around the world. Unfortunately, as in the industries discussed above, rabbits suffer here as well.

Pet rabbits are either bred in small backyard rabbitries or in large commercial operations, some of which could be called rabbit mills. In either case, breeder rabbits are generally kept in solitary cages throughout their lives, being bred and giving birth throughout the year. Large-scale operations with annual profits over $500 must be licensed and thus inspected by the USDA, but these inspections are so infrequent that the facilities might as well be unregulated.

From the rabbitries, rabbits are transported via brokers or wholesalers to pet stores around the country, generally in large crowded trucks, when the babies are four to six weeks old. Many rabbits, perhaps as many as 20–30 percent, die during transport, and many die upon arrival at the pet store, thanks to the stress of the travel, the early age at which they were weaned, and the conditions at the store upon arrival. The Animal Welfare Act does not cover the care of animals at pet stores. Once sold, their fate rarely improves. Most pet stores provide minimal care and behavior information on the animals that they sell, and the pet store industry routinely fights legislation that would force them to give out more comprehensive information, so the new owners are often not equipped with the information and supplies that they need to care for their new pet. In the case of rabbits, the situation is complicated by the fact that the pet industry has long marketed rabbits as starter pets and good pets for children when, in fact, caring for a rabbit in anything other than a cage in the backyard is a complicated proposal, given the rabbit's behavioral and physical needs.

Because of the throwaway mentality in the United States and other countries, rabbits, like other animals, are often discarded when they prove too much a burden to care for. The House Rabbit Society, founded in 1988 as the first American rabbit rescue organization, is now the leading advocate for domestic rabbits around the world, rescuing and rehoming domestic rabbits and educating the public on rabbit care and behavior. Since its founding, hundreds of other rabbit rescue groups have formed, all with the aim of helping the tens or hundreds of thousands of rabbits discarded and brought to animal shelters every year. Unfortunately, because much of the public is still uneducated as to rabbits' unique needs, adoption rates at animal shelters remain terribly low.

Today, the plight of rabbits may be improving, thanks to the work of animal advocates around the world and, in particular, rabbit advocacy organizations like House Rabbit Society, and other groups and individuals working to ensure that domestic rabbits are given a fair shake.

Further Reading

Davis, Susan, and DeMello, Margo. 2003. *Stories rabbits tell: A natural and cultural history of a misunderstood creature.* New York: Lantern Press.

Harriman, Marinell. 1985. *House rabbit handbook: How to live with an urban rabbit.* Alameda: Drollery Press.

USDA, 2002 Rabbit Industry Profile, http://www.aphis.usda.gov/vs/ceah/cei/bi/emerg ingmarketcondition_files/RabbitReport1. pdf. Accessed March 29, 2008.

USDA, Annual Report for Inspection, 2006, http://www.aphis.usda.gov/animal_welfare/

downloads/awreports/awreport2006.pdf.
Accessed March 29, 2008.

Margo DeMello

RATS

The term rat refers not to any one species, for there are more than fifty species of true rats in the world today. However, in Western societies, the rat most commonly referred to is the brown or Norway rat, *Rattus norvegicus*. Because this is the species that is so widely used in laboratory experimentation and testing, this short essay will focus mainly on the Norway rat. However, it should be said at the outset that any ethical considerations that may apply to this species also apply to other rats.

Rat Sentience and Awareness

With the exception of vision, rats have more acute sensory perceptions than humans. They use frequent urine marking to communicate by smell. From these marks, rats can discern each other's individual identity, age, reproductive status, and familiarity. Other rats may even detect the marker's social status and stress level from these cues. Rats also communicate with a variety of vocalizations. Compared to humans' hearing range of from about 16-20,000 Hz, a rat's hearing range is about 200-90,000 Hz. Recent analyses of high-speed video reveal that rats use complex whisker movement patterns to explore their environment, much as humans use their fingertips.

During play, rats produce ultrasonic chirps (around 50 kHz) believed to communicate positive feelings. Subordinates studies by neuroscientists Jaak Panksepp and Jeff Burgdorf show that rats will come running to be tickled by a trusted human handler, and that they will utter many more chirps during these interactions than will rats who are merely stroked on the neck. However, these rats lose their willingness to be tickled if there are cats nearby, or if their handlers have punished them, indicating that the rats' response appears to hinge on their feeling comfortable and safe.

Rats are aware, alert, and intelligent. As early as 1948, it was shown that rats form mental maps. When trained rats were placed in mazes and their optimal pathways to food were blocked, the rats created and remembered new paths. Rats can learn some things faster than human children or chimpanzees, such as the association between a shape or pattern and a food reward. Experiments from the University of Georgia published in 2007 indicate that rats demonstrate metacognition, that is, they know what they know. When presented with an easy discrimination task, rats quickly chose the correct answer for a reward. However, when presented with a difficult discrimination, they usually opted to decline the trial by poking their nose into a cone and proceeding directly to the next trial for a small reward, rather than risking failure and earning no reward. Other experiments show that rats grasp the relationship between seeing and doing, and understand cause and effect.

Highly social mammals, rats have evolved behaviors that can be described as considerate or empathetic. A 1959 study titled "Emotional Reactions of Rats to the Pain of Others" showed that rats would stop pressing a bar to obtain food if doing so delivered an electric shock to a rat next to them. In another study, rats pressed a lever to lower to the floor a squirming, vocalizing rat trapped in a suspended harness, but did not respond to a suspended block of Styrofoam. Possibly the Good Samaritan rats merely wanted to stop a disturbing stimulus and were not concerned for the other rat, but in the very least a form of empathy termed emotional contagion was occurring. Rats become stressed when other rats are suffering or being killed in the same room. Scientists have concluded that rats can feel anticipation, surprise, and disappointment, that they experience joy during rough-and-tumble play, and that they may become optimistic or pessimistic depending on their living conditions.

Rats and Humans

Rats have flourished as human commensals. Wherever man goes, rats are likely to follow. It is estimated that there is about one rat for every human living on Earth today. Native to Japan and possibly eastern Asia, *Rattus norvegicus* arrived in Europe later than its cousin the black rat (*Rattus rattus*). Its first known appearance in Europe was around 1553, and in North America around 1775; both introductions are believed to have happened via ship. Much of humankind's ongoing antipathy toward rats originates in the latter's role as a host to fleas bearing the bubonic plague, which killed an estimated one-third of the human population of Europe during the 1340s. Particularly in Western culture, Norway rats have become popular as companion animals. However, it is as subjects of laboratory tests and experiments that most domesticated rats are used today.

Laboratory Use

Tractable, easily maintained, and readily bred in captivity, rats became popular as subjects in laboratory experiments early in the 20th century. Rats rank second only to mice in frequency of laboratory use. Official estimates are that between 3.4 and 3.7 million rats are killed yearly in American research laboratories, though estimates from other sources are as high as 23.6 million. A search on PubMed, the U.S. National Institutes of Health's online database of biomedical journals dating from 1950, yields over 1.26 million hits for the search term rats as of January 2009.

Much of this use is in product testing. One of the most notorious of tests for which rats are commonly used is the LD50 Test, in which animals are exposed to prescribed amounts of a test substance until 50 percent of the subjects die (hence lethal dose 50%, or LD50). Examples of test substances include drugs, industrial chemicals, household cleaners, and cosmetics. Variations on the LD50 test include the LC50 for assessing lethal concentrations of test substances to which rats may be exposed by air (inhalation LD50) or by applications to the skin (dermal LD50). Rats are also used in standard tests of potential cancer-causing substances (carcinogenicity tests), and standard tests for potential birth defects (teratogenicity tests). Cancer test methods may last up to two years, with chronic exposure to the potential carcinogen. Rats and mice are also routinely used in tests for genetic toxicity, immune toxicity, and skin irritancy.

Rats are used as subjects in a wide range of harmful experiments, including studies of sleep deprivation, noise-induced hearing loss, fracture pain, constriction injury, spinal cord injury, burns, and models of depression and pain, to name a few. Depression and despair models include the forced-swim test, tail-immersion test, inescapable electric shock, and other behavioral, drug- or injury-induced models of depression. One of the criticisms of psychological studies using animals is that the subjects are unable to verbalize their symptoms and feelings in the way that human subjects can do.

For various reasons, and despite their overwhelming use, rodents are poor predictors of human outcomes. The LD50 test, for example, yields wildly varying results that have been attributed to a long list of causes, including strain, age, weight, sex, health, diet, temperature, and housing conditions. A study commissioned by the European Communities found that LD50 tests of the same substance conducted at different laboratories yielded LD50 values that differed by as much as a factor of 12. A follow-up trial with methods carefully standardized across labs still came up with eight-fold differences. These are comparisons of rats with other rats; extrapolating to humans is a far greater leap. In an evaluation of cell toxicity tests conducted at several test facilities, researchers evaluated 68 different methods to predict the toxicity of 50 different chemicals. Rat LD50 tests were only 59 percent accurate (lowest possible accuracy is 50%, or chance), but a combined *in vitro* human cell test was 83 percent accurate in predicting actual human toxicity.

Several investigations have shown that animal carcinogenicity tests are poor predictors of human carcinogenicity. The U.S. Food and Drug Administration determined that the overall failure rate for new drugs is 92 percent after they have passed animal testing and entered clinical (human) trials. This failure rate is at least 95 percent for cancer drugs. There are a variety of causes for the poor human predictivity of studies of rodents and other animals. These include differences across species, strain, and gender; differences in aspects of absorption and metabolism of substances; variable responses of organ systems; and the effects of stress experienced by animals in the laboratory setting. Studies show that rats and other animals have a pronounced stress reaction to routine laboratory events, including injections, blood collections, the forced feedings normally used to deliver test substances, and prolonged confinement in small, uninteresting cages. Changes in hormone levels, blood pressure, heart rate, and other factors accompanying stress can greatly influence how the body responds to drugs and other treatments.

Ethical Considerations

In addition to the question of the possible scientific utility of animal studies is the question of whether it is moral to deliberately harm sentient animals in the name of science. Like all mammals, rats are sensitive to pain and pleasure, and they express a range of emotions. It is often assumed that other animals are not capable of suffering as much as humans are, but this idea is tenuous and there is no rigorous science to support it. Pain is equally adaptive to a rat as to a human. A growing number of scientists are beginning to suggest that animals may be more vulnerable to states of suffering than we are. For instance, we can be told, or can rationalize, that a pain will not last for long, whereas an animal like a rat is presumably unable to do so.

Despite this, the Norway rat, along with house mice and birds, is not covered by the U.S. Animal Welfare Act (AWA). Animal welfare organizations have campaigned vigorously to have them covered by the AWA, but without success. The development of non-animal alternatives has been progressing more rapidly in recent years. Practical advantages shown for many of these methods are that they are quicker, cheaper, and more reliable. Adoption of these alternative methods is currently perhaps the most promising avenue by which rats and other animals will be replaced in laboratory research.

Further Reading

Berdoy, M. 2002. The laboratory rat: A natural history. Film. 27 minutes. www.ratlife.org.

Hanson, A. Rat behavior and biology. Website. http://www.ratbehavior.org/history.htm

Walker, E. P. 1964. *Mammals of the world*, Vol. II. Baltimore: The Johns Hopkins University Press.

Jonathan Balcombe

RELIGION AND ANIMALS

Religion influences our understanding of human and animal relations in three principal ways. The first is the contribution that religion makes to our perception. People sometimes refer to religious vision, and by that they mean that there are ways of seeing that are deeply rooted in religious traditions that can enrich our perspective. The way we view the world is indebted to a range of influences, and religion is one of them. well

What are these religious perceptions? In terms of the animal-human bond, they are both negative and positive. Negatively, some religions tend to exalt human power over animals and exclude animals from the bonds of friendship with humans. Perhaps the most extreme version of this tendency can be found in the writings of Thomas Aquinas (1225–1274), who held that friendship with animals was impossible because they are not rational. Since, according to St. Thomas, friendship was only possible with rational creatures, animals were deemed incapable of "fellowship with man in the rational life" (*Summa Theologica,* Part 1, Question 65.3). This strong emphasis on rationality, which in Western religious traditions was denied to animals, has meant variously that they were largely perceived as being without a mind or an immortal soul, and incapable of having a relationship with God.

Although Judaism, Christianity, and Islam all recognize that animals are creatures of God, that their lives belong to God, even that God loves creatures, it remains true that all have given animals a low status in comparison with human beings. There is little in their religious literature that specifically champions relations with animals. Largely in hagiography, the biographies of saints, and early apocryphal Christian literature are relations with animals recognized and celebrated. St. Francis of Assisi is the obvious example, but there are countless other Christian saints of East and West, such as St. David of Garesja, St. Anthony of Padua, St. Catherine of Siena, St. Guthlac of Crowland, St. Werburgh of Chester, and St. Columba of Iona, who befriended animals and had friendly relations with them. St. Francis's idea that animals are our brothers and sisters has had great symbolic power within the Christian tradition, though it appears to have influenced behavior very little. Within Islam, animals and birds belong to communities (Qur'an 6.38) and give praise to God (Qu'ran 24.41), but animals clearly have an inferior status to that of human beings. Animals may be eaten for food and used for clothing. But the Prophet Muhammad required his followers to be merciful when killing. "Kindness to any living creature will be rewarded," the Prophet said.

Eastern religious traditions have envisaged a much stronger bond between animals and humans. Jainism, Hinduism, and Buddhism all offer cosmologies that explicitly link humans and animals. Chief among the animal-inclusive concepts is the notion of *samsara*, the cycle of death and rebirth, which expresses a radical continuity between all living beings. All life exists as in a chain, and all are linked together. From this perspective, animals and humans are not creatures but subjects; all life is in a state of progress or regress determined by *karma,* understood very simply as a moral law of cause and effect. Animals and humans, thus conceived, are obviously interrelated; each individual soul has not just a biography, but also an ancestry.

The second contribution that religion makes concerns values. In the West, the predominant view of animals is that they exist to serve human interests. The originator of this view, or at least its earliest philosophical exponent, was Aristotle (384–322 BC). He maintained that since "nature makes nothing without some end in view, nothing to no purpose, it must be that nature has made them [animals and plants] for the sake of man" (*The Politics*). Although not specifically religious, this became the predominant lens through which later religious thinkers, including Augustine, Aquinas, Luther, and Calvin, interpreted the place of animals.

What Aristotle held to be the end or *telos* of animals became in later Jewish and Christian thought the God-given purpose of animals as well. Even the Hebrew and Christian scriptures were subsequently interpreted in terms of this instrumentalist model. Thus, for example, dominion in Genesis 1:28 came to be seen as God's validation of human supremacy. The irony is that, in its original context, dominion or *radah* meant something quite different, namely God's commission to humans to care for the rest of creation. Proof that this is the correct reading is given in the subsequent verse (29–30) where humans, like animals, are given a vegetarian diet, a situation that is only reversed after the Fall and the Flood (Genesis 9.3f). God's original will in Genesis 1 was therefore for a peaceful, nonviolent creation. But the idea that animals are given for our use, through either the designs of nature or divine providence, has so caught hold that Western society still principally regards animals as tools, machines, commodities, and resources for human use.

In Eastern religion, the idea of *ahimsa*, meaning non-injury or nonviolence, has a long provenance. Arguably, Jainism taught the concept of nonviolence to the world; it has certainly influenced Hinduism and Buddhism, perhaps more widely. Many believe that it is the noblest of all Indian ethical injunctions, expressed in the incomparable words of the venerable Mahavir: "For there is nothing inaccessible for death. All beings are fond of life, hate pain, like pleasure, shun destruction. To all life is dear" (*Acharanga Sutra*).

These words are the result of a simple but profound spiritual discovery: all life is holy, or sacred, or God-given. Life, therefore, has intrinsic value, and all that lives has an interest in living. It does not follow, of course, that all life is accorded the same value. *Samsara* is not an egalitarian doctrine; on the contrary, those who commit misdeeds, or rather those with bad *karma*, are sent back to live as one of the lower forms of life. While life is an interconnecting chain, humans still represent the apex of the moral hierarchy.

The third contribution that religion makes is in terms of practice. How people perceive the world obviously affects what they do. Religious practices can therefore be seen as the embodiment, that is, the physical shape of religious perceptions of animal-human relations. The obvious example is animal sacrifice. It has been said that the most usual characterization of animals in the Hebrew scriptures is as objects for sacrifice. In fact, there are a wide variety of characterizations. For example, they are perceived as creatures, as covenant partners, as possessors of *nephesh* or God-given life, to take only three examples, but it is the case that animals and birds are most regularly used throughout the Hebrew scriptures as a means of sacrificial offering.

Interpreting what this practice means is less than straightforward. As one might expect of any practice lasting more than a thousand years, various interpretations are possible. Negatively, it can most usually be seen in terms of using animals as a means of reparation for human sin or appeasing the divine.

But it is worth pointing out that this is only one of many views. For example, another view is that sacrifice is to be understood as the returning of an animal to the Creator who made it, so that far from involving the gratuitous destruction of a creature, the practice paradoxically involves its liberation, its final union with God. Whatever interpretation is given, it is significant that within the Hebrew Bible there is a developing criticism of the practice as inefficacious or immoral. Psalm 50 describes the Lord opposing sacrifice on the grounds that creatures belong to him:

I do not reprove you for your sacrifices;

your burnt offerings are continually before me.
I will accept no bull from your house,
nor he-goat from your folds.
For every beast of the forest is mine,
the cattle on a thousand hills.
I know all the birds of the air,
and all that moves in the field is mine. (7–11, *RSV*)

The logic of this protest appears to be that humans should not appropriate what in fact belongs to God. Not only are all creatures his, but he also knows them individually and cares for them.

Eastern religious traditions have, however, firmly set themselves against animal sacrifice, though it is true that Islam retains animal sacrifice for major festivals. And, of course, both Judaism and Islam maintain the practice of religious slaughter, called *shehita* and *halal* respectively. Again, Jainism led the way in rejecting animal sacrifice and in commending the way of peaceable living with all nonhuman creatures. In Mahayana Buddhism, the Bodhisattva postpones his own enlightenment in order to save all living things from the cycle of misery and death:

I have made a vow to save all living beings . . . The whole world of living beings I must rescue from the terrors of birth, old age, of sickness, of death and rebirth . . . I must ferry them across the stream of *samsara* . . . I will help all beings to freedom. (*The Bodhisattva's vow of universal redemption*)

This vision of humanity using its power to save other living creatures, and doing so sacrificially, is characteristic of Jain and Buddhist thought, which seeks *moska,* or liberation, for all. But it is not completely unknown in other religious traditions. In Christianity, the redemptive effects of the death of Christ are understood as inclusive of all beings as, for example, in Colossians, where Christ is described as "the first born of all creation." Through Christ, God has determined "to reconcile to himself all things, whether on earth or in heaven, making peace by the blood of the cross" (1:15–19). In Judaism, there is the vision of a future heaven and earth in which the lion lies down with the lamb, where there is universal peace, and "they shall not hurt nor destroy in all my holy mountain" (Isaiah 11: 6–9).

Oddly, there are no religious rites in Eastern traditions that unite concern for animals and humans, or specifically celebrate the animal-human bond. It may be that because *ahimsa* is such a widely accepted practice that no need was felt for any specific rites. In Western traditions, there are likewise no specific rites, except that Catholicism has always accepted the appropriateness of blessings for animals, presumably mirroring God's own blessing of the creatures recorded in Genesis 1: 20–22. These appear in the *Romanum Rituale,* the priest's service manual, first written in 1614 and left virtually untouched until 1952. This provision has enabled animal blessing services, and latterly animal welfare services, arranged by all Christian denominations in the West. These are usually held on St. Francis Day, October 4th, which is now designated World Day for Animals, and the first Sunday of each October is designated Animal Welfare Sunday.

One of the new, unofficial rites especially concerns the celebration of human relationships with companion animals. The service involves the bringing of animals to the front of the church, where their human companion publicly promises to be faithful in care and love, mirroring God's own covenantal care as shown in Genesis. The priest then says: "May the God of the new covenant of Jesus Christ grant you grace to fulfill your promise and to show mercy to other creatures, as God has shown mercy to you" (*Animal Rites,* 1999). Services of celebration and blessing are held in many cathedrals in Britain and America.

There are resources within almost all the religious traditions of the world for a celebration of the animal-human bond. But it must be said that many of the more positive ideas have been obscured by instrumentalist elements which present animals as wholly separate from humans, or which suppose that they exist only to serve us. There is a need for religious traditions, many think, to respond creatively to the new voices of ethical sensitivity to animals that are now increasingly heard in Western society in particular. At the heart of this sensitivity needs to be a reevaluation of human relations with animals, from one of crude dominance to one of friendship and respect. Ironically, although religion is often seen as an anti-progressive force because of its social conservatism, it contains many subtraditions that offer precisely that vision of filial relations with animals.

Baptist preacher Charles Spurgeon once recounted the view of Rowland Hill that a person "was not a true Christian if his [or her] dog or cat were not the better off for it" and commented: "That witness is true" (*First Things First,* 1885). Many believe that the same should also be said of all world religions.

Further Reading

Birch, Charles and Vischer, Lukas. 1997. *Living with the animals: The community of God's creatures*. Geneva: WCC Publications.

Chapple, Christopher Key 1993. *Nonviolence to animals, earth, and self in Asian traditions*. Albany: State University of New York Press.

Foltz, Richard C. 2006. *Animals in Islamic tradition and Muslim cultures*. Oxford: One World.

Kapleau, Philip 1981. *To cherish all life: A Buddhist case for becoming vegetarian*. Rochester, NY: The Zen Center.

Linzey, Andrew. 1994. *Animal theology*. London: SCM Press and Chicago: University of Illinois Press.

Linzey, Andrew. 1999. *Animal Gospel: Christian faith as if animals mattered*. London: Hodder and Stoughton and Louisville, Kentucky: Westminster John Knox Press.

Linzey, Andrew 1999. *Animal rites: Liturgies of animal care*. London: SCM Press and Cleveland, OH: The Pilgrim Press.

Linzey, Andrew, and Yamamoto, Dorothy, eds. 1998. *Animals on the agenda: Questions about animals for theology and ethics*. London: SCM Press and Chicago: University of Illinois Press.

McDaniel, Jay B. 1989. *Of God and pelicans: A theology of reverence for life*. Louisville, KY: Westminster John Knox Press.

Phelps, Norm 2002. *The dominion of love: Animal rights According to the Bible*. New York: Lantern Books.

Robinson, Neal, ed. 1991. *The sayings of Muhammad*. London: Duckworth.

Sorrell, Roger D. 1988. *St. Francis of Assisi and nature: Tradition and innovation in Western Christian attitudes toward the environment*. New York: Oxford University Press.

Walters, Kerry S., and Portmess, Lisa, eds. 2001. *Religious vegetarianism from Hesiod to the Dalai Lama*. Albany, NY: State University of New York Press.

Webb, Stephen H. 1998. *On God and dogs: A Christian theology of compassion for animals*. New York: Oxford University Press.

Andrew Linzey

RELIGION AND ANIMALS: ANIMAL THEOLOGY

Animal theology relates Christian thinking to contemporary debates about the status and rights of the nonhuman animals. It seeks to address and redress the failure of historical theology to take seriously alternative insights that lie largely silent within the Christian tradition. Systematic theology has largely proceeded on the basis of the virtual nonexistence of animals. Historically, they have been the outcasts of theology, defined as beings with no mind, reason, immortal soul, or moral status. Basic questions about their status and significance have simply not been addressed. The question raised by animal theology is whether Christian doctrine is necessarily speciesist, and whether it can incorporate animal-centered concerns into mainstream thinking. Modern theologians argue variously that even conservative theological understandings can be enhanced and deepened by the adoption and development of these insights.

In terms of traditional doctrine, there are three main areas. The first is creation. Much theological emphasis has been laid on the special creation of humans, to the detriment of nonhumans. But the special place of humanity in creation can be read another way: as support for the special role of humanity in looking after the world, not as the master but as "the servant species" (Linzey, 1994).

The second area is incarnation. Traditional doctrine affirms that God became human in the person of Jesus Christ. While this is frequently taken as a vindication of human uniqueness, some church fathers have argued that the incarnation is the raising up of all fleshly substance (*ousia*) to be with God; the Word becoming flesh affirms all flesh, animal and human.

The third area is redemption. While much traditional interpretation excludes animals directly or indirectly from the sphere of God's redemptive purposes, it can be argued that notions of ultimate justice specifically require animal immortality. Viewed from this threefold perspective, God creates, unites, and redeems all living beings, and the focus of this divine work is not just the human species but specifically sentient, fleshly creatures.

Apart from the plausibility of these reinterpretations, there is one reason why theology needs to take animals more seriously. It lies in the traditional claim that the *Logos* (John 1.3) is the source of all life, because if so, it must follow that a theology based on the *Logos* must be able to render an account not just of the human species but of the entire created universe. In other words, the implicit promise of traditional theology is that it will deliver us from humanocentricity.

See also Blessings of the Animals Rituals; Religion and Animals—Christianity

Further Reading

Linzey, Andrew. 1987. *Christianity and the rights of animals*. London: SPCK and New York: Crossroad.

Linzey, Andrew. 1995. *Animal theology*. London: SCM Press and Urbana: University of Illinois Press

Linzey, Andrew. 2007. *Creatures of the same God: Explorations in animal theology*. Winchester: University of Winchester Press.

McDaniel, Jay B. 1989. *Of god and pelicans: A theology of reverence for life*. Louisville: Westminster/John Knox Press.

Pinches, Charles, and McDaniel, Jay B., eds. 1993. *Good news for Animals? Christian approaches to animal well-being*. Maryknoll, NY: Orbis Books.

Webb, Stephen H. 1998. *On God and dogs: A Christian theology of compassion for animals*. New York: Oxford University Press.

Andrew Linzey

RELIGION AND ANIMALS: BUDDHISM

The Buddhist tradition is, like all of the major religious traditions in the world, a series of diverse and sometimes even contradictory religious phenomena. For this reason, overly simplistic generalizations about large issues can be misleading. Attitudes toward animals, however, are one of the few areas where generalizations can be made. Generally the Buddhist tradition is unconcerned with any systematic exploration of the physical world, including the realities of nonhuman animals. It accepts most of the views of nonhuman animals that are important in the cultures and subcultures where Buddhism has developed.

At its core, Buddhism is a salvation-like concern, usually referred to as liberation, for the individual. Theoretically, each individual Buddhist attempts to discover about himself or herself the basic features of existence experienced by the tradition's founder, Gotama. Referred to often as the historical Buddha (because Buddhists believe there were previous Buddhas many eons ago), Gotama lived in the fifth or sixth century BCE. His core religious teaching was that each living being has, in the end, no lasting self. Similarly, there is no eternal deity or creator of the Earth. Instead, all is in process and subject to change.

The unifying elements in the tradition are reverence of some kind for Gotama and his basic religious insight, and a strong, consistent, hermitlike tradition under which monks, nuns, and others adhere to time-honored rules of conduct. It is this tradition that has provided a relative unity and stability to the moral code.

Buddhist monks, who have been described by scholars as even more important in their own tradition than the church is in the Christian tradition, put together an extensive monastic code known as the *Vinaya* which reveals that early Buddhists accepted the view that all animals other than humans belong to one realm which is lower than that of human beings. Even though early Buddhists claimed that all nonhuman animals, from the simplest of living forms on up to the most complex, such as the large-brained social mammals, form a single kingdom that does not include humans, in a general way the tradition displayed poor awareness of the intimate details of the lives of animals. This may explain in part why Buddhists lumped all nonhuman animals together in a group below humans in the hierarchy of the universe.

In one very important way, however, Buddhism was clearly revolutionary with regard to the moral significance of animals, for Buddhism, along with Jainism, was important in opposing the sacrifice of animals that was part of the brahminical tradition in India which was the forerunner of the Hindu tradition. Similarly, the Buddhist tradition spread important precepts, or moral undertakings, that affirmed that killing other sentient beings was a violation of the most basic moral norms of the universe. The first precept in the tradition, which is also found in Jainism, is "I undertake to abstain from the destruction of life." This is an ethical commitment that the tradition has from its very beginnings identified as part of

Buddha Siddhartha Guataman with a cow, circa 500 BCE. Buddhists believe animals are sentient beings, and thus should not be killed. (Photos.com)

the core of religious living. The idea of our community, then, for a Buddhist, is not to be taken in the narrow sense of human society alone, but in the broader sense of a shared community comprised of all living or sentient beings.

There is another, less favorable side to the Buddhist view of animals, however. The way in which early Buddhists talked about animals reveals that they thought about them in rather negative ways. For Buddhists, any animal other than a human was in an inferior position and could, if it lived a perfect life, be reborn as a human. Similarly, if a human lived immorally, he or she would be punished by being reincarnated as a nonhuman. The tradition relied, as did all of the major religions born in the Indian subcontinent, on reincarnation as an explanation for the justice of any being's present status. Reincarnation explained not only why humans were superior to any nonhuman animal, but also functioned as a justification for many of the social divisions of the day, although Gotama himself resisted the notion that humans in the lower social divisions were less important than high-status individuals. Nonetheless, rich humans were deemed to have been rewarded for past good deeds, and the lame, the mentally disabled, and other unfortunate humans were deemed to be paying for bad acts in past lives. Below even the most unfortunate and morally corrupt humans were all other animals. The Buddhist tradition, through acceptance of these hierarchical notions of life, thus often tolerated some harsh abuses of animals. Elephants, whose natural history was poorly known by Buddhists, were captured from the wild, tamed with painful methods, and used in many different ways. Buddhism did not give approval to all such uses, for example, the use of elephants in war was condemned, but other uses of elephants, such as kings or rajahs using domesticated elephants for transportation, was widely accepted. Early Buddhists consistently spoke as if rich humans were entitled to ride around on elephants, having lived past lives in such a way as to justly deserve this reward. Sadly, though, the Buddhist scriptures also contain many indications that elephants suffered during captivity, being deprived of their naturally complex social lives with other elephants.

Further Reading

Chalmers, R., trans. 1926–1927. *Further dialogues of the Buddha (translated from the Pali of the Majjhima Nikaaya)*, 2 vols. Sacred Books of the Buddhists series, 5 and 6. London: Humphrey Milford/Oxford University.

Gombrich, Richard. 1991. The Buddhist way. In Heinz Bechert and Richard Gombrich, eds., *The world of Buddhism: Buddhist monks and nuns in society and culture*, 9–14. London: Thames and Hudson.

Gombrich, Richard. 1988. *Theravada Buddhism: A social history from ancient benares to modern Colombo*. London and New York: Routledge.

Keown, Damien. 1992. *The Nature of Buddhist Ethics*. London: Macmillan

Keown, Damien. 1995. *Buddhism and bioethics*. London: Macmillan and New York: St. Martin's Press.

Schmithausen, Lambert. 1991. *Buddhism and nature: The lecture delivered on the occasion of the EXPO 1990: An enlarged version with notes*. Tokyo: International Institute for Buddhist Studies.

Story, Francis. 1964. *The place of animals in Buddhism*. Kandy, Ceylon: Buddhist Publication Society.

Waldau, Paul. 1997. Buddhism and animal rights. In Damien Keown, ed., *Buddhism and contemporary issues*. Oxford: Oxford University Press.

Waldau, Paul, and Patton, Kimberley. 2006. *A communion of subjects: Animals in religion, science and ethics*. New York: Columbia University Press.

Williams, Paul. 1994. *Mahayana Buddhism: The doctrinal foundations*. London and New York: Routledge.

Paul Waldau

RELIGION AND ANIMALS: CHRISTIANITY

Many of the important ideas that have governed our understanding and treatment of animals arise from Christian and Jewish sources or from reaction to, development of, or opposition to these sources. Many animal lovers maintain that Christian indifference has been one of the main causes of the low status of animals. Within the Christian tradition in almost every period of history there were both strong negative and positive ideas and attitudes toward animals. Though it is true that largely negative ideas have predominated, it would be false to suppose that sub-traditions have not sustained alternative and sometimes radical viewpoints.

There are three major negative tendencies. The first may be called instrumentalism, the view that animals are here for human use. St. Thomas Aquinas, interpreting Aristotle, held that in hierarchy that God created, animals were the intellectual inferiors of humans and were made essentially for human use. According to this view, the purpose of animals was primarily, if not exclusively, for the service of human subjects. Second, and allied to instrumentalism, there has been a consistent humanocentricity or anthropocentrism that has effectively defined animals out of the moral picture. This has been achieved largely through the emphasis upon certain perceived differences between humans and animals. Animals are judged as beings with no reason or immortal soul who are incapable of friendship with human subjects. From this it has been deduced that humans have no direct duties to animals because they are not moral subjects of worth in themselves. Many contemporary secular theories, for example contractualism, owe their origin to this developing Scholastic view that animals do not form part of a moral community with human beings. The third tendency may be described as dualism, the way Western culture has made distinctions and separations between, for example, the rational and non-rational, flesh and spirit, and mind and matter. Animals are still viewed as being on the wrong side of these desirable attributes, the most important of which is rationality. As Scholastic philosophy

and theology began to stress the centrality of rational intellect, and since it was almost universally accepted that animals had none, it followed that animals had no moral status. Rationality became, and in many ways still is, the key to moral significance.

But in order to see the broader picture, we need to set alongside these negative tendencies a range of positive insights, many of which are clearly biblical in origin. Three are presented here. The first centers on the notion of dominion found in Genesis 1:28. Although dominion has often been interpreted as little less than tyranny, in its original context it meant that humans had a God-given responsibility to care for the Earth, confirmed by the fact that the subsequent verses command a vegetarian diet and envisage a world in Sabbath harmony. A rival interpretation of dominion as stewardship or responsibility can be traced back to the earliest Christian writers, and came to the fore in the emergence of 18th- and 19th-century zoophily or love of animals. The second concerns the notion of covenant found in Genesis 9. Against the prevailing notion that humans and animals are utterly separate, the idea of God's covenant with all living creatures kept alive the sense of a wider kinship. The third positive insight is preserved in the notion of moral generosity, which came to prominence in the emergence of the humanitarian movements of the 19th century. According to this perspective, we owe animals charity, benevolence, and merciful treatment. Cruelty was judged incompatible with Christian discipleship; to act cruelly or to kill wantonly, was ungenerous, a practical sign of ingratitude to the Creator. The Christian tradition, which had in many ways supported, defended, and provided the ideological justification for the abuse of animals in previous centuries, came to spearhead a new movement for animal protection.

See also Dominionism; Moral Standing of Animals; Religion and Animals—Judaism

Further Reading

Gunton, Colin E. 1992. *Christ and creation: The Didsbury lectures*. London: Paternoster Press.

Joranson, Philip N., and Butigan, Ken, eds. 1984. *Cry of the environment: Rebuilding the Christian creation tradition*. Santa Fe: Bear and Company.

Linzey, Andrew. 1987. *Christianity and the rights of animals*. London: SPCK; New York: Crossroad.

Linzey, Andrew and Clarke, Paul Barry, eds. 2004. *Animal rights: A historical anthology*. New York: Columbia University Press.

Linzey, Andrew, and Cohn-Sherbok, Dan. 1997. *Celebrating animals in Judaism and Christianity*. London: Cassell.

Linzey, Andrew, and Regan, Tom, eds. 2008. *Animals and Christianity: A book of readings*. London: SPCK, 1989; New York; Crossroad, 1989; Eugene, Oregon: Wipf and Stock.

Murray, Robert. 1992. *The cosmic covenant: Biblical themes of justice, peace, and the integrity of creation*. London: Sheed and Ward.

Santmire, H. Paul. 1985. *The travail of nature: The ambiguous ecological promise of Christian theology*. Philadelphia: Fortress Press.

Thomas, Keith. 1983. *Man and the natural world: A history of the modern sensibility*. New York: Pantheon Books.

Andrew Linzey

RELIGION AND ANIMALS: DAOISM

Daoism (sometimes written as Taoism), one of the great religions of China, provides people with a rich sampling of core teachings that encourages humans to foster a morality and a lifestyle that will allow nonhuman animals to live freely and peacefully alongside human beings.

The Dao or the Way permeates all that exists, and is therefore present in each creature. Dao, residing in every cow and chicken, offers a measure of perfection to every living being. A contemporary Daoist notes that, in all creatures, "there is the numinous presence of the Dao" (Komjathy).

The great Daoist masters teach that no individual is isolated or enduring; everything that exists is part of a great and ongoing transformation. Daoists therefore acknowledge a link between each entity and every other entity, whether lizard, human, or vulture. We may prefer not to see ourselves in nose-picking apes or scrapping cats, we may prefer to envision ourselves as civilized, educated, or highly intelligent, but Daoism acknowledges humans as mere creatures of the earth, who share critical similarities with other living beings, and who will ultimately decompose and be recycled into other beings and objects in this ever-transforming cosmos. Daoist traditions do not envision a barrier or separation between people and animals. Dao unites humans and animals as common creatures of Planet Earth.

Daoism fosters a sense of humans as an intimate part of a much larger whole; a human is of no greater importance than a turkey or a piglet. Every hen and toad shares in this Great Unity of Being. In the words of Zhuangzi, second only to Lao Tzu: "Although the myriad things are many, their order is one . . . The universe and I exist together, and all things and I are one" (Chan 204, 186). Consequently, harmony pervades the Daoist cosmos.

Daoist philosophy harbors three interrelated moral ideals that are important to understand with regard to animals: *ci* (compassion or deep love), *jian* (restraint or frugality), and *bugan wei tianxia xian* (not daring to be at the forefront of the world). *Ci* is a deep caring and compassion, which requires gentleness and attentiveness to the needs of all species. Early Daoists speak against harming any living being, even the wriggling worm. The second to last sentence in the *Dao de jing*, the primary text of Daoism, reminds readers, "The Way of Heaven is to benefit others and not to injure" (Chan, 176).

Ci and *jian*, when practiced together, encourage people to live simply out of compassion, to live simply so that other creatures might live without being harmed or crowded from the planet. Those who have compassion for other creatures avoid destroying habitat, and do not exploit cattle or hens for the luxuries of eating flesh, reproductive eggs, or nursing milk.

Ci and *jian* are reflected in *bugan wei tianxia xian*. To care about other creatures, to live a life of restraint and frugality, stems from "not daring to be at the forefront of the world." When we place ourselves in the forefront, we push other creatures to the back. If we imagine that we, or our needs, are more important than other creatures or their needs, then our lives become cruel and exploitative. *Bugan wei tianxia xian* teaches people to take their humble place in the universe, allowing other creatures to do the same.

The Daoist concept of *Wu wei,* action as nonaction, cautions humans, highlighting our limitations and noting that we are merely average members of a large and complicated universe. *Wu wei* reminds people that nature requires no human alterations or refinements, and that any such attempts are likely to lead to ruin. The *Dao de jing* notes that "Racing and hunting cause one's mind

to be mad" (#12) and that "Fish should not be taken away from the water" (#36) (Chan, 145, 157). Breeding to acquire fatter cattle, debeaking, artificial insemination, and genetic manipulation are all contrary to *wu wei*. The Dao, which lies behind the smooth functioning of the universe, operates best without human meddling. Daoism teaches people to avoid aggressive and controlling practices such as factory farming or animal experimentation.

Daoism teaches that, if people would leave animals alone, all species will enjoy a golden age of ultimate integrity. In this world, animals will not fear humans, nor will they be domesticated or exploited. Zhuangzi states: "A horse or a cow has four feet. That is Nature. Put a halter around the horse's head and put a string through the cow's nose, that is man." Therefore it is said, "Do not let man destroy Nature" (Chan, 207). Training an animal, in Daoist teaching, is inherently harmful and cruel; training horses turns happy equines into brigands (Mair, 82). Freedom, the ability to live one's life without disturbance or the control of another, is understood to be no less ideal for horses or cattle than for human beings (Anderson, 278). And if taming doesn't turn horses into brigands, Zhuangzi suggests, it will kill them.

Daoism teaches that all things natural are preferable to human contrivance. For example, humans often imagine that animals are better off in human care, where food and water are abundant. Zhuangzi disagrees: The "marsh pheasant has to take ten steps before it finds something to pick at and has to take a hundred steps before it gets a drink. But the pheasant would prefer not to be raised in a cage where, though you treat it like a king, its spirit would not thrive" (Mair, 27).

Nonhuman animals are best left free, in their natural state.

Animals are explicitly protected by an array of Daoist precepts, the first of which is almost always an injunction not to kill. *The 180 Precepts of Lord Lao* (*Yibaibashijie,* fifth century), one of the oldest Daoist compositions, warns against killing animals, eating flesh, and harming animals (insects, birds, and mammals) by disrupting their homes, destroying their families, or through abuse and overwork. Other Daoist precepts specifically denounce slaughtering domestic animals, shooting wild animals including birds, setting traps to catch fish, capturing animals including birds, imprisoning animals in cages, digging creatures out of the earth, or even startling animals. *The Great Precepts of the Highest Ranks* (fifth century) offers a list of affirmative precepts, three out of six of which focus on the protection and benefit of other species:

> Give wisely to the birds and beasts, to all species of living creatures. Take from your own mouth to feed them, let there be none left unloved or not cherished. May they be full and satisfied generation after generation. May they always be born in the realm of blessedness.
>
> Save all that wriggles and runs, all the multitude of living beings. Allow them all to reach fulfillment and prevent them from suffering an early death. May they all have lives in prosperity and plenty. May they never step into the multiple adversities.
>
> Always practice compassion in your heart, commiserating with all. Liberate living beings from captivity and rescue them from danger. (Kohn, 175)

Daoist monastic practice forbids violence of any kind, including the taking of flesh; for centuries, monastery meals have consisted of rice, wheat, and barley, combined with various vegetables and tofu. Meat is not included in the five main food groups.

Daoism, which fosters compassion, the simple life, and humility, which discourages arrogance, exploitation, or manipulation of any kind, provides the basis for a remarkably animal-friendly religion.

Further Reading

Anderson, E. N., and Raphals, Lisa. 2006. Daoism and animals. In Paul Waldau and Kimberley Patton, eds., *A communion of subjects: Animals in religion, science, and ethics*, 275–290. New York: Columbia University Press.

Chan Wing-tsit, ed. and trans. 1963. *A source book in Chinese philosophy.* Princeton: Princeton University Press.

Henricks, Robert G., trans. 1989. *Lao-Tzu Te-Tao Ching: A new translation based on the recently discovered Ma-wang-tui texts.* New York: Ballantine.

Kirkland, Russell. 2001. "Responsible non-action" in a natural world: Perspectives from the Neiye, Zhuangzi, and Daode Jing. In N. J. Girardot et al., eds. *Daoism and ecology: Ways within a cosmic landscape*, 283–304. Cambridge: Harvard University Press.

Kohn, Livia, ed. 2000. *Daoism handbook.* Leiden: Brill.

Komjathy, Louis. 2008. Meat avoidance in Daoism. In Lisa Kemmerer and Anthony J. Nocella II, eds., *Call to compassion: World religions and animal advocacy.* New York: Lantern.

Mair, Victor H., ed. 1994. *Wandering on the way: Early Taoist tales and parables of Chuang Tzu.* New York: Bantam.

Merton, Thomas. 1965. *The way of Chuang Tzu.* New York: New Directions.

Tu, Wei-ming. 1989. The Continuity of Being: Chinese Visions of Nature. In J. Baird Callicott and Roger T. Ames, eds., *Nature in Asian traditions of thought: Essays in environmental philosophy*, 67–78. Albany: State University of New York.

Wing-Tsit, Chan, trans. 1973. *Tao te jing.* Attributed to Lao Tzu. In *A source book in Chinese philosophy.* Princeton, NJ: Princeton University Press.

Lisa Kemmerer

RELIGION AND ANIMALS: DISENSOULMENT

Disensoulment is the stripping away of the spirit powers or souls of animals and of the sanctity of the living world. This process occurred over the centuries, as early herders and farmers intensively exploited animals and nature and needed new myths and other psychic levers to resolve their very old beliefs in animals as First Beings, teachers, tribal ancestors, and the souls of the living world.

In the ancient Middle East, the cradle of Western culture, where animal husbandry was the key to nation and wealth building, agrarian societies invented misothery and other ideas that aided in the debasement of animals. There, the builders of the bustling city-states preached misothery in their arts and in their rising new agrarian religions. In these, the essential message was to debase animals and nature and to elevate human beings over them. The effect, spiritually speaking, was to turn the world upside down. Before domestication, the powerful souls or supernaturals or gods were animals, and primal people looked up to them; after domestication, the gods were humanoid, and people looked down on animals. In primal culture, all beings had souls, of which the greatest was the tribe's totem animal; in agriculture, humans alone have souls, and god is in human form. Animal-using agrarians stripped animals of their souls and pow-

ers and put them in what they perceived to be their proper place: far beneath, and in the service of, humankind.

See also Dominionism; Misothery

Further Reading

Campbell, Joseph. 1988. *The way of the seeded Earth,* vol. 2 of *Historical atlas of world mythology.* San Francisco: Harper and Row.
Eisler, Riane. 1987. *The chalice and the blade.* San Francisco: Harper and Row.
Fisher, Elizabeth. 1979. *Woman's creation.* Garden City, NY: Anchor Press/Doubleday.
Lerner, Gerda. 1986. *The creation of patriarchy.* New York: Oxford University Press.

Jim Mason

RELIGION AND ANIMALS: HINDUISM

Hinduism, the oldest of the major religious traditions, is not a single religion, but an umbrella under which one finds very different kinds of beliefs. These include, among others, Vaishnavism, Shaivism, Shaaktism, and Tantrism, each of which in turn is a complex religious tradition that has many forms of its own. The term Hinduism was coined by European scholars in the 19th century as a description of native beliefs, other than Buddhism and Islam, which occurred in the Indian subcontinent. Hindus' beliefs are startlingly diverse, such that nontheistic beliefs sometimes coexist with theistic and devotional beliefs.

In this diverse tradition we call Hinduism, there is no single view of animals. However, the many different views one finds in Hinduism are dominated by two general beliefs that govern the ways in which nonhuman animals are conceived. First, human beings, though recognized to be in a continuum with other animals, are considered the model of what biological life should be. A corollary of this first belief is the claim that the status of human is far above the status of any other animal. The second general belief in various forms of Hinduism is that any living being's current position in the cycle of life, created by repeated incarnations, is determined by the strict law of karma. Belief in reincarnation is the hallmark of most, though not all, Hindu beliefs. These two beliefs have resulted in other animals being viewed with uncertainty. In a positive sense, animals have been understood to have souls just as do humans. In a negative sense, they have been understood to be inferior to any human, a corollary of which is the belief that the lives of animals must be particularly unhappy, at least compared to human existence.

Importantly, humans are by no means considered equal to one another in classical Hinduism, for according to the *sanatana dharma*, the eternal law or moral structure of the universe, human beings are not born equal to one another. Each human is born into that station in life for which their past karma has fitted them. Inequalities within the social system, then, are not viewed as unjust; rather, they are seen as merely the result of good or bad deeds performed in former lives. A common claim is that those who act morally are assured of a good rebirth in higher social classes, while wrongdoers are assured of being reborn into the wombs of outcasts or, worse yet, as nonhuman animal.

Despite all this, the tradition has often exhibited great sensitivity to animals. In the Srima Bhagavantam, the believer is told, "One should treat animals such as deer, camels, asses, monkeys, snakes, birds, and flies exactly like one's own children" (7.14.9, Prime, 51). A con-

temporary Hindu environmental ethicist argues, "All lives, human or nonhuman, are of equal value and all have the same right to existence" (Dwivedi, 203). More generally, the economics of village life in India provide many examples of coexistence with animals and environmentally sensitive ways of living.

The tradition has truly vast sources. Hindu scriptures, for example, are more than ten times the length of the Bible, and some do support the view that humans have no special privilege or authority over other creatures, but instead have moral obligations to protect other living beings. Arguments in favor of an obligation to protect other living beings rely on the widespread belief that many Hindu deities, such as Rama and Krishna, closely associated with monkeys and cows, respectively, have been incarnated as animals. In addition, the deities worshipped in India include Ganesh, an elephant-headed god, and Hanuman, the monkey god.

This sensitive side of the Hindu awareness of animals is often symbolized by the image of sacred cows wandering the streets of India unmolested and free; yet the realities for animals in Hindu societies have been and continue to be far more complicated. The traditional respect for animals has been affected greatly by economic factors that inhibit transmission of ancient values encouraging respect for animals. Nowadays, the pace of India's development as one of the leading industrialized nations is leaving behind the strong emphasis that almost all Hindu scriptures place on the innate sacredness of animals. Thus, while there is throughout the Hindu tradition a culturally significant sense of the continuity of all life, the already-pronounced sense of discontinuity between humans and animals threatens to change for the worse.

One important ancient form of the tradition, sometimes known as brahminical religion, was challenged by the Buddhist and Jain traditions because it was, as were so many ancient religions, characterized by a heavy emphasis on animal sacrifice. This practice stemmed from the ancient scriptures known as the Vedas. The Jains and Buddhists challenged these sacrifices as cruel and unethical, and thereby had a great effect on the later Hindu views of the decency of intentionally sacrificing animals. *Ahimsa,* the historically important emphasis on nonviolence, has now become a central feature of the Hindu tradition, and some Hindu groups even advance vegetarianism as essential for a morally upright life.

Hindu social codes, embodied in the ancient Laws of Manu, continue in some ways to support a one-dimensional view of animals as completely inferior to humans. This belief that all animals are qualitatively inferior to any human is also reflected in some of the myths of the origin of animals. For example, one important myth, the Purusa Sukta in the Rig Veda, attributes the origin of all nonhuman animals to the leftover parts of a primal male (*purusa*) sacrificed by the gods. Thus, in the Hindu tradition, as with the Buddhists and with Plato (*Timaeus*) in the West, animals are seen by many as having their origin in, and thus being a degenerate form of, elevated humanity.

See also Religion and Animals—Buddhism; Religion and Animals—Jainism

Further Reading

Basham, A. L. 1990. *The sacred cow: The evolution of classical Hinduism,* ed. Kenneth G. Zisk. London: Rider.
Chapple, Christopher Key, 1993. *Nonviolence to animals, earth, and self in Asian traditions.* Albany: State University of New York Press.
Dwivedi, O. P. 1990. *Satyagraha* for Conservation: Awakening the Spirit of Hinduism. In J. Ronald Engel and Joan Gibb Engel, eds., *Ethics of environment and development: Global challenge, international response,* 202–212. London: Bellhaven Press.
Hardy, Friedhelm. 1994. *The religious culture of India: Power, love, and wisdom.* Cambridge: Cambridge University Press.
Prime, Ranchor. 1992. *Hinduism and ecology: Seeds of truth.* London: Cassell.
Waldau, Paul, and Patton, Kimberley. 2006. *A communion of subjects: Animals in religion, science and ethics.* New York: Columbia University Press.
Zaehner, R. C. 1966. *Hinduism.* New York: Oxford University Press.

Paul Waldau

RELIGION AND ANIMALS: ISLAM

Based on a reading of the standard textual sources for the Islamic tradition, including the Qur'an, reports about the prophet Muhammad (*haidīth*s), and the classical legal texts (*fiqh*), several general points emerge in terms of animal rights. The Islamic textual tradition takes the relationship between humans and other animal species quite seriously, in contrast to Christianity, where this relationship is scarcely mentioned. Nonhuman animals are seen as having feelings and interests of their own, and the overriding ethos enjoined upon humans is one of compassionate consideration. Humans are seen as occupying a special place in Creation, that of Allah's deputies (*khalīfa*), but they are to exercise this role responsibly. Based on textual sources, it would seem that the Islamic ethical system extends moral consideration to nonhuman animals, although not on the same level as humans.

Ritual Slaughter

Ritual slaughter of animals for food (*dhabh*) is said to follow the principle of compassion for the animal being killed. According to a hadith, Muhammad enjoined his followers to "kill in a good way," stating that "every one of you should sharpen his knife, and let the slaughtered animal die comfortably." Yet, on another occasion, when Muhammad saw a man sharpening his knife while an animal waited nearby, he reprimanded him, "Do you wish to slaughter this animal twice, once by sharpening your blade in front of it and another time by cutting its throat?"

Ritual sacrifice, such as that customarily performed by Muslims on the occasion of 'Īd al-Adha, is not prescribed as a duty in the Qur'an, but a hadith is sometimes cited to provide the sense that it is an obligation. Whether or not Muslims are obligated to perform a blood sacrifice during 'Īd al-Adha has recently become a matter of debate.

The Qur'an and the hadiths are the main sources, along with analogical reasoning and consensus among scholars, for the body of Islamic law known as the *sharī'a*. Shairī'a law assumes without question that humans will make use of animals and eat them. The legal questions therefore center on how to define and circumscribe the limits of these behaviors. The issues are which animals to eat, how to kill them properly in preparation for eating and, to a lesser extent, what responsibilities humans have to the animals that serve them. Questions about whether humans have the innate right to do these things do not arise.

Islamic laws pertaining to animals are included under categories such as their treatment, their sale, how to include them in *zakāt* calculations, their lawfulness as food, prescriptions for slaughter, and restrictions on hunting. Thus, animals are discussed in terms of both their use by humans and, less extensively, the obligations humans have toward them.

The various schools of law each classified all known animals in terms of whether eating them was *halāl* (permissible), *harām* (forbidden), or *makrūh* (discouraged). All schools placed the vast majority of animals in the first, permitted category. Some animals presented special cases; frogs, for example, which would normally meet the conditions for a *halāl* designation, were determined to be *harām* on the basis of a hadith in which Muhammad forbade the eating of frogs.

Differences among the schools regarding these classifications occur mainly in cases of reasoning by analogy, such as whether or not to forbid the eating of animals that have similar names to those of forbidden animals, for example "dogfish." Another kind of ambiguity arises when an animal that would normally be considered *halāl* (such as an eel, which is a kind of fish) resembles an animal which is *harām* (for example, the snake, to which eels appear similar). The Maliki and Shafi'i schools allow the eating of fish found floating dead in the water, whereas other schools forbid it. Various schools disagree over the lawfulness of eating crustaceans and insects. Carnivores, which are *harām*, are identified in the legal tradition by their possession of fangs or claws; thus, there is disagreement over the lawfulness of eating elephants, because, although they are herbivores, their tusks resemble fangs.

Human Obligations to Domestic Animals

The Shafi'i jurist 'Izz al-din ibn 'Abd al-salam al-Sulami (d. 1262), in his legal treatise *Rules for Judgment in the Cases of Living Beings (Qawā'id al-ahkām fī masālih al-anām)*, has the following to say about a person's obligations toward his domestic animals:

- He should spend [time, money or effort] on it, even if the animal is aged or diseased in such a way that no benefit is expected from it. His spending should be equal to that on a similar animal useful to him

- He should not overburden it

- He should not place with it anything that might cause it harm, whether of the same kind or a different species

- He should kill it properly and with consideration; he should not cut its skin or bones until its body has become cold and its life has passed fully away

- He should not kill an animal's young within its sight

- He should give his animals different resting shelters and watering places, which should all be cleaned regularly

- He should put the male and female in the same place during their mating season

- He should not hunt a wild animal with a tool that breaks bones, which would render it unlawful for eating (cited in Izzi Dien, 2000, pp. 45–46)

Although the rights of nonhuman animals are guaranteed in the legal tradition, their interests are ultimately subordinate to those of humans. As Sulami argues, "The unbeliever who prohibits the slaughtering of an animal [for no reason but] to achieve the interest of the animal is incorrect because in so doing he gives preference to a lower, *khasīs,* animal over a higher, *nafīs,* animal" (cited in Izzi Dien, 2000, p. 146).

Sport Hunting Despite its prohibition in Islamic law, sport hunting remained a major form of entertainment in Muslim societies, especially among the elite. In Arabia, the oryx was hunted to near extinction, and only recently have measures been taken to preserve the species. In Iran, species such as the lion, tiger, and cheetah were hunted into oblivion before modern times, and leopards have become exceedingly rare. Even gazelles, which were the favored game at royal hunting preserves up until recently, are now generally found only on government lands where private individuals may not enter without special permission.

Historically the most egregious violations of the proscription against sport hunting were in India, where hundreds or thousands of creatures at a time would be indiscriminately slaughtered in bloody orgies of killing for the amusement of the rich and powerful. The favored method, a Central Asian technique called the *qamargha,* was to go out into the wilderness and create a wide circle of beaters who would make as much noise as possible as they slowly closed the circle, forcing huge numbers of terrified creatures toward the center. When the circle was almost closed, the royal hunters would fire at will into the throng of panic-stricken animals. So horrific was the resulting bloodbath that at one point the Mughal emperor Akbar the Great (r. 1555–1605) decided enough was enough and banned the sport, though apparently only for a time.

Wildlife Preservation The Islamic legal tradition contains two institutions that some contemporary scholars have argued could be considered forms of wildlife preserves. They are the *himā*, protected area or sanctuary, and the *harīm*, which was a greenbelt or easement around settled areas intended mainly to ensure a safe water supply. A related institution, the *harām*, refers to areas around the sacred cities of Mecca and Medina (called the *harāmayn*, the two forbidden areas) where hunting is outlawed.

The *harāmayn* were apparently established in the Prophet Muhammad's time when, according to the hadiths, he declared Mecca "sacred by virtue of the sanctity conferred on it by God until the day of resurrection. Its thorn trees shall not be cut down, its game shall not be disturbed." He also made a sanctuary of Medina, whose "trees shall not be cut and its game shall not be hunted."

The prohibition on hunting while on pilgrimage comes from the Qur'an, which states that the penalty for killing game is to offer a comparable domestic animal in sacrifice, that is, to God, by way of compensation (5:96). It would seem from this verse that killing wild animals when one is supposed to be in a state of purity is wrong because it is a crime against God, not against the animals in question. One must atone for this by paying the equivalent in one's own domestic livestock back to God. This atonement for the killing of wild animals by killing yet more domestic animals can hardly be seen to benefit the animals themselves.

Some traditional *himā*s still exist in Saudi Arabia, but they are much diminished from former times and continue to disappear. Most of these preserves are aimed at excluding sheep and goats from grazing lands in preference to cattle, camels, and donkeys, but others exist to control the cutting of firewood or to keep flowering meadows intact for honeybees.

Even in the *harām*s around the holy cities, species such as the ibex and gazelle are no longer found. In fact the laws pertaining to these preserves have been generally ignored, on the basis that development, geared largely toward the millions of pilgrims who now descend on the holy sites, is a need that overrides that of preserving nature.

What is important to note is that these areas were restricted primarily so that they might benefit humans. The *himā*, which in pre-Islamic times was an institution that allowed powerful landowners to keep others off their grazing lands, was transformed in the Prophet Muhammad's time into a means for preserving certain tracts of land for the public benefit. Significantly, the preserved areas were not to be too large, so as not to take too much land out of circulation.

In short, the institutions of *himā*, *harīm*, and *harām* are all clearly meant to preserve resources for human needs, not those of animals. If animals are preserved, or if they benefit from the preservation of water and vegetation, this is a secondary benefit, because they themselves are seen in the law as existing for the good of humans. Thus, in order for the institution of *himā* to be revived in Muslim regions today in a form that would actually serve to protect wildlife for the sake of biodiversity and ecosystem balance, the traditional rationale for its existence would have to be reinterpreted in light of contemporary scientific understanding. To date such an effort has not been undertaken, as few if any Islamic legal scholars seem to have ventured into the works of specialists in biodiversity.

Nevertheless, Islamic jurisprudence has entered a dynamic period in its history, and it may be hoped that in the years to come Muslims will increasingly ask their legal scholars for rulings on wildlife preservation and other issues connected with the world's ecosystems, which Islam states were created by God and belong to Him alone.

Further Reading

Foltz, R. C. (2005). *Animals in Islamic tradition and Muslim cultures.* Oxford: Oneworld.
Izzi Dien, M. (2000). *The environmental dimensions of Islam.* Cambridge: Lutterworth.
Masri, A.B.A. (1989). *Animals in Islam.* Petersfield, UK: The Athene Trust.
Pellat, C. (1971). Hayawān. In *Encyclopedia of Islam* (new ed.) (*3*, 305). Leiden: Brill.

Richard C. Foltz

RELIGION AND ANIMALS: JAINISM

One of the world's oldest religions, Jainism, is also distinguished as one of the faiths that cares the most about non-human animals. Nonetheless, animals receive scant mention in most books on Jainism. The Jains practice a religion without God that yet holds that our souls can become gods through liberation or *moksa*. It is said that our souls accumulate karmic particles through both good and bad actions, which make good or bad things, respectively, happen to us in turn. The goal is to eliminate all passions and actions that generate good and bad karma, as these literally make us too heavy to leave the realm of rebirth. The soul that has escaped the cycle of rebirth ascends to a permanent resting place at the very apex of the universe. The key to achieving divine liberation is to practice *ahimsa*, or avoiding injury to all life. The positive side of this is a reverence for all life or a universal and unconditional love for all creatures. Mohandas Gandhi was a Hindu, but adopted the Jains' principle of *ahimsa*, becoming its most famous champion.

If one acts badly in a lifetime, one might be reborn as a primitive being. There are simple one-sense beings with only a sense of touch, for example, plants and microscopic *nagodas*, which come in the form of earth bodies, water bodies, fire bodies, and wind bodies, two-sense beings which also have taste, for example, worms and leeches, three-sense beings which can also see, for example ants and moths, four-sense beings that can smell things as well, for example, bees, flies, mosquitoes, and five-sense beings that hear in addition to the other senses, for example, fish, dolphins, elephants, or any being born in a womb. There are rational and nonrational five-sense beings, which include humans, gods, hellbeings, and animals, presumably those other than the ones listed with fewer senses. A human can be reborn as a microbe, and a microbe can eventually be born human, ascending the Jains' evolutionary scale.

Inflicting injury on these creatures is wrong because of the suffering caused, and also because it produces passions in the killer leading to karma and rebirth. The Jains condemn all animal sacrifices. They build animal shelters, and never hunt or fish. They avoid any professions causing harm to animals. A Jain named Acarya Hemancandra once convinced King Kumrapala to forbid animal slaughter during the nine-day Paryusan festival in India. During that time, ordinary householders are expected to conform in part to the strictures of the Jain monks. Farming, which injures insects, is permitted because the harm is unintentional, but Jain monks beg with a bowl so that crumbs will not attract insects that would be crushed underfoot. Monks brush the path before them to sweep away small life forms they might otherwise step on. It is prohibited to breed destructive animals, and considered noble to allow oneself to be bitten by a snake rather than kill it. Jains are vegetarians, but consume milk. According to the Jain cosmic wheel of time theory, we are now in a fifth downward cycle, meaning a decline in morality, a craving for material things and success, and increased violence and cruelty. The advent of factory farming and vivisection is viewed to be a part of this downward trend. However, Jainism holds out hope for the eventual liberation of all if even the lowly *nagodas* can eventually be born human and then achieve liberation.

Further Reading

Dundas, Paul. 1992, 2002. *The Jains*. New York: Routledge.

Gopalan, S. 1973. *Outlines of Jainism*. New York: Halsted Press.

Jain, Jyotiprasad. 1975. *Religion and culture of the Jains*. New Delhi: Bharatiya Jnanpith.

Jain, Sagarmal, and Pandey, Shriprakash, eds. 1998. *Jainism in global perspective*. Varanasi: Parsvanatha Vidyapitha.

Jaini, Padmanabh S. 1979. *The Jaina path of purification*. Berkeley: University of California Press.

Mardia, K. V. 1990. *The scientific foundations of Jainism*. Delhi: Motilal Banarsidass.

David Sztybel

RELIGION AND ANIMALS: JUDAISM

Judaism has developed across thousands of years and under a great variety of different cultural, social, geographical, political, and technological circumstances, each of which has left its mark on the role of animals in Jewish tradition and society. According to Jewish tradition, the Written Torah, the first five books of the bible, may be understood as containing 613 commandments, which form the outline of Jewish law. The commandments are further expounded upon and extended by the Oral Torah, the living tradition of Jewish law that was first codified in the Mishnah, circa 200 CE, and further developed and expounded on in the Talmud and many other works. According to one recent count, some 138 of the commandments have some connection with animals.

Judaism has always valued the preservation of conflicting voices within the tradition, and countless references to animals are found throughout Jewish legal, philosophical, mystical, ethical, exegetical, liturgical, and homiletic literature. Furthermore, since the break-up of the Sanhedrin or High Court nearly 2,000 years ago, Judaism has lacked institutions authorized to make universally binding legal decrees and interpretations. These two factors make it difficult to formulate statements that are universally true of Judaism in all of its varied manifestations. The goal of this article is merely to outline some of the major Jewish themes, ideas, and practices relating to animal rights and welfare.

The Status of Animals According to Judaism

According to the first chapter of Genesis, after creating the animals, God created a male and a female human in the divine image. They were meant to "rule the fish of the sea, the birds of the sky, the cattle, the whole earth, and all creeping things of the earth" (Verse 26) and they were told, "Be fruitful and increase, fill the earth and master it; and rule the fish of the sea, the birds of the sky, and all the living things that creep on the earth" (Verse 28). Some recent writers have claimed that these statements support the right of human beings to treat animals as they please. This impression is immediately tempered, however, by the next verse's call for vegetarianism: "God said, 'See, I give you every seed-bearing plant that is upon all the earth, and every tree that has seed-bearing fruit; they shall be yours for food'" (Verse 29).

On the one hand, humans are thought of as created in the image of God and fundamentally superior to animals, which they are permitted to use for their own purposes. On the other hand, humans must take the wellbeing of animals into account, and the exploitation of animals for human ends must be regulated by moral considerations. While the Written Torah contains no explicit general principle concerning animal welfare, many individual laws are concerned with particular aspects of it. These are understood to supply examples of a general prohibition against *tza'ar ba'alei hayyim* (Hebrew for "suffering caused to animals"), which is usually considered to have the legal force of an explicit biblical commandment.

Judaism's self-understanding of its concern for animals has developed in ways that parallel developments in Western moral philosophy. Some thinkers, such as Moses Maimonides (1135–1204), believe that the wellbeing of animals is of intrinsic moral importance, while others, such as Moses Nachmanides (1194–1270), believe that while only humans are intrinsically deserving of moral consideration, people must treat animals humanely in order to properly cultivate their own moral virtue.

The Limits and Applications of Tza'ar ba'alei Hayyim

The general idea of *tza'ar ba'alei hayyim* is that people should not inflict needless suffering on animals. Almost every parameter of the application of *tza'ar ba'alei hayyim* has been subject to multiple interpretations in the Jewish legal tradition. Some authorities exclude pests and insects from the rule's purview, and there are those who say it only applies to domesticated animals, a position that would seem to be contradicted by Jewish prohibitions against hunting for sport. There is also disagreement regarding the minimum intensity of suffering that is prohibited, and about what kinds of human benefits gained from animal suffering are sufficient to keep it from being considered needless. It is unclear whether the otherwise painless death of an animal constitutes *tza'ar ba'alei hayyim*, and to what extent a person must prevent suffering inflicted by one animal upon another. Beyond all of these strictly legal debates and considerations, Jewish discussions of animal welfare make constant mention of *midat hahassidut*, the virtue of piety, that is, the expectation that people should go beyond the letter of the law in demonstrating compassion toward animals.

Human Obligations toward Working Animals

While wanton cruelty towards any animal is forbidden by Jewish law, anyone who owns or works with an animal has many additional obligations towards it. For instance, Jews are required to make sure that their animals have been fed before sitting down to eat themselves.

The Torah contains a number of commandments which specifically deal with the working conditions of animals. According to Deuteronomy 25:4, one is not allowed to muzzle an ox while it is threshing grain. This commandment is understood to prohibit people from stopping any kind of animal from eating any kind of food with which it is presently working. One corollary of this rule is that a pack animal must be allowed to nibble from whatever it is carrying (Maimonides' *Mishneh Torah, Laws of Hiring* 13, pp. 1–2). Deuteronomy 22:10 prohibits people from using a mixed team consisting of an ox and an ass to plow a field. This verse eventually gave rise to rabbinic legislation prohibiting people from using any combination of animals belonging to different species to pull the same vehicle or object. One explanation for these laws is that animals often find it stressful to be forced into close contact with members of other species (*Sefer HaHinukh,* Commandment 550); another possibility is that an animal from a weaker species will have trouble keeping up with a stronger work-partner. Other work-related laws include the obligation upon humans to assist in the unloading of a pack animal that has collapsed under its burden (Exodus 23:5) and the obligation to help a fallen animal get back on its feet (Deuteronomy 22:4).

A vast section of Jewish law deals with the prohibition of work on the Sabbath and festivals. The Torah makes it clear that one's animals must also be allowed to rest on those days: "The seventh day is a Sabbath of the Lord your God; you shall not do any work—you . . . your ox or your ass, or any of your animals" (Deuteronomy 5:14). "On the seventh day you shall cease from labor, in order that your ox and your ass may rest" (Exodus 23:12). Many laws derive from these verses; for instance, an entire chapter of the Mishnah (Shabbat 5) is devoted to the question of which items one may have one's animal carry into a public area on the Sabbath. A Jew is also not allowed to lend or rent an animal to a gentile who might force it to work on the Sabbath (Maimonides' *Mishneh Torah,* Laws of the Sabbath 20, p. 3).

Interestingly, Jewish law permits humans to perform certain kinds of work necessary for their animals' wellbeing on the Sabbath, even though those tasks would otherwise be prohibited by rabbinical edicts. For instance, Jews are allowed to milk cows on the Sabbath in order to alleviate the pain caused them by swollen udders. In Israel some milking parlors are fitted out with specially designed systems so that religiously observant Jewish dairy farmers can milk their herds on the Sabbath in a manner permitted by Jewish Law.

Laws Respecting the Parent-Child Relationship Among Animals

Judaism places great stress on the importance of the human parent-child relationship, and this concern extends to parent-child relationships among animals as well. In his *Guide for the Perplexed,* which is usually considered to be the most important work of medieval Jewish philosophy, Maimonides writes that when animals see their offspring die, they

> feel very great pain, there being no difference regarding this pain between man and the animals. For the love and tenderness of a mother for her child is not consequent upon reason, but upon the activity of the imaginative faculty, which is found

in most animals as it is found in man. (*Guide* III, p. 48; Pines, p. 599)

Several laws reflect concern for the human-parent relationship among animals. Leviticus 22:28 prohibits the slaughter of an animal together with its offspring on the same day. Maimonides (loc. cit.) states that this law is intended to prevent situations in which the parent might witness the slaughter of its offspring. Similarly, Leviticus 22:27 states that a newborn animal "shall stay seven days with its mother, and from the eighth day on it shall be acceptable as an offering by fire to the Lord." Deuteronomy 22:6–7 states that a mother bird and her eggs should not be taken together, and that the mother bird must be shooed away before the eggs are taken from her nest. Nachmanides argues that this last law is intended to preserve bird species by making sure that the mother bird will survive to produce a new future generation. This interpretation offers a foundation for the value of biodiversity in Jewish law.

Two Contemporary Applications

Sports and entertainment involving animal suffering do not jibe well with the restrictions of *tza'ar ba'alei hayyim,* and as a result Jewish law has generally taken a quite negative view of hunting for sport, bullfighting, and the like. Israeli Chief Rabbi Shlomo Moshe Amar has supported a recent ruling by Rabbi David Bardugo extending the prohibition to include horse-racing. It states that, "one ought to instruct every God-fearing person . . . not to participate in horse-races—neither in establishing them, nor by watching them: because of the pain to animals caused thereby . . ."

The use of animals in medical and biological research is another question that has generated considerable interest and activism in recent decades. In the conclusion to his comprehensive review of Jewish legal attitudes towards this question, Rabbi David Bleich writes:

there is significant authority for the position that animal pain may be sanctioned only for medical purposes, including direct therapeutic benefit, medical experimentation of potential value and the training of medical personnel. *A fortiori,* those who eschew . . . [this] . . . position would not sanction painful procedures for the purpose of testing or perfecting cosmetics. An even larger body of authority refuses to sanction the infliction of pain upon animals when the desired benefit can be acquired in an alternative manner, when the procedure involves "great pain," when the benefit does not serve to satisfy a "great need," when the same profit can be obtained in another manner, or when the benefit derived is not commensurate with the measure of pain to which the animal is subjected.

See also Religion and Animals—Judaism and Animal Sacrifice

Further Reading

Bardugo, David. 2006. *Responsa on horse-racing.* An English translation of the original Hebrew may be found at: http://www.chai-online.org/en/campaigns/racing/e_racing_psak.pdf.

Bleich, J. David. 1989. "Animal experimentation" and "Vegetarianism and Judaism." In *Contemporary Halakhic problems,* Vol. III, 194–236 and 237–250b. New York: Ktav Publishing/Yeshiva University Press.

Cohen, Norman J. 1976. *Tsa'ar Ba'ale Hayim: The prevention of cruelty to animals, its bases, development and legislation in Hebrew literature.* Jerusalem and New York: Feldheim Publishers.

HaLevi, Aharon (traditional attribution) 1988–9. *Sefer haHinnuch* (5 volumes, translated with notes by Charles Wengrov). Jerusalem and New York: Feldheim Publishers.

Kalechofsky, Roberta, ed. 1992. *Judaism and animal rights: Classical and contemporary responses.* Marblehead, MA: Micah Publications

Maimonides, Moses. 1963. *A guide for the perplexed*, S. Pines, trans. Chicago: University of Chicago Press.

Maimonides, Moses (volumes published at various dates). Maimonides' Code *(Mishneh Torah).* (Various trans.). New Haven: Yale University Press.

Schochet, Elijah Judah. 1984. *Animal life in Jewish tradition: Attitudes and relationships.* New York: Ktav Publishing.

Berel Dov Lerner

RELIGION AND ANIMALS: JUDAISM AND ANIMAL SACRIFICE

During biblical times animal sacrifice or *zebach* was practiced as part of Jewish religious observance. As happened in so many other religions at the time, domesticated animals were offered to God as an institutionalized means of relief from the impurity generated by human violations of moral rules or purity taboos. The animals selected for sacrifice were those that were deemed useful to humans, and both anthropomorphism and anthropocentrism can be seen in the description of these animals and not others as pleasing to God. The well-known "Thou shall not kill" was not thereby violated because, in the Hebrew tradition, this moral rule is interpreted as "Thou shall not kill unlawfully." Methods for lawful killing are defined by the Torah, which contains a written code with 613 laws of ethical human behavior, and by the later oral tradition and rabbinical commentary. The practice of animal sacrifice was discontinued after the destruction of the second temple by the Romans in 70 CE, although Orthodox Jewish prayer books to this day ask for a reestablishment of the temple sacrifices.

Another view of sacrifice appears in the criticism of the tradition, although in this criticism of sacrifice there was little emphasis on the obvious point that it was cruel to the individual animal. Maimonides, a 12th-century Jewish philosopher, argued that sacrifices were a concession to barbarism. Some modern theologians continue to argue that sacrifice in its way represented respect for animal life. A more balanced observation is that sacrifice does not necessarily involve a low view of the sacrificed animals' lives (Linzey, *Christianity and the Rights of Animals,* p. 41). This is plausible, given that the tradition contains powerful passages recognizing that the blood of humans and animals is sacred (for example, Leviticus 17:10). Ultimately, Judaism moved away from the practice of animal sacrifice, although there remain rules governing ritual slaughter or *shechita* by a specially trained religious functionary called a *shochet* when an animal is killed for food purposes.

The occurrence of these instrumental uses of animals and the ultimate rejection of the old sacrificial practices are of limited value in assessing Judaism's views of animals, as they deal with only a few domestic animals. Far more helpful in assessing Jewish views of animals is an evaluation of the ways in which Jews in their diverse communities have treated and continue to treat the living beings in their care.

See also Religion and Animals—Judaism

Further Reading

Clark, Bill. 1990. "The range of the mountains is His pasture": Environmental ethics in Israel." In J. Ronald Engel and Joan Gibb Engel, eds., *Ethics of environment and development: Global challenge, international response*, 183–188. London: Bellhaven Press.

Kalechofsky, Roberta. 1992. *Judaism and animal rights: Classical and contemporary responses*. Marblehead, MA: Micah Publications.

Linzey, Andrew. 1987. *Christianity and the rights of animals*. New York: Crossroad.

Maimonides. 1956. *A Guide for the perplexed*. M. Friedlander, trans. New York: Dover Publications.

Murray, Robert. 1992. *The cosmic covenant: Biblical themes of justice, peace, and the integrity of creation*. London: Sheed and Ward.

Schwartz, Richard H. 1998. *Judaism and vegetarianism*. Marblehead, MA: Micah Publications.

Waldau, Paul, and Patton, Kimberley. 2006. *A communion of subjects: Animals in religion, science and ethics*. New York: Columbia University Press.

Paul Waldau

RELIGION AND ANIMALS: PANTHEISM AND PANENTHEISM

Pantheism and panentheism exist in nearly every religious tradition, especially among mystics, who hope and strive for unity with the divine.

The word pantheism stems from two Greek words, *pantos*, meaning "all," and *theos*, meaning "God." Pantheists believe that the divine and the natural world are one and the same. Whatever exists is God, and God is all that exists. The pantheist's world is divine; from lizards to piglets, from rocks with flowers to fish; God is all, and all is God.

Panentheists believe that the divine permeates the natural world, but the divine is yet more than what we see and experience. Pantheists identify ultimate reality directly and solely with the physical world, whereas panentheists view ultimate reality as within, but also more than, the natural world.

Hinduism, the dominant faith of India, expresses both pantheism and panentheism in sacred writings such as the *Upanishads* and the *Mahabharata*. In Hinduism, *Brahman* is the divine, the greatest principle of the universe. Some authors translate *Brahman* as "God." *Brahman* is the substratum underlying the universe, the unknowable, undefinable power behind and within all that exists. The Hindu *Upanishads,* composed about 2,500 years ago, teach that each individual is *Brahman:* "This Great Being . . . forever dwells in the heart of all creatures as their innermost Self . . . [and] pervades everything in the universe" (*Svetasvatara*, pp. 122–23). *Brahman* is identified with nature and nonhuman animals:

> Thou art the fire,
>
> Thou art the sun,
>
> Thou art the air . . .
>
> Thou art the dark butterfly,
>
> Thou art the green parrot with red eyes,
>
> Thou art the thunder cloud, the seasons, the seas. (*Svetasvatara*, pp.123–24)

Brahman pervades every living being. Every creature shares this ultimate reality; the ground of each individual's being is identical with that of *Brahman.* Turkey, wombat, human being, Hindu pantheists find "in all creation the presence of God" (Dwivedi, p. 5).

With Hindu pantheism and panentheism, to understand what it means to be human is also to understand what it is to be a sparrow or a Leghorn chicken. For the Hindu, what is important about the existence of coho salmon or black angus cattle, the divine within, is the same in both human and nonhuman. The foundation and fundamental core of all beings is *Brahman.* As *Brahman* is essential to human essence, so this divine force is also essential to a pollywog wiggling in a mud puddle, or a fish struggling to escape the net of a fisherman. As a pinch of salt dissolved in a glass of water cannot be seen or touched but turns the contents to salt water, so the subtle essence of *Brahman* runs through all beings, creating their essential essence, yet this divine element cannot be perceived or touched. This subtle essence makes each living being, all that exists, holy. As all rivers are temporarily distinct but ultimately join one great sea, so all living beings appear in separate bodies. The indigo bunting sitting on your neighbor's fence, the tuna fish darting through the sea, the sow brimming with piglets, and the blue heron stepping gingerly through shallow pond waters, all are ultimately united by *Brahman.* "[A]s by one clod of clay all that is made of clay is known," so all things are one in essence, and that essence is sacred (*Chandogya*, p. 92).

The Hindu epic the *Mahabharata* teaches that those who are spiritually learned behold all beings in Self, Self in all beings, and *Brahman* in both. Hindus understand that all living beings have *atman* (usually translated as "soul"), and that *Brahman,* or God, is that soul.

Panentheism is one of the core teachings in the most famous portion of the *Mahabharata,* the *Bhagavad Gita,* in which the beloved god Krishna explains what it means to be divine: "I am the life of all living beings . . . All beings have their rest in me . . . In all living beings I am the light of consciousness" (*Bhagavad,* pp. 74, 80, 86). The divine, in this case Krishna, is not only a great deity, but is also indwelling in the cockroach and the elephant. The *Bhagavad Gita* presents the divine as an essential part of who we are, as an essential part of every aspect of every creature, and of nature: "I am not lost to one who sees me in all things and sees all things in me."

Panentheism and pantheism teach that all beings share in the divine. What does this mean about the relationship between white-tailed deer and Buff Orpington hens? What does this mean for ethics? The *Bhagavad Gita* notes that we exist in the heart of all other beings and the heart of all other beings exists within our own self. Not only is the divine in all beings, but we, as part of the divine, are also part of all other living beings. In this way Hindus come to love all beings, and the pleasure and pain of other creatures becomes personal (*Bhagavad,* pp. 71–72). Those who love God also love the ladybug and the anteater, the tulip and the turkey.

This rich and pervasive pantheistic and panentheistic vision of the universe affects Hindu ethics, as it does all religions that honor the divine in nature. Consequently, *ahimsa* is central to Hinduism. *Ahimsa,* often translated as nonviolence, is more literally translated as not to harm. Practicing contemporary Hindus strive to avoid harming to any creature or to the natural world because the divine is all that exists. Devout Hindus must extend their caring and compassion not only to other human beings, but to dogs and halibut, turkeys and hogs. As a result, many Hindus have been vegetarians for centuries, eschewing flesh in their diet, and also abstaining from reproductive eggs, such as chicken eggs.

Pantheistic and panentheistic religions, which teach that the divine is indwelling in the world around us, in all that exists in this great universe, also teaches respect for nature. Pantheism and panentheism discourage human arrogance and pride, greed and dominion, which might otherwise lead people to believe that we are superior to nonhuman animals, that we are somehow separate and more important than other earthbound creatures. Hinduism, like all great religions of the world, teaches people that every aspect of the natural world shares in the divine, is divine. For pantheists and panentheists, the spiritual life demands respect and reverence for all living beings and for the natural world in general.

Hinduism provides but one example of earth-centered beliefs and their accompanying ethics. Every religion teaches pantheism or panentheism in one form or another; each religion teaches that the world is sacred, and that every calf and garter snake holds some measure of the divine.

Further Reading

Buck, William, trans. 1973. *Mahabharata.* Berkeley: University of California.

Dwivedi, O. P. 2000. Dharmic ecology. In Christopher Key Chapple and Mary Evelyn Tucker, eds., *Hinduism and ecology: The intersection of earth, sky, and water*, 3–22. Cambridge: Harvard University.

Embree, Ainslee T., ed. 1972. *The Hindu tradition: Readings in Oriental thought.* New York: Vintage.

Harrison, Paul A. 2004. *Elements of pantheism.* Tamarac, FL: Llumina Press.

Nelson, Lance E. 2000. Reading the *Bhagavadgita* from an Ecological Perspective. In Christopher Key Chapple and Mary Evelyn Tucker, eds., *Hinduism and ecology: The intersection of earth, sky, and water*, 127–64. Cambridge: Harvard University Press.

Prabhavananda, Swami, and Manchester, Frederick, trans. 1948. *Svetasvatara Upanishad: The wisdom of the Hindu mystics: The Upanishads: Breath of the eternal.* New York: Mentor.

Lisa Kemmerer

RELIGION AND ANIMALS: REVERENCE FOR LIFE

Reverence for life is a concept pioneered by the Alsatian theologian and philosopher Albert Schweitzer in 1922. According to Schweitzer, ethics consists in experiencing a "compulsion to show to all will-to-live the same basic reverence as I do to my own." The relevance of Schweitzer's thought to modern debates about animals is significant. According to Schweitzer, other life forms have a value independent of humans, and our moral obligation follows from the experience and apprehension of this value. This insight is essentially religious in character and therefore basic and nonnegotiable. Schweitzer was undoubtedly prophetic. "The time is coming," he wrote, "when people will be astonished that mankind needed so long a time to learn to regard thoughtless injury to life as incompatible with ethics."

Further Reading

Barsam, Ara Paul. 2008. *Reverence for life: Albert Schweitzer's great contribution to ethical thought.* Oxford: Oxford University Press.

Linzey, Andrew. 1981. Moral education and reverence for life. In David A. Paterson, ed., *Humane education: A symposium*, 117–125. London: Humane Education Council.

Linzey, Andrew. 1995. *Animal theology*. London: SCM Press; Urbana: University of Illinois Press.

Schweitzer, Albert. 1970. *Reverence for life*, R. H. Fuller, trans., foreword by D. E. Trueblood. London: SPCK.

Schweitzer, Albert. 2008. The ethics of reverence for life. In Andrew Linzey and Tom Regan, eds., *Animals and Christianity: A book of readings*, 118–120, 121–133. London: SPCK, 1989; New York: Crossroad, 1989; Eugene: Oregon: Wipf and Stock.

Andrew Linzey

RELIGION AND ANIMALS: SAINTS

There is a remarkable range of material linking Christian saints with animals. The stories of St. Francis of Assisi preaching to the birds and St. Anthony of Padua preaching to the fishes are well known. Much less well known are the stories, for example, of St. Columba and the crane or St. Brendan and the sea monster. Most scholars and theologians have dismissed this wealth of material as legend or folklore, but its significance, historically and theologically, can be noted.

First, it is testimony to a widespread positive tradition within Christianity that has linked spirituality with a benevolent and sensitive regard for animals. The underlying rationale for this study of saints appears to be that, as individuals grow in love and communion with their Creator, so too ought they to grow in union and respect for animals as God's creatures. Something like two-thirds of canonized saints East and West apparently befriended animals, healed them from

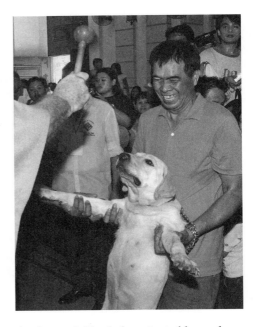

A priest sprinkles holy water to bless a dog during the celebration of the feast of Saint Francis of Assisi, at a church in Manila. Animal lovers brought their pets to celebrate the annual celebration of the known animal-lover saint. (AP Photo/ Pat Roque)

suffering, assisted them in difficulty, and celebrated their life through prayer and preaching. Second, despite the negative tradition within Christianity that has frequently downgraded animals, regarding them, at worst, as irrational instruments of the Devil, the literature on these saints makes clear God's benevolent concern for nonhuman creatures and the common origin of all life in God. Third, because of this common origin in God, it necessarily follows that there is a relatedness, a kinship between humans and nonhumans. According to St. Bonaventure, St. Francis was able to call creatures "by the name of brother or sister because he knew they had the same source as himself." Fourth, many of these stories prefigure a world of peaceful relations between humans and animals where human activity is no longer injurious or detrimental to other creatures. St. Brendan's voyage, for example, culminates in the discovery of a new Eden-like land characterized by widespread vegetarianism and the absence of predation. Such stories are testimonies to a substratum within Christianity that is inclusive of concern for animal life. The ideas they embody of respect, generosity, and kinship between species reflect the themes that mainstream Scholastic tradition has almost entirely failed to incorporate into its thinking.

Further Reading

Butler, Alban. 1946. *Lives of the saints,* revised by Herbert Thurston and Donald Attwater, 4 vols. New York: P. J. Kennedy and Sons.

Hobgood-Osler, Laura. 2008. *Holy dogs and asses.* Chicago: University of Illinois Press, 2008.

Linzey, Andrew, and Cohn-Sherbok, Dan. 1997. *Celebrating Animals in Judaism and Christianity.* London: Cassell, 1997.

Low, Mary. 1996. *Celtic Christianity and nature: Early Irish and Hebridean traditions.* Edinburgh: Edinburgh University Press, 1996.

Sorrell, Roger D. 1988. *St. Francis of Assisi and nature: Tradition and innovation in Western Christian attitudes toward the environment.* New York: Oxford University Press.

Waddell, Helen. 1995. *Beasts and saints,* rev. ed. by Esther De Vaal. London: Darton, Longman and Todd.

Andrew Linzey

RELIGION AND ANIMALS: THEODICY

Theodicy comes from the Greek words *theos* (God) and *dike* (justice), and is a branch of theology concerned with exploring and defending the justice of God in relation to physical and moral evil. Theodical issues are frequently at the heart of debates about animal rights and animal welfare, and are used both positively and negatively in encouraging or discouraging concern for animal suffering. A great deal of historical theology has utilized theodical arguments negatively, in ways that seem to satisfy the claim that God is just and good, but at the expense of animals. The first negative type solves the problem of animal pain by effectively denying its existence. Historically, Cartesianism has played a vital part in the development of this argument, but it has not lacked modern adherents. For example, as late as 1927, Charles Raven argued that "it may be doubted whether there is any real pain without a frontal cortex, a fore-plan in mind, and a love which can put itself in the place of another; and these are the attributes of humanity." Clearly there can be no problem of animal pain to solve if such pain is illusory.

The second negative type admits of some animal pain but minimizes its significance morally. For example, John Hick holds that animal pain is necessarily different from human pain because animals cannot anticipate death. "Death is not a problem to the animals . . . We may indeed say of them 'Death is not an injury rather life a privilege.'" Clearly, if death is not a problem to animals, then the moral significance of killing is necessarily reduced.

The third negative type also admits of the existence of animal pain but denies its significance theologically. For example, Peter Geach holds that God is essentially indifferent to animal pain. "The Creator's mind, as manifest in the living world, seems to be characterized by mere indifference to the pain that the elaborate interlocking teleologies of life involve." This appeal to the world as it now exists has historically been one of the major theodical arguments against animal welfare. In the crisp summary of Samuel Pufendorf (1632–92): "For it is a safe conclusion from the fact that the Creator established no common right between man and brutes that no injury is done brutes if they are hurt by man, since God himself made such a state to exist between man and brutes." Such an argument finds its contemporary and largely secular expression in an ecological form of theodicy that maintains that since nature is essentially predatory, we should abide by nature's rules. Nature's perceived law is baptized into natural or moral law.

Alongside these negative types, there are positive ones, too. Here are three examples. The first is that animal pain and predation, far from being the Creator's will, are actually contrary to it. C. S. Lewis, for example, held that both animal pain and carnivorousness were the result of Satanic corruption of the earth before the emergence of human beings. It follows that humans therefore have a duty not to imitate such malevolent distortion, but to fight against it. The second is that while the Creator allows pain in creation, both animal and human, as an inevitable corollary to the freedom allowed to creation itself, such pain will eventually be transformed by a greater joy beyond death. Keith Ward, for example, holds that "immortality, for animals as well as humans, is a necessary condition of any acceptable theodicy," and that "necessity, together with all the other arguments for God, is one of the main reasons for believing in immortality." Such a prospect both maintains the ultimate justice of God and justifies the alleviation of pain, as an anticipation of God's final will, in the present. The third form of positive theodicy maintains that the God revealed in the suffering of Jesus suffers with all innocents, whether human or animal, in this world, and will redeem all such suffering. From this perspective, Andrew Linzey concludes that the "uniqueness of humanity consists in its ability to become the servant species," that is, "co-participants and co-workers with God in the redemption of the world" (Linzey, 1994). Far from being indifferent to suffering, God is seen as manifest within it, beckoning human creatures to active compassion to remove its causes.

How ever we may judge the satisfactoriness of these negative or positive theodicies, it is inevitable that ethical concern for animals will continue to be influenced by one or more of them in one form or another. Concern for animal suffering rarely stands by itself as a philosophical position, and requires the support of some form of meta-ethical framework in which the problem of a specific injustice can be properly recognized and addressed only within the context of a sufficiently comprehensive vision of ultimate justice for all.

Further Reading

Geach, Peter. 1977. *Providence and evil.* Cambridge: Cambridge University Press.

Hick, John. 1967. *Evil and the god of love* (1966), Fontana, ed. London: Collins.

Kingston, A. Richard. 1967. Theodicy and Animal Welfare. *Theology* 70 (569): 482–488.

Lewis, C. S. 1940. *The problem of pain.* London: Geoffrey Bles.

Linzey, Andrew. 1995. *Animal theology.* London: SCM Press; Urbana: University of Illinois Press.

Linzey, Andrew, and Regan, Tom, eds. 2008. *Animals and Christianity: A book of readings.* London: SPCK; New York: Crossroad; Eugene, OR: Wipf and Stock.

Raven, Charles E.. 1927. *The creator spirit.* London: M. Hopkinson.

Ward, Keith. 1982. *Rational theology and the creativity of God.* Oxford: Blackwell.

Andrew Linzey

RELIGION AND ANIMALS: THEOS RIGHTS

Theos rights denotes God's own rights as Creator to have what is created treated with respect. According to this perspective, rights are not awarded, negotiated, or granted, but recognized as something God-given. Comparatively little attention has been devoted to the theological basis of animal rights, though it offers a coherent theoretical basis for the intrinsic value of, especially, sentient beings. Whereas in secular ethics, rights are usually correlative of duties, for example, if A has a duty toward B, it usually follows that B has a right against A, in theological ethics the reverse may be claimed. For example, Dietrich Bonhoeffer maintains that "we must speak first of the rights of natural life, in other words of what is given to life and only later of what is demanded of life." Rights thus may be characterized as what are given to creatures by their Creator, to whom humans owe a primary obligation. The value of theos rights lies conceptually in the way in which it frees ethical thinking from humanocentricity. As Andrew Linzey writes:

According to theos rights what we do to animals is not simply a matter of taste or convenience or philanthropy. When we speak of animal rights we conceptualize what is objectively owed to animals as a matter of justice by virtue of their Creator's right. Animals can be wronged because their Creator can be wronged in his creation.

Although some Christians oppose the language of rights altogether as unbiblical or contrary to creation construed as grace, the notion of rights has a long history in theological ethics. Thomas Tryon was probably the first to use it in a specifically theological context relating to animals in 1688, but it continues to be used in modern contexts as well. For example, John Cardinal Heenan stressed that "animals have very positive rights because they are God's creatures . . . God has the right to have all creatures treated with proper respect."

Further Reading

Bonhoeffer, Dietrich. 1971. *Ethics,* 2nd ed. London: SCM Press; New York: Macmillan.

Heenan, John. 1970. Foreword to Ambrose Agius, *God's animals.* London: Catholic Study Circle for Animal Welfare.

Linzey, Andrew. 1987. *Christianity and the rights of animals.* London: SPCK; New York: Crossroad.

Linzey, Andrew. 1995. *Animal theology.* London: SCM Press; Urbana: University of Illinois Press.

Linzey, Andrew, and Clarke, Paul Barry, eds. 2004. *Animal rights: A historical anthology.* New York; Columbia University Press,.

Tryon, Thomas. 1688. Complaints of the birds and fowls of Heaven to their creator. In *The country-man's companion.* London: Andrew Sowle.

Andrew Linzey

RELIGION AND ANIMALS: VEGANISM AND THE BIBLE

Many vegans, that is, those who eat no food made from animals, including dairy products, and who do not use products made from animals, believe that the Bible, including the story of creation and the life of Jesus, presents a moral imperative for a vegan way of life.

In the many books of the Bible, only Genesis 1 and 2, honored by Jews, Christians, and Muslims, reveal the world as the creator preferred and intended creation to be. After the fall, which occurs in Genesis 3, God's perfect creation is changed. To many believers, in these first two chapters of Genesis, the divine being creates and sanctifies what is in essence a vegan world.

In Genesis 1, God grants humans rulership over the creatures, over everything that moves. God also creates people in the image of himself. Genesis 2 defines this divinely ordained rulership as God "took the man and placed him in the Garden of Eden, to till it and tend it," (Gen. 2:15). The Hebrew word most frequently translated in Genesis 2 as tend (*shamar*) also appears in Numbers 6:24, translated as protect ("The Lord bless you and protect you"). The Hebrew word most often translated as till (*'abad*) in Genesis 2, is translated as serve in other portions of the Bible, such as Joshua 24:15: "choose this day which ones you are going to serve— the Gods that your forefathers served . . . or those of the Amorites." Humans are therefore placed in the Garden of Eden to protect and serve creation, including animals.

Immediately after humans are granted rulership, God instructs the first humans as to what they may eat: "I give you every seed-bearing plant that is upon all the earth, and every tree that has seed-bearing fruit; they shall be yours for food" (Gen. 1: 29). To many readers, this means that food for humans should come from plants.

Jesus, the quintessential Christian moral exemplar, was devoted to weak and imperfect beings; he was deeply concerned for the oppressed and downtrodden. His life and teachings speak of compassion and service of the strong for the weak. Jesus provides an exemplar of Genesis 2 in action, of serving and tending creation.

Fundamental among Christian moral teachings is the commandment to love. Love is "the paramount scripture . . . essential to the Christian way of life" (Allen, 1971, p. 214). In the Christian worldview, love is limitless. Any understanding of Christian love or of God's love that limits care and affection "is spiritually impoverished" (Linzey, 1997, p. 131). The Catholic catechism notes that God surrounds animals with "providential care," that the creatures of the earth bless God and bring glory to the creator, and so we "owe them kindness" (1994, p. 2416): "It is contrary to human dignity to cause animals to suffer or die needlessly" (1994, p. 2418). Vegans believe that eating animals and exploiting them for their reproductive eggs and nursing milk causes them to suffer and die needlessly.

Scripture also notes that the human body is "a temple of the Holy Spirit" (1 Cor. 6:19). A flesh-based diet has been linked to heart disease, cancers, obesity, and numerous other serious health problems. Steve Kaufman, founder of the Christian Vegetarian Association, comments that a vegan diet is neither a burden nor a self-sacrifice, but part of a broader spiritual life that shows respect for creation, including the human body, "manifesting core values such as love, compassion, and peace."

Most Christians believe that Christianity holds the dream of universal peace and a vision of a future world devoid of violence, and that they are to work toward this great peace. A number of Christians believe that killing animals to consume their bodies, or keeping them in cramped cages to obtain their eggs and milk, is a failure of Christian love, and cannot bring about the Peaceable Kingdom of all creation, which includes both humans and animals. The creation story, the life of Jesus, and primary moral ideals such as the peaceable kingdom and the sanctity of our bodies provide a vegan Biblical imperative.

See also Veganism

Further Reading

Allen, Clifton J. 1971. *Broadman Bible commentary,* 12 vols. Nashville: Broadman.
Catechism of the Catholic church. 1994. Liguori, MO: Liguori.
Holy Bible: New revised standard version. 1989. New York: American Bible Society.
Linzey, Andrew, and Cohn-Sherbok, Dan. 1997. *After Noah: Animals and the liberation of theology.* London: Mowbray.
Phelps, Norm. 2002. *The dominion of love.* New York: Lantern.
Phelps, Norm. 2003. *Love for all creatures: Frequently asked questions about the Bible and animal rights.* New York: Fund For Animals.

Lisa Kemmerer

RELIGION, HISTORY, AND THE ANIMAL PROTECTION MOVEMENT

From ancient times, religion has played the same contradictory roles in shaping human relationships with animals that it has played in other areas of human life. On the one hand, religion has been a powerful force for the advancement of humanity out of our fearful, benighted past toward an open, generous, enlightened future. It was morality codified as religion that introduced concepts like compassion, altruism, and nonviolence into the human dialogue, and religion created the ethical values that enable people to live in relative peace and harmony with one another.

At the same time, religious institutions have been among the fiercest opponents of human progress, willing, even eager at times, to use violence to defend the status quo and halt the extension of compassion and love to groups considered other. All too often, religion has taught people to hate in the name of love and kill in the name of God.

Sacrifice

The origins of religion are lost in the darkness of prehistory. What we do know is that when religion was first practiced by human beings, it was organized around the cult of sacrifice. Most ancient religions were based upon sacrifices to appease angry gods or curry favor with helpful deities. Ancient temples were first and foremost slaughterhouses.

But it was also religious leaders who first called for the abolition of sacrifice.

The ideas of good and evil that have guided human thinking about ethics for more than two millennia, epitomized in the Hebrew Scriptures as "You shall love your neighbor as yourself" (Leviticus 19:18), were created almost simultaneously along a band stretching some four thousand miles from China through India, Persia, and Israel to Greece during the most remarkable period of spiritual, intellectual, and social progress in human history. During this Axial Age, as it was dubbed by philosopher Karl Jaspers, which lasted roughly from 800 to 200 BCE, sages like Confucius, LaoZi, the founder of Taoism, Mahavira, the founder of Jainism, the Buddha, the great Hindu reformer Vyasa, a name usually taken to stand for several teachers whose names have been lost, Zoroaster, the Latter Prophets, and Pythagoras, revolutionized religion and ethics by introducing the idea that our lives should be based upon helping others rather than entirely upon self-interest.

The Origins of Animal Rights and Animal Welfare

From the animals' standpoint, what is most important here, and all too often overlooked, is that several of these teachers: Mahavira, the Buddha, and Vyasa in India, Pythagoras in Greece, and the Latter Prophets in Israel, counted animals among the neighbors whom we should love as we love ourselves. They recognized that the suffering and death of a chicken or a lobster is as urgent to the chicken or the lobster as your suffering and death are to you and my suffering and death are to me. Therefore, they taught that the chicken and the lobster are entitled to the same moral consideration to which human beings are entitled. They taught the moral equality of all sentient beings.

Thus, animal protection began as animal liberation, not as animal welfare, and was part and parcel of the same movement that pioneered human liberation. In the minds of these thinkers, animals were not second-class citizens. They were co-equal beneficiaries, along with the humans of the Axial Age movement, to end violence and oppression.

Mahavira, the Buddha, Vyasa, and Pythagoras explicitly forbade the raising and slaughter of animals for food or sacrifice. Down to the present time, the vast majority of Jains are ethical vegetarians, as are a large percentage of Buddhists and Hindus. Neither Jainism nor Buddhism has ever indulged in animal sacrifice, and in Hinduism it became a marginalized vestigial practice.

In Israel, the Latter Prophets, radical religious and social reformers who included such familiar names as Isaiah, Jeremiah, Amos, and Hosea, condemned the oppression of the poor by the rich and powerful, and also called for an end to animal sacrifice. In Isaiah, God tells the people,

I take no pleasure in blood of bulls, lambs, and goats. When you come to appear before Me who requires of you this trampling of My courts? . . . [E]ven though you multiply prayers, I will not listen. Your hands are covered with blood. (excerpted from Isaiah 1:11–17, *New American Standard Bible*).

In Hosea, God speaks just as clearly. "For I desire mercy and not sacrifice, and acknowledgement of God rather than burnt offerings" (Hosea 6:6, *New International Version*). Condemnations of sacrifice are also found in Isaiah 66:3–4, Jeremiah 7:21–23, Hosea 8:11–13, and Amos 5:21–25.

In very ancient times, Jews were permitted to eat meat only from an animal that had been offered as a sacrifice (I Samuel 14:31–35). But gradually this law was relaxed to allow Jews to eat meat as long as sacrifices were offered at the Temple in Jerusalem, the only place that sacrifice was permitted. When the Temple was destroyed by the Romans in 70 CE, bringing animal sacrifice to an abrupt halt, the leading rabbis of the day debated whether meat eating was still allowed. As so often in human history, appetite triumphed. And so when the prophets condemned sacrifice, they were also condemning meat eating, as their contemporaries would have understood perfectly well.

Animal welfare, the belief that we may enslave and slaughter animals for our own benefit as long as we spare them any suffering that is not inherent in their use, was a compromise between unrestricted animal exploitation and abuse and the call of the Latter Prophets for an end to animal abuse and slaughter. Over time, this compromise became the normative view of Judaism, and is enshrined in the Hebrew Scriptures, for example, in Proverbs 12:10, Deuteronomy 5:14, and Exodus 23:12. This Biblical compromise holds that we may exploit and slaughter animals for our own benefit, including for food and sacrifice, as long as we spare them any suffering that is not essential to the purpose for which they are being exploited and killed. It establishes two levels of morality: one to guide our treatment of human beings, and another, lower level to guide our treatment of animals. In recent years, Jewish animal advocates, notably Dr. Richard Schwartz and Dr. Roberta Kalechofsky, have sought to move beyond the compromise and reclaim the original call of the Latter Prophets for a single standard of treatment for both human beings and animals.

Christianity and Islam

Jesus appears to have endorsed the tradition of the Latter Prophets which condemned animal sacrifice and meat eating. There is no record that he ever sponsored a sacrifice at the Temple; twice he quoted with approval the passage above from the prophet Hosea denouncing sacrifice (Matthew 9:13, 12:7), and in history's first recorded animal liberation, he freed animals being held in the Temple precincts to be killed as sacrifices (John 2:14–16). The only animal product that Jesus is said to have consumed is fish, and that only once, following the crucifixion, when he ate a small morsel of fish to prove to his disciples that he had been resurrected in the flesh, leading some scholars to suspect that this story was a legend added to the gospel for theological reasons (Luke 24:36–43). The gospels describe bread, not lamb, as the main course at the Last Supper. According to ancient Christian sources, Jesus' brother James and several of the other Apostles appear to have been vegan, and the original Jewish Christians, who learned their traditions directly from Jesus and his immediate disciples, remained vegan until their movement died out sometime around the fourth century.

Christianity never practiced animal sacrifice, but rather taught that Jesus's sacrifice of himself for the sins of humankind rendered it obsolete. And when Christianity triumphed in the Roman world, the Empire's pagan religions were forcibly abolished, bringing animal sacrifice to a permanent end in the West.

Christianity as a gentile religion owes much of its theology and practice to Paul of Tarsus, a Greek-speaking Jew who spread the new faith to the gentile populations of the eastern Mediterranean in the decades following Jesus' crucifixion. Paul rejected the Biblical Compromise, as well as the single morality taught by the Latter Prophets, in favor of a Greek philosophical tradition derived from Aristotle and the Stoics which held that animals exist solely for human benefit and we may exploit and slaughter them as we like (I Corinthians 9:9–14; 10:25–31) In this he was followed by the leading theologians of the Middle Ages, St. Augustine and St. Thomas Aquinas, who taught the Aristotelian doctrine that only beings with rational souls, that is, human beings, are entitled to ethical treatment, and that we have no direct moral duties to animals.

From the conversion of the Roman Empire to Christianity in the fourth century, when Pythagoreanism was eradicated, until the Protestant Reformation more than a millennium later, there was no animal advocacy in Christian Europe. To the extent that kindness to animals was encouraged at all, it was on the grounds that it predisposed people to kindness toward other humans. The Catholic Church did not fully endorse animal welfare until a new universal catechism issued by Pope John Paul II in 1992 embraced both elements of the Biblical Compromise.

Judaism's other daughter religion, Islam, took the opposite tack. From the beginning, Islam incorporated animal welfare into its core ethical teachings. But Islam also continued the ancient practice of animal sacrifice, which to this day takes place once a year, at the festival of Eid-al-Adha, which celebrates the Hajj, the pilgrimage to Mecca that every Muslim whose circumstances permit is obligated to make at least once in his or her lifetime. In Mecca and around the world, millions of animals, primarily sheep and goats less than a year old, are slaughtered as a token of the believers' submission to the will of God.

The Protestant Reformation Revives Animal Welfare

As the modern age arrived in Europe, Protestant theologians like John Calvin and John Wesley discovered the Biblical Compromise in the Hebrew Scriptures and taught both of its elements: that we may exploit and slaughter animals for human benefit, and that we must spare them any suffering that is not essential to their use. In 1641, a Puritan clergyman named Nathaniel Ward wrote the Western world's first animal welfare law into the legal code of the Massachusetts Bay Colony, the so-called *Massachusetts Body of Liberties*: "No man shall exercise any cruelty or tyranny toward any brute creature which is usually kept for man's use."

It was an Anglican priest, Rev. Dr. Humphrey Primatt who was largely responsible for bringing animal welfare to the attention of the general public. In 1776, he published a small book which achieved a broad readership among England's liberal social reformers, *The Duty of Mercy and the Sin of Cruelty to Brute Animals.*

Inspired by *The Duty of Mercy,* another Anglican priest, Rev. Arthur Broome, came to see animal welfare as his Christian ministry. In 1824, Rev. Broome convened a meeting in London of leading British abolitionists and social reformers, including Richard Martin, a member of Parliament who two years earlier had sponsored the second animal welfare act in the modern world following the Massachusetts Body of Liberties. These liberal opinion leaders created an organization to educate the public about animal cruelty and to bring prosecutions against abusers under Martin's Act. The group did not challenge the exploitation and slaughter of animals for human benefit, but vigorously opposed cruelty that was not intrinsic to their use. In 1840, it received the sponsorship of Queen Victoria and became known by its present name, The Royal Society for the Prevention of Cruelty to Animals. From this beginning, animal welfare, *tsar ba'ale hayyim* in modern dress, spread to North America, Australia, continental Europe, and around the world.

Christianity and Animal Rights

There matters rested until 1976, when yet another Anglican priest, Rev. Dr. Andrew Linzey, published *Animal Rights: a Christian Perspective.* In this and subsequent books, Professor Linzey brought the Christian view of animals full circle by moving beyond the Biblical Compromise, back to the original view of Jesus and the Latter Prophets, that non-human animals have moral equality with human beings. In fact, Professor Linzey goes farther by arguing that the essence of Christian ethics, as expressed in the life and teachings of Jesus, is to serve those who suffer and are powerless to end their suffering. And since, as a generality, animals suffer more grievously and have less power than human beings, they actually have a kind of moral priority. "The uniqueness of humanity," Professor Linzey tells readers in his book *Animal Theology,* "consists in its ability to become the servant species."

Further Reading

Phelps, Norm. 2002. *The dominion of love: Animal rights according to the Bible.* Brooklyn, NY: Lantern Books.

Phelps, Norm. 2004. *The great compassion: Buddhism and animal rights.* Brooklyn, NY: Lantern Books.

Norm Phelps

REPTILES

Reptiles are poorly understood by most people, which leads to their mistreatment. In their natural environments, they may be killed by fearful humans or by those seeking their skins. Reptiles are becoming more popular as pets, but in this role they may also suffer due to human ignorance about reptilian physical needs. The traditional classification of the class *Reptilia* includes turtles, squamates (lizards, snakes, and relatives), crocodilians, and the two recognized species of tuatara. The later are relic and highly protected species found on several islands off the coast of New Zealand. All reptiles share several traits, including being ectothermic (dependent on external sources of heat) and covered with hard plates, scales, or bony shells. Reptiles live in almost all habitats, except for year-round subfreezing or deep sea environments. Within these limits, reptiles have adapted to many conditions, exploit a wide range of food items with diverse foraging methods, and have evolved diverse social systems. All tuataras, turtles, and crocodilians lay eggs, with the last group also showing a highly developed system of nest guarding and post-hatching parental care. Squamate reptiles, which constitute about 95 percent of all reptile species, have egg-laying, egg retention (ovoviviparity), and viviparous reproduction, the first two sometimes occurring in the same species. Egg brooding and postnatal parental solicitude also occur in a number of squamate species, and complex social and multigenerational groups have been documented in some lizards. Social, foraging, and anti-predator defensive behavior can differ greatly within and between closely related forms, especially in squamates. Thus it is very difficult to generalize across species, which raises problems in maintaining many species in captivity, developing effective conservation plans, studying their behavior, and understanding the way they experience their lives.

The ability of reptiles to learn, suffer, communicate, play, and socialize is generally underestimated, even by many herpetologists. Data are accumulating on many species which indicate that reptiles are not the robot-like, insensitive, simple, and stupid animals many think they are. This mistake is fostered because while reptiles do not have complex facial or vocal repertoires, tactile, chemical, and whole body visual displays are common and important in communication. The metabolic rate of reptiles is about 10 percent that of mammals and birds, and thus their behavior is often slow, for example, in the movements of land turtles, or sporadic, although there are many exceptions. Furthermore, reptiles are often ecologically specialized and critically dependent upon having proper temperature, humidity, diets, lighting, substrates, perches, retreats, and other captive arrangements to stimulate normal activity. Knowing their natural behavior aids greatly in providing appropriate captive conditions for reptiles. An indication of the bias against this constricted view of reptilian abilities is seen in the hot-blooded dinosaur controversy which elevates dinosaurs above mere reptiles by willful ignorance of the documented complexity of reptile behavior (Burghardt, 1977).

Reptiles are fascinating both in how they look and in their behavior. They are now highly popular as pets, especially rat snakes, leopard geckos, bearded dragons, and boas. A major problem is that the behavioral, nutritional, environmental, medical, and psychological needs of reptiles are rather different, indeed alien to, ours and those of our common companion animals, namely dogs, cats, birds, and rodents. However, many people keep reptiles because they seem to need less care than traditional mammalian and avian species. This leads to many problems and the premature deaths of literally thousands of animals each year. For example, reptiles can go much longer without food than other vertebrates, and many slowly starve to death or succumb to poor nutrition, insufficient temperatures for digesting food, or inadequate lighting with insufficient ultraviolet radiation.

Reptiles possess many traits that are useful for answering important questions about animal biology and behavior (Greenberg et al., 1989). Snakes possess chemosensory abilities more acute than most other terrestrial vertebrates. Reptiles can be both short- and long-lived, have behavior patterns that can be measured and recorded easily, and are important ecological components of many habitats where they occur. Many species are affected by habitat loss or changes due to human activity. Many reptiles are also killed directly by ignorant, fearful people. Others are exploited for food, skins, and the pet trade, in numbers that threaten the survival of many species, including once-common species of turtles in North America.

There are many sources of accurate information on reptiles. Several organizations and publications in the United States and Europe are devoted to investigation and dissemination of accurate information on captive reptiles, including books and pamphlets on selected species or groups. More scholarly sources are now also available. The multivolume *Biology of the Reptilia* series founded and edited by Carl Gans and subsequent coeditors has been published since the 1960s and now contains over 20 volumes covering almost every aspect of anatomy, physiology, ecology, and behavior. The Society for the Study of Amphibians and Reptiles is the largest organization devoted to reptiles, and publishes many important publications including the *Journal of Herpetology* and the *Herpetological Review.* There are also many books published at the state, regional, country, and continent-wide level devoted to reptiles as a whole, or to various subgroups such as lizards, that contain a wealth of information on exotic or less popular species.

See also Amphibians

Further Reading

Burghardt, G. M. 1977. Of iguanas and dinosaurs: Social behavior and communication in neonate reptiles. *Amer. Zool.,* 17, 177–190.

Burghardt, G. M. 2005. *The genesis of animal play: Testing the limits.* Cambridge, MA: MIT Press.

Greenberg, N., Burghardt, G. M., Crews, D., Font, E., Jones, R., & Vaughan, G. 1989. Reptile models for biomedical research. In A. Woodhead, ed., *Nonmammalian models for biomedical research,* 290–308. Boca Raton, FL: CRC Press.

Halliday, T. R., & Adler, K., eds. 2002. *New encyclopedia of reptiles and amphibians.* Oxford: Oxford University Press.

Schaeffer, D. O., Kleinow, K. M., & Krulisch, L., eds. 1992. *The care and use of amphibians, reptiles, and fish in research.* Bethesda, MD: Scientists Center for Animal Welfare.

Warwick, C., Frye, F. L., & Murphy, J. L., eds. 1995. *Health and welfare of captive reptiles.* London: Chapman Hall.

Gordon M. Burghardt

RESCUE GROUPS

Animal rescue groups are typically privately funded groups, often made up of volunteers, which rescue domesticated animals and place them up for adoption. The animals may be surplus animals from public or private animal shelters, unwanted pets from the general public, or strays.

While individuals have been finding and keeping stray animals for centuries, organized animal rescue groups are a relatively recent invention, dating to the early 19th century in England. Here, in 1824, a handful of animal lovers formed the Society for the Prevention of Cruelty to Animals, the first SPCA in the world, later renamed the Royal Society for the Prevention of Cruelty to Animals (RSPCA).

In the United States, the first formal animal welfare group was formed by Henry Bergh, who founded the American Society for the Prevention of Cruelty to Animals (ASPCA) in 1866. The group was formed not to rescue individual animals but, initially, to protect animals like carriage horses in New York City and to fight other forms of cruelty. While Bergh's efforts were primarily focused on anticruelty campaigns, resulting in the passage of the nation's first anticruelty law in New York in 1866, the ASPCA, which took over New York City's animal control contract in 1894, becoming one of the first privately-run animal shelters in the country, became the model for the country's first animal rescue groups, many of whom still use SPCA in their names today.

The 19th century saw the formation of other animal rescue groups, in the United States and England, some of which still exist today. The methods of these groups varied, but all were founded in order to help alleviate the suffering of companion animals, usually cats and dogs. Often focused on picking up stray cats and dogs, providing them with medical care, and attempting to find them homes, many animal rescue groups operated and, in many areas of the world today, still operate in the absence of any formalized, municipally-run animal control agency.

These small, often volunteer-run organizations are often the only source of help for stray, sick, abused, or starving domestic animals.

Today, in most locations in the developed world, animal rescue groups operate alongside city-and county-run shelters. Many groups are species- or breed-specific, rescuing only rabbits, Great Danes, or Chihuahuas, for example. As nontraditional pets become mainstream, and as the public purchases and then discards these animals, rescue groups are popping up to handle every species from turtle to parrot to rat.

Those groups with a relationship with their local shelter are generally contacted by staff at the shelter when an animal meeting the breed or species requirement is brought in, and representatives from that group will then pick the animal up. These groups both aid their local shelters by cutting down on the volume of animals the shelter must deal with, and are often better able to find a suitable adoptive home for animals which, because of their breed, species, or temperament, may be difficult for the shelter to place.

Rescue groups that focus on rescuing and placing nontraditional pets are faced with some unique challenges. Many so-called exotic pets, for example, are in fact wild animals that are not at all domesticated, and should not even be kept as pets, because of the damage that the exotic pet industry does to wild habitats and

Marc Bekoff (editor of this encyclopedia) shares a moment with a rescued dairy cow, Bessie, at the Farm Sanctuary in California. The Farm Sanctuary rescues and cares for animals freed from factory farms, slaughterhouses, and stockyards. (Marc Bekoff)

species, and because of these animals' unique behavioral and physical needs. These groups must attempt to place birds, reptiles and other non-domesticated animals into new homes, while at the same time discouraging people who can't adequately provide for their needs from obtaining these animals in the first place. In addition, animal shelters often have to rely on these rescue groups to take their nontraditional animals because they are so overwhelmed in the first place, are often ill-equipped to deal with the specific needs of exotic animals, and the shelter adoption rates of these animals are consequently much lower than those of cats and dogs.

Some animal rescue groups specialize in rescuing animals, both domestic and wild, from disasters. In the United States, Hurricane Katrina in 2005 demonstrated the need for comprehensive disaster plans that included provisions for animals to be rescued along with people. By the devastating Southern California wildfires in 2007, local and national officials had recognized the necessity of providing for animals, and joined with numerous animal rescue groups to provide for hundreds if not thousands of companion animals during that disaster.

Animal rescue groups are funded primarily by private donations. Those which have charitable status can offer tax benefits to their donors, but many groups are operated by well-meaning individuals who have not taken the steps to incorporate or obtain tax-deductible status. Many groups have newsletters, and most engage in fundraising efforts such as walkathons, merchandise sales, or services such as boarding, veterinary care, or grooming. Groups that are staffed entirely by volunteers and that operate out of a network of foster homes will have lower operating costs than those groups with paid staff and/or a permanent facility such as an animal shelter. These groups must do additional fundraising in order to meet their expenses. Some groups, like the ASPCA in New York, operate their city's animal control contracts, and thus are paid in part by the city. Other groups may run their own private shelters, taking in animals from the public, often for a fee, but not collecting stray animals or responding to cruelty calls. While most animal rescue groups do not euthanize animals except for health reasons, some, in particular those that operate shelters, do.

Many animal rescue groups make use of foster homes that provide permanent sanctuary care to animals that, by virtue of their age, health, or temperament, are deemed unadoptable. Other groups specialize in certain kinds of animals, such as disabled animals or seniors, often keeping them as sanctuary animals, but also often offering them for adoption.

Before the age of the Internet, animal rescue groups were primarily local operations with a network of local volunteers, a relationship with their local shelters, and a list of local supporters to provide funding. In the 1980s in the United States, Project BREED (Breed Rescue Efforts and Education) was founded to provide a resource for animal shelters, the public, and rescue groups. The Project BREED directory, still published today, listed thousands of breed-specific, for dogs, and species-specific rescue groups, as well as specific information on the breed or species, to aid people who are interested in adopting a particular kind of animal.

Today, with the Internet, not only are there numerous websites that provide such information, but rescue groups are able to operate nationwide, and even worldwide. The House Rabbit Society (HRS), a rabbit rescue organization, is one such group. Originally based in the San Francisco Bay area, the group developed a nationwide mailing list, attracting members and new volunteers from around the country, leading to chapters or volunteers in almost every state. With the Internet, HRS has gone international, with representatives in Europe, Asia, Canada, and Australia.

Most rescue groups today, in fact, use the Internet for public outreach, to advertise their adoptable animals, to fundraise, and to attract supporters. One important website for many groups is Petfinder, which allows those groups without their own websites or the technology to easily update a website to quickly and easily list their adoptable animals for the public to see, along with photos and, now, videos.

Animal rescue groups have, especially compared to breeders and pet stores, strict adoption procedures that typically include an adoption application, an interview, and often a home visit. These requirements are often stricter than those found at animal shelters, and are put in place to ensure that animals end up in a safe, loving, permanent home.

Animal rescue groups can be found around the world today, although developing nations, which often have greater problems with stray animals, tend to have fewer groups with fewer resources.

Further Reading

Best Friends Animal Society. 2006. *Not left behind: Rescuing the pets of New Orleans.* New York: Yorkville Press.

Goodman, Susan. 2001. *Animal rescue: The best job there is.* New York: Aladdin Press.

Petfinders. www.petfinder.com. Discovery Communications. More than 10,000 animal shelters and adoption organizations in the United States, Canada, and Mexico.

Stallwood, Kim. 2001. *Speaking out for animals: True stories about people who rescue animals.* New York: Lantern Press.

Margo DeMello

ROYAL SOCIETY FOR THE PREVENTION OF CRUELTY TO ANIMALS (RSPCA), HISTORY

At the beginning of the 19th century, the English would have been surprised to hear themselves praised for special kindness to animals. City streets were crowded with horses and dogs that served as draft animals and beasts of burden, as well as with herds of cattle and sheep being driven to slaughter. Many of these animals were obviously exhausted or in pain, as were many of the horses and donkeys used for riding. Popular amusements included cockfighting, dog fighting, rat killing, bull running, and the baiting of wild animals. Elsewhere in Europe, England was generally known as the hell of dumb animals, and early 19th-century English humanitarian crusaders sadly agreed with this criticism. By the end of the century, however, officials of such organizations as the Royal Society for the Prevention of Cruelty to Animals routinely claimed that kindness to animals was a native English trait and that, within Europe at least, cruelty was to be associated with foreigners, especially those from southern, Catholic countries.

This shift in opinion reflected real changes. The 19th century saw a series of administrative and legal breakthroughs with regard to the humane treatment of animals, as well as steadily widening public support for animal welfare, and for the laws and societies dedicated to protecting animals from cruelty and abuse. Although the first animal protection bill to be introduced in Parliament failed miserably in 1800, in 1822 a pioneering piece of legislation was enacted. Known as Martin's Act, after its originator and chief advocate Richard Martin, it aimed to prevent cruel and improper treatment of cattle, which included most farm and draft animals, but not bulls or pets. Later legislation was passed in 1835, 1849, and 1854, and periodically extended protection until all domesticated mammals, as well as some wild mammals in captivity, were covered.

These extensions did not inspire universal rejoicing. For example, the first extension of the provisions of Martin's Act, in 1835, specifically prohibited the keeping of places dedicated to fighting or baiting bulls, bears, badgers, dogs, and cocks, which had become local institutions in many rural communities. Bull baiting or bull running, where people and dogs chased an animal through village streets before cornering and killing him, were particularly cherished traditions of this type and, perhaps for that reason, particularly obnoxious to humanitarian reformers. When, supported by the 1835 act, they attempted to suppress such observances, they often encountered physical as well as moral resistance. For example, the bull running at Stamford, Lincolnshire survived several attempts at suppression. Finally, in 1838, the local magistrates and the RSPCA successfully enforced the ban by calling in twelve London policemen and a troop of dragoons.

When the Society for the Prevention of Cruelty to Animals was founded in 1824, one of its primary goals was to ensure that the provisions of the new legislation actually took effect. The SPCA funded its own special corps of constables, and instructed civilian sympathizers in how to arrest the variety of aggressive wrongdoers who might be encountered in the streets, including livestock drovers and recreational sadists, as well as cabmen and wagoneers. Despite the initial obstacles it faced, the SPCA (RSPCA beginning in 1840, when Queen Victoria granted the Society permission to add the prefix "Royal" to its name) was successful on every front. As legal protections for animals expanded, so did the Society's membership, in both numbers and social prestige. It boasted a series of royal patrons, and the aristocracy was heavily represented on its governing board.

By the 1900s, the RSPCA epitomized respectable philanthropy, the kind of charity routinely remembered in the wills of the prosperous. With such powerful backing, the size of the RSPCA increased from its initial complement of only a few men, to eight officers by 1855, 48 by 1878, and 120 by 1897. In its first year of operation, the society conducted 147 successful prosecutions under Martin's Act. By the end of the century successful prosecutions peaked at over 8,000 per year, before horses, the most frequent victims of prosecuted offenses, were replaced by motor vehicles.

One reason that cab horses and draft horses figured so prominently in RSPCA prosecutions was that there were many them, and they were abused in plain sight on the public streets. But another was that their abusers were apt to belong to the part of human society where the middle and upper-class members of the RSPCA expected to encounter depraved behavior. Indeed, it is likely that some humanitarians viewed the animal protection laws as a useful supplement to existing legal and social mechanisms for controlling unruly humans. When animals suffered at the hands of the genteel, the RSPCA and kindred organizations found it more difficult to prosecute or often even to acknowledge that a problem existed. For this reason, such sports as steeplechasing, foxhunting and, indeed, hunting of all kinds, were subjects of contention within the mainstream Victorian humane movement. The hardest case of all in these terms was posed by vivisection, an exclusively middle- and upper middle-class pursuit. Although John Colam, then the Secretary of the RSPCA, offered strong testimony against the use of vivisection in teaching when he testified before a Royal Commission on vivisection in 1876, few of his constituents shared his strong views. As a consequence, committed antivivisectionists withdrew from the mainstream humane movement and, for at least several generations, they languished while it prospered.

Further Reading

Fairholme, Edward G., and Pain, Wellesley. 1924. *A century of work for animals: The history of the R.S.P.C.A., 1824–1924*. New York: E. P. Dutton.

Kean, Hilda. 1998. *Animal rights: Political and social change in Britain since 1800*. Chicago: University of Chicago Press.

Ritvo, Harriet. 1987. *The animal estate: The English and other creatures in the Victorian age*. Cambridge, MA: Harvard University Press.

Salt, Henry. 1980. *Animals' rights in relation to social progress* (1892; rpt.). Clark's Summit, PA: Society for Animal Rights.

Shevelow, Kathryn. 2008. *For the love of animals: The rise of the animal protection movement*. New York: Henry Holt.

Thomas, Keith. 1983. *Man and the natural world: A history of the modern sensibility*. New York: Pantheon.

Turner, James. 1980. *Reckoning with the beast: Animals, pain, and humanity in the Victorian mind*. Baltimore: Johns Hopkins University Press.

Harriet Ritvo

ROYAL SOCIETY FOR THE PREVENTION OF CRUELTY TO ANIMALS (RSPCA) REFORM GROUP

The Royal Society for the Prevention of Cruelty to Animals (RSPCA) is the oldest, largest and most influential animal protection organization in the world, and so its vigor and radicalism or lack of same are of great importance for the whole movement internationally.

Frustrated by the ineffectiveness of the RSPCA in dealing with the modern cruelties of factory farming, animal experimentation, and the increasingly internationalized abuse of wildlife, some members of the RSPCA, led by Brian Seager, John Bryant, and Stanley Cover, formed the RSPCA Reform Group in 1970. They supported the attempt by Vera Sheppard to persuade the RSPCA to oppose foxhunting and other cruel sports, and succeeded in 1972 in securing the election to the RSPCA Council of five Reform Group supporters, including Bryant, Seager, Andrew Linzey, and Richard Ryder. Over the next eight years, until the end of the decade, the Reform Group faction succeeded in changing the world's oldest and largest animal welfare organization beyond recognition. In 1976, Ryder was made Vice Chairman, and was then Chairman of the RSPCA Council from 1977 until 1979. During these years of reform, the Society not only came out against cruel sports but, for the first time, developed comprehensive animal welfare policies across the board, elevating the welfare of farm, laboratory, and wild animals to a priority status equal with the welfare of domestic species. Against stiff opposition, the reformers set up expert staff departments to deal with these areas of abuse, and revived the Society's campaigning function, which had been allowed to lapse since the Edwardian era. Publicity, parliamentary, and scientific facilities were established, and the Society even gave its support to Douglas, Lord Houghton's successful initiative, the General Election Coordinating Committee for Animal Welfare, to persuade all major British political parties to include officially, for the first time, animal welfare policies in their election platforms in 1979. Before the end of Ryder's first term of office, which was followed by a temporary reversal of the Society's performance initiated by conservatives, an undercover plainclothes section of the RSPCA's Inspectorate was also established and perhaps most important, the Society initiated the establishment and funding of a powerful political lobby for animals in the European Community, subsequently to be named Eurogroup for Animal Welfare.

In 1906, pro-foxhunting members had changed the constitution of the Society's ruling Council so as to bring in rural representatives, thus strengthening their position.

Tensions persist to this day between, on the one hand, the nationally and internationally-minded campaigners and, on the other, the dogs and cats rural conservatives. During the 1990s, some five thousand pro-foxhunting people infiltrated the Society in a vain attempt to prevent the banning of hunting with hounds by the Labour government in 2004. The modernists continued to resist this attempted takeover of the RSPCA and, with the support of the High Court, several hundred members were ejected. Supported by the RSPCA, much new European and British legislation has been passed, culminating in the Animal Welfare Act of 2006, which marked the end of an era.

Further Reading

Ryder, Richard D. 2000. *Animal revolution: Changing attitudes towards speciesism.* Oxford: Basil Blackwell Ltd.; rev. ed. Berg.

Richard D Ryder

S

SANCTUARIES

There are thousands of sanctuaries for animals around the world, havens offering protection to individuals rescued from circuses and other forms of entertainment, from medical research laboratories, from factory farming, from the pet trade, and indeed from all situations in which they are shamefully neglected or horribly abused, as well as orphan animals whose mothers have been killed for food, for sport, or for any other reason. There are sanctuaries for animals of all kinds, from hens to elephants, and while some are small enterprises in backyards, others are big operations that require a large staff and considerable funding. Unfortunately, once a refuge has been created it tends to grow, an indication of the number of desperate creatures needing care. Fortunately, there are also a growing number of people around the globe who are not only aware of, but care, sometimes desperately, about animal suffering.

Chimpanzees are classed as an endangered species and it is illegal to hunt and sell them; however, these laws are seldom understood, let alone enforced. I became involved with sanctuaries for orphan chimpanzees in Burundi in 1990. This was when The Jane Goodall Institute (JGI) initiated a conservation project there, and people began telling us about the pet chimpanzees in the country, many

of them held in appalling conditions. One of these, whom I went to see for myself, was Whiskey. His owner came to greet me and led me through his noisy garage to a cement-floored 6-foot-by-6 foot space that had once been a lavatory. The only light came through a hole in a corner of the corrugated iron roof. A five- or six-year-old male chimpanzee with a collar around his neck was chained to a pipe in the wall. Whiskey held his hand towards us, stretching as far as he could, but his chain was only two foot long and we were out of reach, so he turned and stretched back with one foot. When I went in and crouched down beside him, he put his arms around my neck.

Whiskey's mother had been shot for the illegal live animal trade so that her infant could be stolen and sold as a pet or to attract visitors to a hotel or bar. He had been captured in neighboring Zaire (now the Democratic Republic of the Congo), then smuggled over the border and sold in Bujumbura. At first he had been part of a human family, sitting with them at table, riding in their car, and playing with the children, until he was about four years old, when they realized how strong and potentially dangerous he was. Then he was banished to his prison cell. Eventually we (formerly JGI-Burundi) persuaded his owner to hand him over to what we called The Half Way House. This was a small backyard facility where other ex-pets were waiting until we could raise the money for a permanent sanctuary. But in 1994, because of the ethnic violence in Burundi, we had to move all of them, 20 by then, to Kenya, where a new sanctuary, Sweetwaters, had been built for them.

That is how it always starts. An individual chimpanzee looks, from his place of fear and confusion and pain, into your eyes, and reaches out to touch you. The very first African sanctuary began when one tiny and badly wounded infant was confiscated from a hunter, who had shot her mother in neighboring Zaire, and taken to a British couple, Dave and Sheila Siddle, who run a cattle ranch in Zambia. They nursed her back to health and were given a permit to keep her. And so, of course, government officials brought them the next confiscated infant. And the next, and the next. And when people realized that, at last, there was a place where young chimpanzees would be properly cared for and loved, youngsters began arriving from other parts of the world. Of course, as the Siddles' chimpanzee family grew, so did their expenses. They had to fence in a large area of their land, and build strong cages for night quarters and where the chimpanzees could be cared for if they were sick or injured. The Chimfunshi Animal Orphanage is now home to more than 100 chimpanzees.

There are now 13 sanctuaries in Africa that care for orphan chimpanzees, including the Tchimpounga Chimpanzee Rehabilitation Center in the Republic of the Congo, which JGI manages. In Zambia, Kenya, and South Africa, where there are no wild chimpanzees, the orphans are considered refugees from neighboring countries. This gives a total of over 600 chimpanzees in sanctuaries. Of course, the number is constantly changing as new orphans arrive and, in-

evitably, a few die. The best place to check on the African chimpanzee sanctuaries is the Pan African Sanctuary Alliance (PASA), www.pasaprimates.org. There is an additional colony of chimpanzees belonging to the New York Blood Center located in Liberia. Dr Fred Prince is working to move these ex-experimental chimpanzees to a safe sanctuary. It must be mentioned that there are also five sanctuaries that care for orphan gorillas, two in Cameroon, one in Gabon, one in the Republic of the Congo, and one in DRC. These sanctuaries care for over 78 gorillas. Another sanctuary in the DRC cares for over 50 bonobos.

Often I am asked why we do not return our orphans to the wild. The answer is that we would if we could, but it is a very difficult process. First, it is necessary to find an area of suitable chimpanzee habitat where there are few wild chimpanzees, who are territorial and typically kill strangers, especially males, and no people, for our orphans have no fear of humans and would wander into a village and either be hurt or hurt someone. We are actively searching for ideal places for reintroduction in Congo-Brazzaville, for our sanctuary is currently operating over capacity. If we are successful we shall then have to ensure that our youngsters acquire the skills they need to survive in the wild. One sanctuary that has successfully released chimpanzees into the wild is H.E.L.P. in the Republic of Congo-Brazzaville.

Conservationists often accuse us of wasting money by caring for captive individuals rather than spending our precious dollars on trying to save wild chimpanzees and their vanishing habitat; however, I feel we have no choice. After all, ever since I began my research into chimpanzee behavior at Gombe,

I have stressed the importance of individuality. Each chimpanzee has his or her own, unique personality, and each plays an important role in his or her society. This thinking was not fashionable among scientists back in 1960, but is widely accepted today. From the beginning I insisted that the chimpanzees had feelings and emotions similar to ours. (After all, I had learned during my childhood that this was true for my dog.) Thus to abandon an orphaned chimpanzee would be, for me, as unethical as abandoning a small human child. However, I know also that it is desperately important to do everything we can to protect the remaining wild chimpanzees and their habitat. And so JGI struggles to achieve both of these goals.

A final and important point is that our orphans serve as ambassadors for the wild chimpanzees. Most people, even if they live near a forest, have little or no opportunity to observe chimpanzees. When visitors from the villages or from a nearby town come to see the sanctuary chimpanzees, they are typically amazed to see how like humans they are. A number of people, after watching the youngsters kissing and embracing, using tools, playing, and so on, have said that they will never purchase, trade, or eat chimpanzees or other apes again, and never go to a restaurant that serves ape meat. We especially encourage children to visit. And we aim to provide research opportunities for students from universities to study chimpanzee behavior.

Another criticism often leveled at those working to save orphan chimpanzees or other animals in Africa is that we have got our priorities wrong. Surely, in view of the poverty and suffering of the people of Africa, we should not be wasting money on animals. We realize only too well the desperate need of hundreds of thousands of Africans, and JGI is working hard to improve the lives of the people living around our sanctuaries. We are modeling these efforts on our highly successful TACARE (Lake Tanganyika Catchment Reforestation and Education) program, which continues to improve the lives of over 150,000 people in 24 villages around the Gombe national park. This program, in addition to facilitating the introduction of tree nurseries, agroforestry, the most suitable environmentally sustainable farming techniques, and conservation education, also provides primary health care through the regional health authority, AIDS education, and family planning. Micro-credit banks enable women to start their own environmentally sustainable projects, thus earning some money for themselves, often for the first time. Gifted girls can apply for scholarships to go from primary to secondary school. Around the world, it has been shown that as women's education and self-esteem improves, family size drops. In Uganda there is also a developing community outreach program. And in the Congo, in the villages adjacent to the Tchimpounga Chimpanzee Rehabilitation Center, a similar program is in its early stages. We have built a health dispensary and a school, introduced our education program Roots & Shoots, and are in the planning stages of many other projects to help people to help themselves. Of course, in and of themselves, our sanctuaries provide ongoing jobs for local people, help to boost the local economy and, when tourism is possible, bring foreign exchange into the country.

Unfortunately, it is not only in Africa that chimpanzees desperately need the help of dedicated people. In the Americas,

Europe and Asia chimpanzees have been mistreated, often shockingly, in zoos, circuses and other forms of entertainment, and in medical research laboratories. Many of these were taken from Africa, snatched from the dead bodies of their mothers as infants. Others were born in captivity. We owe it to these unfortunate individuals to provide them with safe havens where they can live out their lives in relative freedom once they have been rescued.

In the UK, Jim Cronin founded the Monkey World Ape Rescue Centre, which he runs with his wife, Alison. Originally this center was built to provide a home for the infant chimpanzees smuggled into Spain from West Africa and used as photographers' props in tourist resorts. Jim worked with a British couple who lived in Spain, the Templars, and with the police, to stop the illegal trafficking, and also with tourist agencies, persuading them to warn visitors of the cruel practice. Jim has now rescued chimpanzees and other primates from many parts of the world. In America, Wally Swett began taking in abused animals, mostly primates, discarded by the pet and entertainment industries. His Primarily Primates is situated in San Antonio, and now provides sanctuary for several groups of chimpanzees. Patti Regan, at the Center for Orangutan and Chimpanzee Conservation, Vachula, Florida, and April Truit, at the Primate Rescue Center, Inc., Nicholasville, Kentucky, have both built small sanctuaries for ex-pet and ex-entertainment chimpanzees.

A very difficult challenge is to create sanctuaries for chimpanzees who have been used and abused in medical research laboratories. These individuals are typically full grown, and often they have been housed alone for most of their lives, so that it can take years to re-socialize some of them. The very first rescue of a group of ex-lab chimps, released onto a manmade island at Lion Country Safaris in Florida, is described by Linda Koebner in her moving book *From Cage to Freedom.* They are still there.

Years later, the Chimpanzee Health Improvement, Maintenance and Protection Act or CHIMP Act, H.R. 3514, sponsored by U.S. Representative James Greenwood, was passed by both House and Senate in 2000, and signifies the U.S. government's commitment to partner with the private sector to provide sanctuaries for chimpanzees retired from medical research. In December 2006, President George W. Bush signed the Chimp Haven is Home Act into law, which prohibits the removal of or research on retired chimpanzees living in federal sanctuaries. Chimp Haven (http://www.chimphaven.org/index.cfm), a nonprofit organization, has received $24 million from the National Institutes of Health (NIH) to build and manage a sanctuary on a 200-acre site of forested land donated by the citizens of Caddo Parish in Louisiana. Chimp Haven must raise funds themselves equal to 10 percent of the government grant.

There are other sanctuaries in North America. Richard Allen and Gloria Grow of the Fauna Foundation have built a sanctuary for 15 chimpanzees near Montreal in Canada. It was the first sanctuary of its kind, built to house chimpanzees infected with AIDS and hepatitis as well as clean individuals. It serves as a precedent, inspiring others to make the same commitment. The next sanctuary for ex-lab chimps was built by Carole Noon in Florida. The first group to be housed there comprised 21 of the so-called Air Force chimpanzees from the Holloman Air Force Base in New Mexico. Some of these are descendents of the original group that was captured in the wild for the space research that culminated in sending the first astronauts to the moon.

In 2002, Carole Noon took on the biggest chimpanzee rescue in history when the Coulston laboratory in New Mexico was finally closed down after accumulating countless violations of the animal welfare act for years, and the whole facility was bought with an incredibly generous grant from the Arcus Foundation. Dr. Noon's first task was to make immediate improvements to the existing facility. The chimpanzees were sleeping on concrete or metal flooring with no blankets or straw, given only one piece of fruit each per week, and many had no contact with each other even visually. Gradually the Florida facility will be enlarged, and more and more of these chimpanzees will move to the relative freedom of grassy islands with shade and climbing structures, and be cared for by humans who understand and love them.

Wherever a sanctuary is located, the chimpanzees rescued from abuse have so much to teach us. Many have lived alone for years, deprived of everything that a chimpanzee needs to enjoy life. Often they have acquired psychotic behaviors, such as rocking from side to side, banging their heads on the wall, mutilating themselves, showing sudden violent outbursts of rage, or huddling alone in a corner for hours on end. Many can never fully recover psychologically. But it is inspirational to watch how they can gradually manage to lose some of their abnormal behavior, and learn to live in chimpanzee society, and there is much they can teach us. Not so long ago, psychiatrists and psychologists used to raise chimpanzees in conditions designed to replicate the abnormal early experiences of psychologically disturbed humans. It was argued that this would be helpful to scientists seeking to better understand mental illness in people and thus help human patients. Now there are hundreds of chimpanzees who have been exposed

Mariah, a Siberian tiger, right, is penned next to two timberwolves, Apache and Noshoba, at Noah's Lost Ark animal sanctuary in Berlin Center, Ohio. Mariah was raised by an elderly woman who also took care of wolves. When brought to the sanctuary, Mariah began losing weight and was not adjusting well to her new environment. The original owner suggested her enclosure be placed next to the wolf enclosure. She immediately began gaining weight. Ellen Whitehouse, who has run the sanctuary with her husband for the past five years, calls the wolves "the best tiger baby sitters." (AP Photo/Tony Dejak)

to all manner of abnormal conditions. It is important that sanctuaries open their doors to scientific observation of a strictly noninvasive, non-disruptive nature. Surely there are lessons we can learn from the rehabilitation of our closest relatives that will benefit the thousands of humans who, like the chimpanzees, carry deep psychological scarring from past traumatic experiences.

Further Reading

Editor's Note: For general information about animal sanctuaries see:

http://www.pasaprimates.org

The Pan African Sanctuary Alliance, or PASA, is an alliance of sixteen primate sanctuaries from all over Africa. The Web site lists the sanctuaries and provides information on most of them. It also provides an extensive list of resources about animals, endangered species and conservation, sanctuaries, animal behavior, and related topics.

http://www.taosanctuaries.org/sanctuaries/species.htm

The Association of Sanctuaries, TAOS, was founded in 1992 as a not-for-profit organization to assist sanctuaries in providing rescue and care for displaced animals. It accredits superior sanctuaries for wild, farmed, and companion animals. Provides a worldwide list of accredited sanctuaries.

http://www.cwu.edu/~cwuchci/index.html

The Chimpanzee and Human Communication Institute at Central Washington University in Ellensburg, Washington, provides sanctuary to a group of adult chimpanzees who communicate with humans and each other using American Sign Language (ASL).

Jane Goodall

SANCTUARIES, ETHICS OF KEEPING CHIMPANZEES IN

The mandate of sanctuaries is to provide for both the physical and psychological requirements of chimpanzees, a new life in which to heal and recover from previous abuses, and a life rich providing for their complex needs, striving to replace and fulfill a chimpanzee's natural requirements. Sanctuaries consist of a team of committed humans who tend to the chimpanzees; a board of directors, donors, management, care staff, veterinarians, and volunteers, all of whom support a new life for the chimpanzees.

Chimpanzees do not live in sanctuaries by choice. Their native environment and natural daily lives are often in sharp contrast to what sanctuaries can provide. In even greater contrast is the emotional and mental status of chimpanzees in sanctuaries. In addition, previous psychological damage from being subjects of research, animal actors, roadside zoo attractions, or household pets exacerbates these chimps' emotional maladies.

An ever-challenging mission of sanctuaries is to mitigate the damage done to chimpanzees in captivity, while going far beyond to provide an environment in which they can heal and thrive emotionally. Sanctuaries must also provide a physically stimulating and enriching environment to give their charges a healthy life. Captive chimpanzees are completely dependent on our intelligence and compassion.

Chimpanzees are extraordinary and complex, emotional, and sensitive. They are fully aware of themselves as captives. Key to restoring chimpanzees' sense of self and confidence is freedom of choice. Environmental designs and enrichment programs within sanctuaries must provide that freedom of choice through creative and constructive methods to help heal and nurture chimpanzees. Although seemingly simplistic, the value of freedom of choice cannot be overemphasized.

The physical strength and athletic ability of chimpanzees is remarkable. Sanctuaries are responsible for the safety of their charges, as well as that of their human caregivers. Careful planning and sound design of enclosures which prevent harm and escape yet provide for the chimpanzees' physical and emotional needs have proven successful at many model sanctuaries. The use of open chimp islands with mature trees and large climbing structures provides for the basic needs of exercise, fresh air, enrichment, and natural behaviors.

Sanctuaries have the obligation to provide enclosures and enrichment that serve the chimpanzees, with little regard for how this may perceived by humans. In contrast, traditional zoos have a perceived obligation to provide a living photographic image for the pleasure of their admission-paying customers. Zoo exhibits are often designed by architects to appear as if the display is natural and the chimps are content. They are designed to keep chimps in front of the public, even though chimpanzees require solitude and privacy, variety and change. Sanctuaries exist for the chimp's approval, not the public's. With this mandate, sanctuaries can design enclosures and provide enrichment that maximizes choice, stimulation, and interest for their charges.

Some of the more progressive zoos understand the emotional needs of the chimpanzees in their care, designing spacious exhibits and providing a wide variety of enrichment elements on a daily basis. Although a cardboard box would not be found in a chimpanzee's natural habitat, a chimpanzee's natural habitat would not be limited to a large rock and plastic tree, either. Progressive zoos and responsible sanctuaries are able to reconcile the difference between a clean and sterile chimp environment which may be attractive to humans, but deprives intelligent chimpanzees with emotional and physical stimulation. The most content chimps in the best sanctuaries may have the messiest enclosures. Responsible guardianship also mandates the cleanliness of the chimps' environments, achieved through routine daily cleaning and maintenance.

Providing for comfort and natural behaviors is also critical for chimpanzees' wellbeing. Although chimpanzees have coarse hair covering most of their bodies, it is not dense. Some sanctuary individuals may have very thin hair coverage or be completely bald, which may be attributable to age, poor nutrition, or often to emotional disorders resulting in hair-plucking behaviors.

Their native equatorial Africa remains warm throughout the year, as it has throughout their evolution. Therefore strong consideration must be given to chimpanzees' low tolerance and emotional discomfort when chilled or cold. Healthy chimpanzees are muscularly dense, usually with few fat reserves for warmth or protection from the elements. Proper sanctuaries will ensure proper ambient heating, as well as providing further options for warmth and comfort, such as blankets and other nesting materials. Even in the heat of summer, chimps will make a new bed or nest every night in which to sleep. Sanctuaries should provide a variety of materials to encourage this activity, with items such as blankets, sheets, leaves, newspaper, or cardboard, for example.

Progressive sanctuaries exercise sound judgment when providing enrichment for their chimpanzees. There is an inexhaustible list of safe foods, toys, natural browse, and so on which can be used to enhance and stimulate a chimpanzee's daily life. However, even the most benign objects have the potential to be a hazard. Caretakers are well informed on safe and proper use of enrichment by the chimps, and whether or not an individual's toys must be limited for their own safety.

If new residents in a sanctuary are not familiar with enrichment, there are safe protocols established to introduce enrichment and allow for timely introduction of new items. Chimpanzees cannot live in a bubble, because environmental enrichment may involve accidents. Caretakers' experience and intelligence play an important role in maintaining a margin of safety, just as experience and intelligence play a role in our safety every time we drive a car or allow our children to play at the beach.

Nutrition and food opportunities play a critical role in the health and wellbeing of chimpanzees. Wild chimpanzees forage for food six to eight hours every day. Their natural diet is varied and mostly vegetarian, measuring approximately 60 percent fruit, 30 percent other vegetation, and 10 percent animal matter (Nowak, 1999, p. 183). Food may take on greater importance in a captive chimp's life, in part due to boredom. Nonetheless, captive chimps are exceedingly motivated and excited by food. A sanctuary's responsibility with regard to feeding chimps is important both nutritionally and psychologically. There must be a balance between feeding opportunities and nutrition. It is not unusual for former pets or circus performers to arrive at a sanctuary with poor eating habits and medical sequela, such as diabetes. Some chimps from research studies may arrive with a preference for pre-packaged chow biscuit and reject fresh produce. Optimal physical and emotional health can be attained by providing a varied and well-rounded diet daily. Regardless of convenience and cost, to deny chimpanzees a variety of produce, natural foods, and freedom of choice is unacceptable.

Sanctuaries are responsible for forming social groups of chimpanzees. This responsibility is not for the inexperienced or unintuitive. The desired result is a cohesive and dynamic social group while minimizing the risk of injury or possible death to an individual unprotected and unable to defend himself. Although there are no guarantees of harmony, careful planning and consideration, enclosure design and introduction techniques must be properly administered. It is the sanctuary's responsibility to oversee all aspects upon which the final outcome relies.

The greatest responsibility of a sanctuary is to those individuals in their immediate care for whom the sanctuary has accepted lifetime guardianship. Sanctuaries are often faced with the difficult choice of denying a home to yet another chimp in need. These facilities make an effort to help place the chimp in another sanctuary or at least offer advice to benefit the chimp, but sound sanctuaries know their capacity. Decline in the overall success of a sanctuary can occur when accepting another chimp compromises the care of those to whom the sanctuary is already committed. However, reasonable circumstances may allow for stretching resources on an emergency or temporary basis. But a chronically overpopulated sanctuary ceases to be a refuge and becomes a place that chimps may need to be rescued from.

Philosophically, how humans can and should interact with chimpanzees greatly determines the safety and contentedness of both humans and nonhuman primates. True sanctuaries treat their charges with respect and dignity. For the sentient and intelligent chimpanzee, a life behind bars is hell. It is the sanctuary's responsibility to make every effort to equalize the power and offer pride and dignity to the powerless. An angry chimpanzee is a very dangerous chimpanzee, to both his human caregivers and the social group. Gaining trust and respect is a two-way street, and how both parties interact, chimps with humans and humans with chimps. is generally indistinguishable. The humans, however, must take the initiative and have inexhaustible patience and compassion.

Daily, sanctuary staff plays a critical role in the emotional recovery and stability of the chimpanzees in their care. Staff's arrival should be met with excited pant-hoots in anticipation of food and the arrival of a trusted friend with whom the chimps have a bond. Once a friendship is developed, it is sacred, and is held in the highest regard. A chimp will not befriend anyone just for food, which alone indicates a more complex and sophisticated emotional capacity. Befriending a chimpanzee is a significant accomplishment. There are many chimps who may forever refuse the friendship of a human, and in their rebuffs they will attempt to physically injure or assault the very caregiver whose patience and olive branch remain proffered. The role of the caregiver often means ever remaining an honorable and true friend even to the most jaded of chimpanzees, because often it is these individuals who've suffered most at the hands of other humans.

In 2007, the National Center for Research Resources of the National Institutes of Health permanently adopted a temporary 1995moratorium on breeding chimpanzees. Although this was met with criticism from many in the biomedical research community, it was applauded by animal welfare and animal rights advocates as an ethically and morally sound decision. The greatest number of captive chimpanzees in the United States live in research institutes. Lesser numbers are those from the entertainment industry and zoos, or pets. The current and future need for sanctuaries is far greater than the number of sanctuary homes available. For this reason and the obvious wrongness of breeding into captivity, sanctuaries do not breed chimpanzees. Failed contraception may result in a birth, which realistically enhances and stimulates chimpanzee group dynamics, but at the expense of yet another chimp living an incarcerated life.

Regarding euthanasia, the Chimpanzee Health Improvement, Maintenance, and Protection (CHIMP) Act signed by President Clinton in 2000 states that "The CHIMP Act prohibits routine euthanasia. No chimpanzee can be killed simply because they are no longer of "use," the facility is overpopulated, or they are too costly to maintain. Euthanasia as a humane option during an intractable illness is permitted."

This is a significant moral and ethical advancement in a country where millions of dogs and cats are euthanized annually due to overpopulation. There is hope for a county's moral conduct when the responsibility to preserve lives transcends speciesism to include great apes.

Most animal rights advocates believe that regardless of how a chimp comes to a sanctuary, whether it is research, entertainment, a roadside zoo or the pet trade, fundamentally humans have failed. Chimpanzees should not be kept captive for the use of humans. It is impossible to provide an environment that exactly mimics their natural habitat. The goal for sanctuaries is that they eventually become unnecessary, to serve chimpanzees in need until there is no more need.

Reintroductions of chimpanzees in Africa to their native habitat have been marginally successful. However, sheer logistics all but eliminates this as an option for non-African chimpanzees. The ultimate goal is for chimpanzees to thrive in the wild, and although the era has passed where no human intervention is necessary to protect wild populations, most animal advocates believe that strict effective protection and respect for chimpanzees, other primates, and all natural wildlife must become realized if they are not to perish before our very eyes.

Sanctuaries are the self-appointed guardians of some of the most complex and remarkable beings with whom we share this earth. Chimpanzees' exquisite evolutionary achievement and their human-like familiarity account for both their intrigue and the cause of their demise. Humans reign supreme over this world, and many believe that with that dominion comes not the privilege to exploit and discard at will, but rather the responsibility to protect and preserve. The restorative power and potential of sanctuaries have been proven and must continue to advance, but ultimately to preserve chimpanzees in their wild habitat is essential.

Further Reading

Wise, S. 2000. Chimpanzee and Bonobo minds. In *Rattling the cage toward legal rights for animals*, 179–237, Cambridge, MA: Perseus Publishing.

Bradshaw, G. A., Capaldo, T., Lindner, L., & Grow, G. 2008. Building an inner sanctuary: Complex PTSD in chimpanzees. *Journal of Trauma and Dissociation*, 9 (1). http://www.haworthpress.com/web/JTD.

Lee Theisen-Watt and Chance French

SCHOLARSHIP AND ADVOCACY

Traditionally, a strong distinction has been drawn between scholarship and advocacy. The scholar's job, it was held, was limited to providing an understanding of a problem or issue, independent of any advocated position on it. In the positivist philosophy of science supporting that distinction, understanding can be independent or neutral, and can provide credible information of use to advocates on all sides of the issue.

This contrasts with the legal system, in which a lawyer explicitly advocates for a particular party, namely the client. The lawyer's brief is a presentation of facts and their application to relevant case law on one side of the issue, that is, either the innocence or guilt of a defendant. Historically, the term advocate was used in a legal context and the word derives from the Latin *ad* (for or toward) and *vocare* (to call), and later from the French *l'avocat* which means legal counsel or lawyer. In contemporary use it is broadened beyond the legal context to refer to taking up the cause of another. In a free society, any individual can advocate for any cause.

Scholarship is a researcher's day job, and he or she can advocate for any issue in her free time. However, in practice, when a scholar acts or speaks as a private citizen, the public interprets the pronouncement as being that of a scholar, an expert who works in a space outside of the fray of callings, causes, and partisanship. Scientists as individual citizens have been powerful after-hour advocates both outside of and within their primary areas of scholarship. Noam Chomsky, the seminal thinker on modern linguistics, for example, was a major advocate for the Left; after inventing the atomic bomb, Albert Einstein was a strong advocate against its use.

The more difficult question is the role of advocacy within the actual enterprise of scholarship. Contemporary philosophers and sociologists of science challenge the received view that it is possible for a scholar to take no position on the material he or she investigates. One major idea in this challenge is field theory. Both the object of investigation and its investigator are embedded in the same field or system. With regard to the object of investigation, the strategy of breaking it down into small bits, isolating them, and controlling for all other variables is thrown into question. If the targets of study, particularly involving social or cultural topics, are inherently embedded in a field or system, investigations must include, rather than control for, those contexts.

Far from being independent or outside of the scholarly enterprise, the investigator is embedded within that field, and so necessarily has a position (attitudes, values, biases) on it. Although presented as facts or findings, scientific findings are actually social constructs, products of political, social, economic, and even personal forces. Although science can take as its regulating ideal the goal of understanding, independent of these forces and treating them as contaminants, in practice science is messier. Independence of view, the view from nowhere, is a fiction. For example, although scientists strive to maintain the same external, objective non-relationship to animate objects as they do to inanimate objects of study, they do form a relationship even with the likes of mice and rats, as described in Davis and Balfour's *The Inevitable Bond: Examining scientist-animal interactions*.

A bond or relationship implies an evaluative view of the another being, including recognition of his or her interests and the pull to advocate for those interests.

Although a possible and even admirable ideal, there is no value-free inquiry. Scholarship occurs in an enterprise that is value-laden. Scholars bring their values to it, those values are changed through the research, and the results of the research influence society's values and practices.

Is an enterprise that is value-laden in these ways distinguishable from advocacy? Are scholars necessarily advocates?

To look more closely at the advocacy side of the question, it's necessary to distinguish advocacy from activism. Activism is one form of advocacy, emphasizing vigorous action for a cause—protests, strikes, sabotage, boycotts, and sit-ins. But advocacy has a quieter, more slow-burning side as well. Much of the work of contemporary animal protection organizations involves exposing the public to animal abuse and exploitation through an array of printed and other media—leafleting in the mall or showing a video on a truck jerry-rigged for that purpose.

This quiet activism makes no claim to scholarship, as typically no new knowledge or understanding is developed. But is it even education? Although an exposé may be factual and the information it provides is often new to the targeted audience, it presents only one side of an issue and explicitly advocates for that side. Although we loosely refer to it as educational, it is closer to propaganda, in that it propagates or promotes a particular view or practice.

From advocacy as activism and advocacy as exposé, consider classroom education. A teacher may or may not be a researcher, but the curriculum he or she presents relies more on research findings than on material developed by advocates. So is it proper for a teacher to be an advocate in the classroom or, the other extreme, is a teacher necessarily, even unwittingly, an advocate? Is she required to and can she, actually and metaphorically, take off her Obama button before entering the classroom? The earlier argument regarding scholarship applies roughly to pedagogy. Ideally, a teacher presents all sides of an issue or, better still, fosters critical thinking applicable to understanding any issue. But, in practice, teaching, like scholarship, is value-laden. Choice of curriculum, questions raised, even style of teaching all occur within a context of personal and professional assumptions and values that have leanings, that are evaluative and that, at least implicitly, are advocatory.

In response to the recognition that both pedagogy and scholarship are advocatory platforms, many critics have suggested that the scholar, the main concern here, present her position on issues raised by the research. An investigator of the history of meat-eating would, in effect, indicate what she typically has for dinner and why. The investigator explicitly presents her perspective and reflects on the way in which that perspective informed the investigation. The effort to be transparent by indicating how the issue under study is informed by the author's values inoculates the reader against this personal bias. Whether subtracting that perspective or, better, critically adding it to others, the reader can take it into account.

The recognition that all scholars are advocates even while on the job legitimizes wearing those two hats, scholar and advocate. This enriches both the academy and the animal protection movement as it fosters reciprocity between the two. Animal protection organizations currently employ scholars from many fields as part of their own research and policy development efforts. Other scholars, remaining in universities, can be recognized as advocates without losing their credibility as independent investigators of the issues for which they advocate.

This is particularly the case for fields that investigate issues that are necessarily evaluative. For examples, a philosopher of ethics develops a theory that nonhuman animals should be taken into consideration when we evaluate whether an act is good or bad. The thesis itself is necessarily advocatory, and therefore no heads turn when that philosopher joins forces with those who work to implement it in treatment of animals. A second example, that of a political scientist studying social justice movements, uses the animal protection movement as an example. It is difficult for this investigator to not be influenced by her views on animals and the movement in their behalf, whether negatively or positively. Again, no criticism of the scholar would follow if she ended the study with recommendations intended either to assist or constrain that movement.

Another example is research that establishes that individuals of a particular species have sophisticated intellectual and social capabilities. An investigator in the natural sciences is more likely to maintain the traditional posture of neutral or independent observation, contemporary challenges to that claim notwithstanding. However, a number of such investigators eventually become advocates. Some do so because they feel the need to compensate animals for using them purely as objects of study, particularly when that study involved deprivation or harm to those animals. Arguably, a scholar whose use of animals had a cost to them is obligated to become an advocate for their benefit.

In some instances, this advocacy is part of a collective action within the particular field of scholarship. Psychologists, ethologists, and veterinarians have established advocacy organizations to advance the interests of animals used as objects of study within their respective fields.

In addition, scholars have developed a multi-disciplinary field devoted to the study of human-animal relationships. In fact, scholars within this burgeoning field of Human-Animal Studies (also known as Animal Studies and Anthrozoology) debate the issue of the relation between scholarship and advocacy. Many HAS scholars are after-hours advocates, having self-selected the field to integrate their professional and personal lives. Their research and teaching varies from unabashed advocacy to the declaration of bias, as discussed earlier, to overcorrection to avoid even the appearance of bias. It appears that the overcorrection approach, motivated by the need for this new field to gain credibility, is giving way in the second generation of HAS scholars to the declaration of bias approach.

Further Reading

Davis, H. and Balfour, Dianne, eds. 1992. *The inevitable bond: Examining scientist-animal interactions.* New York: Cambridge University.

Latour, Bruno, and Woolgar, Steve. 1979. *Laboratory life: The construction of scientific facts.* Princeton, NJ: Princeton University.

Shapiro, Kenneth. 2008. *Human-animal-studies: Growing the field, applying the field.* Ann Arbor: Animals and Society Institute.

Kenneth J. Shapiro

SENTIENCE AND ANIMAL PROTECTION

Why is it important for humans to understand the nature of sentience in the animals under our protection? Put simply, a sentient animal has feelings that matter. Sensation is interpreted as emotion; the strength of emotion determines the strength of the motivation to seek satisfaction and avoid suffering. Moreover, the emotional reactions of a sentient animal are governed by experience. If it learns that it can cope with the challenges of life, then it can achieve a state of wellbeing. If it learns that it cannot cope, then it will suffer. The human duty of care to sentient animals is, at the least, to protect them from suffering. Ideally our aim should be to give them a life worth living.

Animal Protection: Our Responsibility

Sentient animals deserve more than our protection; they deserve our respect. This moral principle derives from the recognition that the animals humans use for their own purposes on the farm, in the laboratory, or in the home are able to experience emotions ranging from comfort and pleasure to pain and suffering. Our aim should be to keep them fit and happy: to create within reasonable limits a physical and social environment wherein they can achieve a sense of wellbeing, defined in terms of both their physical and emotional state. This applies whatever our intentions for the animal may be: to love it, eat it, to use it in scientific procedures to find a cure for cancer, or to establish the safety of a detergent.

Within the European Union, farm animals have, since the signing of the Treaty of Amsterdam in 1997, been reclassified not as commodities but as sentient creatures, and this has generated new legislation that takes their sentience and capacity to suffer into consideration. In the UK, the Animals (Scientific Procedures) Act of 1986 creates an obligation to minimize cruelty to laboratory animals resulting from pain, suffering, distress or lasting harm. On the other hand, at the time of writing (2008), federal anti-cruelty laws do not yet apply to farm animals kept for commercial purposes, although a proposed Farm Animals Anti Cruelty Act is under consideration.

Those with a direct responsibility for animal care need skills that can only be acquired through education, understanding, and experience. They include, of course, the promotion of animals' physical welfare through the provision of appropriate food, shelter, and protection from disease. They must also recognize that the welfare of a sentient animal is also determined by how it feels as it seeks to achieve a sense of wellbeing, that is, meet its physiological and behavioral needs, when faced by the challenges of life. To this end, we need to explore the nature of sentience itself.

Sentience

Most dictionary definitions of sentience resort to apparent synonyms such as "feeling and sensation, consciousness and awareness" that have little biological meaning. "Sensation" is too broad, since all animals from the simple amoeba are responsive to stimuli. "Conscious" and "aware" are terms used by most biologists only in the context of human perception. To understand animal sentience we need to explore, without preconceptions, the nature of stimulus and response. To give two examples, simple orders of vertebrates such as reptiles and fish respond to, and seek to avoid, stimuli likely to cause harm, but do they suffer? Many dogs display extreme distress when separated from their owners. Indeed, most vets will treat more dogs for separation

anxiety than for all the varieties of infectious disease. We may conclude that dogs suffer from separation anxiety, but that may be just because we think we understand dogs better than fish.

Animal sentience implies much more than simple response to sensation. A frog with its head removed but spinal cord intact will respond to a nociceptive stimulus to its foot (a pinch) by withdrawing its leg. Nociceptive is a term used by physiologists to describe reflex or conscious evidence of response to a painful stimulus, but not its complex consequences for a sentient animal. A sentient, conscious rat will respond similarly to a nociceptive stimulus (an electric shock) from the floor of its cage. If these shocks are repeated, the rat will learn to associate them not only with the physical sensation of pain but also an emotional sense of distress. This physical and emotional impact will motivate the rat to seek ways to avoid receiving further shocks. If it discovers a way to escape the source of the shock, it will learn that it can cope and feel better. If it is helpless to avoid repetition of the stimulus, it may develop extreme anxiety or learned helplessness, that is, it will feel progressively worse.

Animal sentience therefore involves not just feelings but feelings that matter. The behavior of animals is motivated by the emotional need to seek satisfaction and avoid suffering (Fraser and Duncan, 1998). Marian Dawkins (1980, 1990) has pioneered the study of motivation in animals by seeking to measure their strength of motivation, that is, how hard they will work to obtain a resource or stimulus that makes them feel good or avoid one that makes them feel bad. Many of these emotions are associated with primitive sensations such as hunger, pain, and anxiety. Some species may also experience higher feelings such as friendship and grief at the loss of a relative, and this may expand the range of their sentience. However, we humans should not underestimate the emotional distress caused by hunger, pain, and anxiety. These emotions may be primitive, but that does not make them any less intense.

Sentient animals perceive their environment and this motivates their behavior. Control centers in the central nervous system constantly process information from the external and internal environment. Most of this information, for example our perception of how we stand and move in space, is processed at a subconscious level. However, any stimulus that calls for a conscious decision as to action must involve some degree of interpretation. Scientists define these stimuli as positive, aversive, or neutral. In effect, when presented with the stimulus, the animal will ask itself "do I feel good, bad or indifferent?" This is an emotional, that is, sentient, response. The sentient animal, within which category we must include *Homo sapiens,* may also interpret the incoming information in a cognitive fashion, that is, apply reason. However they, and we, are usually and most powerfully motivated by how we feel.

To illustrate this point, consider the primitive sensation of hunger. Central nervous system centers responsible for control of appetite and satiety receive a variety of internal and external stimuli, for example, low blood glucose, the sight or smell of food, or a conditioning stimulus such as the bell that preceded the meal for Pavlov's dogs. This information is categorized and integrated in the form of an emotion. If the animal feels hungry, it will be motivated to seek food. If a good meal arrives, it will feel pleasure. If no food is available, it will feel bad.

This psychological model of mind makes a clear distinction between the reception, categorization, and interpretation of incoming stimuli. Moreover, it is consistent with new research in neurobiology. Keith Kendrick (1998), for example, has made recordings from single neurons within the brains of sheep presented with external stimuli, or photographic images of external stimuli. A wide range of images, for example, sacks of grain, bales of hay, trigger signals in a family of neurons that convey the generic information "food." A second set of stimuli or images, for example, dogs and men, forms another category of information that we may call "predator." These categories of information then proceed to a second control center for emotional interpretation. "Food" alone is interpreted as a positive emotion: good. "Predator" becomes a negative emotion: bad. However, when the sheep is presented with a picture of a human carrying a sack of food, two categories of information, food and predator, are evaluated together and interpreted as a single, unconfused emotional message, namely good. The animal's decision as to how or indeed whether to respond is therefore determined by how it feels at the time, good or bad. Moreover, this is not a simple yes/no decision. The intensity of its feelings will vary. It will, for example, feel more or less hungry, more or less afraid, and this will determine the strength of its motivation to respond in a positive or negative fashion.

The traditional stimulus/response concept of animal psychology proposed by Pavlov and Skinner held that the behavior of most nonhuman animals involved no more than simple reactions to stimuli that directly or indirectly predict a reward or punishment, for example, a bell that presages the arrival of food or an electric shock. This hypothesis can accommodate sentience, just, but precludes cognition. Moreover, it struggles with the concept of strength of motivation, that is, the emotional measure of how much feelings matter. There is now abundant evidence that mammals and birds can employ cognition to interpret incoming sensation in a reasoned fashion. One of the first and best proofs of animal cognition was the classic experiment of Edward Tolman (1948). He introduced rats to mazes with two exits. In one group, a food reward was provided at one exit only. After an average of 12 trials, almost all rats unerringly took the route to the exit where food was provided. In the other group, no reward was offered, in the first instance, at either exit. Unsurprisingly the rats showed no consistent preference as to route. However, when these rats were subsequently offered food at one exit only, they learned the correct route after only three to four trials. During the first stage of the trial they had, in the absence of any reward, been acquiring an education: gathering spatial information for interpretation and use at such time as they might need it.

The study of animal cognition is a necessary guide to our understanding of and respect for animals under our protection (see Shettleworth, 1998). However, we should not infer that the capacity of an animal to suffer is proportional to the extent of its cognitive ability, still less to its apparent similarity to humans. Pain, for example, is a physical and emotional phenomenon. Cognitive interpretation of the sensation of pain can make things either better or worse. The emotional response of a woman to severe abdominal pain will differ according to whether the pain arises from normal childbirth or stomach cancer.

The first big message to be taken from the story so far is that animal sentience involves feelings that matter. The second message is that sentient animals do not just live in the present. Table 3 first describes the sequence of events involved in perception, categorization, and interpretation of incoming sensation and how this motivates a sentient animal to respond. It then lists what happens next. Having evaluated incoming sensation in emotional, and possibly cognitive fashion, the animal makes a measured response designed to make it feel better. Having acted, the animal then assesses, emotionally and possibly cognitively, the effectiveness of its response. If it judges that its response has been effective, then it is likely to feel better when a similar event occurs in the future. It has learned to cope. If it judges that its response was ineffective, or if it was prevented by environmental or other constraints from behaving in a way designed to improve how it feels, then it is likely to feel worse.

Stress and Suffering

The importance of sentience to evolutionary fitness was recognized by Charles Darwin. The fact that the emotional response of an animal to stimuli is governed by its past experience carries obvious survival advantages in a challenging environment, whether the challenges be wild or domestic. To illustrate this point, consider the difference between fear and anxiety. Fear is an adaptive emotional response to a perceived threat, which motivates action designed to deal with that threat. It is also an educational experience, since the memory of previous threats, actions taken in response to those threats, and the consequences thereof will obviously affect how the animal feels the next time it

TABLE 3 A sentient perception of stimulus and response: Sequence of events

1. Perception of incoming stimuli as categories of information
2. Interpretation of information categories
   positive and negative emotions
   stored information
3. Motivation or aversion: (the measure of behavioral need)
4. Measured response from repertoire of available behaviors
5. Emotional (and possibly cognitive) assessment of effectiveness of action
6. Modification of mood and understanding in light of experience

encounters such a threat. If it learns it can cope, then it will acquire confidence, if it discovers it cannot cope, then the adaptive sensation of fear can proceed to a non-adaptive state of suffering from chronic anxiety or learned helplessness.

Thus, stress and suffering are not the same. Animals are equipped to respond and adapt to challenge in circumstances that permit them to make an effective response. If so, then they learn that they can cope. An animal is likely to suffer when it fails to cope or has extreme difficulty in coping with stress:

- Because the stress itself is too severe, too complex or too prolonged
- Because the animal is prevented from taking the constructive action necessary to relieve the stress

Care of the Sentient Animal

Animals under human protection deserve a fair deal, a sense of wellbeing in life and a humane death. This does not mean that their lives should be entirely free of stress. Our responsibility is to protect them from suffering. Suffering can certainly result from failure to cope with

primitive stresses such as hunger and thirst, heat and cold, pain, fear, and exhaustion. It may also involve higher emotions such as frustration and boredom, loneliness and depression. However, sentience implies the capacity not just for suffering but also for pleasure. Our duty to sentient animals should therefore include the possibility of promoting elements of positive welfare within the reasonable constraints of, for example, viable livestock farming. At the very least, our aim should be to give them a life worth living.

See also Whales and Dolphins, Sentience and Suffering in.

Further Reading
Dawkins, M. S. (1980). *Animal suffering: The science of animal welfare*. London, UK: Chapman and Hall.
Dawkins M. S. (1990). From an animal's point of view: motivation, fitness and animal welfare. *Behavioural and Brain Sciences*. 13, 1–61.
Fraser D., and Duncan, I.J.H. (1998). Pleasures, pains and animal welfare: Toward a natural history of affect. *Animal Welfare* 7, 383–396.
Kendrick, K. M. (1998). Intelligent perception. *Applied Animal Behaviour Science*. 57, 213–231.
Shettleworth S. J. (1998). *Cognition, evolution and behaviour*. Oxford, UK: Oxford University Press.
Tolman, Edward. (1948). Cognitive maps in rats and men. *The Psychological Review* 55, 189–208.
Webster, John. (2005). *Animal welfare: Limping towards Eden*. Oxford, UK: Blackwell Publishing.

John Webster

SENTIENTISM

Sentientism, a term coined by Andrew Linzey in 1980, denotes an attitude that arbitrarily favors sentient beings over the nonsentient. The term is historically parallel to that of speciesism, coined by Richard

Ryder in 1970. Although Linzey was one of the early advocates of sentiency as the basis of rights, he subsequently warned against claiming too much for any one form of classification as the basis of moral standing or rights. Raymond Frey specifically argues that sentiency as the basis of rights "condemns the whole of nonsentient creation, including the lower animals, at best to a much inferior status or . . . at worst possibly to a status completely beyond the moral pale." The issue is how to recognize the value and moral relevance of sentiency as a criterion, while avoiding falling into the error of previous generations who have isolated one characteristic or ability—for example, reason, language, culture, or friendship—and used it as a barrier to wider moral sensibility. There is a need to be aware that, as our own moral sensibilities develop and our scientific understanding increases, all moral categories and distinctions are themselves liable to change.

Further Reading

Frey, R. G. 1979. What has sentiency to do with the possession of rights? In David A. Paterson and Richard D. Ryder, eds., *Animals' rights: A symposium*, 106–111. London: Centaur Press.

Linzey, Andrew. 1976. *Animal rights: A Christian assessment*. London: SCM Press.

Linzey, Andrew. 1981. Moral education and reverence for life. In David A. Paterson, ed., *Humane education: A symposium*, 117–125. London: Humane Education Council.

Schweitzer, Albert. 1967. *Civilization and ethics* (1923), trans. C. T. Campion. London: Unwin Books.

Andrew Linzey

SHELTERS, NO-KILL

The animal rights movement has been steadily gaining converts in the United States, and its scope and influence continue to grow. To those who work for animal rights, while there is still a long way to go to persuade the majority of consumers to make more ethical decisions in their daily lives around what they eat, wear, and purchase, there is no question that the acceptance of animal rights is greater now than at any time in our history. It is hardly surprising then that the issue is taking center stage in the area of companion animals, animals with which millions of people have a deep, personal relationship.

Unlike animals on factory farms, in research laboratories, or in circuses, dogs, cats, and other domestic companion animals enjoy very high esteem in the psyche of the public. In the United States, for example, Americans share their homes with ninety million cats and seventy-five million dogs. Every year they spend more than forty billion dollars on their care, and they donate hundreds of millions of dollars more to charities that promise to help companion animals, with the largest of these having an annual budget in excess of one hundred million dollars (Winograd, 2007). However, the agencies that the public expects to protect animals are instead killing millions yearly.

Today, shelter killing of companion animals remains the leading cause of death for healthy dogs and cats in the United States; between four and five million are killed in our nation's shelters every year (Merritt, 2007; Winograd, 2007). These numbers are staggering. Increasingly, however, animal advocates are working to oppose this life ending. The growing No-Kill movement in the United States is not only calling into question the shelter killing of animals, but is moving to end the practice altogether.

In the Beginning

The modern humane movement began in earnest in the United States with the 1866 founding of the nation's first humane society in New York City. Today we know it as the American Society for the Prevention of Cruelty to Animals.

While the ASPCA focused much of its effort on trying to protect working horses, abolishing vivisection, and outlawing hunting and other conduct it viewed as exploitive, it labored equally hard to protect the city's stray dogs, particularly against the cruel practices of city dog-catchers. As in most American cities of the 19th century, dogs were kept in rough sheds, with no food or water for several days, until they were killed by drowning, beating, or shooting.

The ASPCA worked to outlaw and reform these conditions, succeeding in forcing city dogcatchers to provide food and water, advocating that strays be treated kindly, replacing existing methods of killing with more humane ones, and forcing the city to build a more modern dog pound. Its efforts were highly successful and influential.

In a very short period of time, Canada and twenty-five states and territories across North America had used the ASPCA as a model for their own independent humane societies and SPCAs, and the numbers continued to grow. By the end of the first decade of the 20th century, virtually every major city in the United States had an SPCA or humane society (Winograd, 2007).

Unintended Consequences

While most of these agencies initially focused on oversight of dog pounds, advocacy to increase the status of animals, direct action to assist animals in need, and cruelty prosecutions, most ultimately moved toward direct administration of animal shelters. The guaranteed source of income provided by municipal contracts helped sway many of them to abandon their traditional platforms around horses in favor of administering dog control for cities and counties. In many American cities, pound work was placed in the hands of the SPCA. Within a decade or two, most mainstream humane societies and SPCAs did little more than kill dogs and cats. In 1910, for example, the Animal Rescue League of Boston adopted the following policy, more or less identical in practice to most shelters of the time:

> We keep all dogs we receive, unless very sick or vicious, five days; then those unclaimed are humanely put to death except a limited number of desirable ones for which we can find good homes. We keep from twenty to thirty of the best of the cats and kittens to place in homes and the rest are put to death. . . We do not keep a large number of animals alive. (Winograd, 2007)

From the ASPCA in New York City to humane societies throughout California, the 20th century saw a national shift away from a tenacious focus on saving lives to pound work that resulted in high rates of killing. A critic of this shift summarized it as follows:

> Historically, SPCAs made the tragic mistake of moving from compassionate oversight of animal control agencies to operating the majority of kill shelters. The consequences in terms of resource allocation

and sacrificing a coherent moral foundation have been devastating. (Duvin, 1989)

Today, key changes in society's attitudes towards animals, as well as other technical and demographic changes, have increased the criticism of current sheltering practices, called them into question and, most important, provided a solution to the problem.

Demographics for Change

In the United States, people hold the humane treatment of dogs and cats as a personal value, reflected in laws, the proliferation of organizations founded for their protection, increased per capita spending on animal care, and great advancements in veterinary medicine.

In addition to the integration of companion animals in people's lives, three other key changes in our relationship with dogs, cats, and other companion animals since World War II have become evident. The first is technical. Veterinarians have gained the ability to perform widespread and high volume sterilization of animals easily, safely, and at relatively low cost. By partnering with veterinarians, shelters are able to dramatically reduce births and thus the number of animals surrendered, and subsequently killed, in shelters.

The second change is economic. The growth of the middle class after World War II meant a spread of America's wealth across a wider range of people. This wealth, combined with an unfolding humane ethic, meant donations and bequests to animal welfare organizations increased on a scale previously unimaginable. The wealth made available to these agencies, combined with a prospering economy, resulted in shelters with very

significant annual budgets. By the 1980s, top organizations had assets ranging from forty million to one hundred million dollars, a net worth which continues to grow. Today, giving to animal-related charities is the fastest growing segment in American philanthropy (Duvin, 1989).

The third and perhaps most important change is suburbanization. People moved from farms into cities, and eventually out of cities into suburbs. These households had yards, nearby parks, and open space. Since animals were no longer seen as needed for farm-related work, suburban households became homes for animal companions, and often homes for multiple animals. Americans began to view animals, particularly dogs and cats, very differently and opened their hearts and homes as never before, vastly increasing the number of homes available for animal companions.

These moral, technical, economic, and demographic changes offer the ability to end the era of mass killing in American shelters. And that is exactly what one city, under the leadership of its local SPCA, sought to accomplish.

San Francisco Achieves Success

In 1994, San Francisco became the first community in the nation to end the killing of healthy dogs and cats in its animal shelter system (Clifton, 1995). By the turn of the millennium, roughly eight out of ten dogs and cats in city shelters were being released alive, either back to their caretakers or to new homes. At a time when shelters were killing the majority of animals entering their facilities, this citywide achievement was over twice that of any other major urban area and approximately three times the national average (Clifton, 1995).

The success of San Francisco involved a paradigm shift from a reactive and traditional public health orientation to a proactive and community-based adoption and rescue agency. This involved putting in place programs and services that had a measurable lifesaving impact, rather than basing shelter responses and operations on tradition or longstanding practices.

The mandatory programs and services, collectively called the No-Kill Equation, developed in San Francisco include the following, which must be implemented rigorously enough so that they replace killing in their entirety:

a feral cat trap-neuter-release program

high-volume, low- and no-cost spay/neuter

working with rescue groups

foster care

comprehensive adoption programs including evening and weekend hours and offsite adoption venues

animal retention efforts

medical and behavior socialization, prevention and rehabilitation programs

proactive stray redemption efforts

public relations/community involvement, and

volunteerism

The model has since been used with great success by other communities, many of which have even surpassed San Francisco's rate of lifesaving.

The Current State of Sheltering

Unfortunately, this success has not been met with universal celebration but,

in many cases, by an entrenched defeatism. Traditional shelter proponents blame pet overpopulation caused by public irresponsibility for the continued killing in many shelters, and suggest that the success of San Francisco had more to do with the particular demographics of a city described as progressive, educated, and affluent, than with program implementation.

Without denying public irresponsibility, four important factors weigh heavily against this interpretation as the cause of shelter killing. First, over the past five years, by embracing not only the no-kill philosophy but the programs and services which make it possible, several animal control shelters in communities across the United States have achieved unprecedented lifesaving success, saving in excess of 90 percent of all impounded animals. Not only are death rates plummeting and adoptions increasing in these communities, but these results have been achieved in a very short period of time, virtually overnight, underscoring that saving lives is less a function of any perceived pet overpopulation and more a function of a shelter's leadership and practices.

Second, current estimates from a wide range of groups indicate that between four million and five million dogs and cats are killed in shelters every year (Clifton, 2007). Of these, given data on the prevalence of aggression in dogs in society based on dog bite extrapolation (Bradley, 2005), and rates of lifesaving at the best-performing shelters in the country from diverse regions and demographics, about 90 percent of all shelter animals are salvageable (No Kill Advocacy Center, 2008). The remainder are either hopelessly ill or injured, or vicious dogs whose prognosis for rehabilitation is poor or grave. That would put the number of salvageable dogs and cats at roughly 3.6 million on the low end and 4.5 million on the high end of the spectrum.

But even at the high end this means that, nationally, shelters only need to increase their adoption market by 2–3 percent in order to eliminate all population control killing. Today, there are about 165 million dogs and cats in homes. Of those, about 20 percent come from shelters. Three percent of 165 million is 4.9 million, more than all the salvageable animals being killed in shelters (Keith, 2007). This is a combination of what statisticians call stock and flow. In layman's terms, some of the market will be replacement life, that is, a companion animal dies or runs away and the owner wants another one, some will be expanding markets, that is, someone doesn't have an animal companion but wants one, or they have one but want another. But it all comes down to increasing market share, that is, where they get their companion animals from.

No-kill advocates believe that these same demographics show that every year about twice as many people are looking to bring a new dog into their home as the total number of dogs entering shelters, and every year more people are looking to bring a new cat into their home than the total number of cats entering shelters (Winograd, 2007; Merritt, 2007). Moreover, not all animals entering shelters need adoption; some will be lost strays that will be reclaimed. Some cats are feral or wild and need sterilization and return to their habitats. Vicious dogs, and animals that are irremediably suffering or hopelessly ill/injured will not be eligible for adoption. From the perspective of achievability, no-kill advocates point out, the prognosis is very good.

Third, many downplayed the significance of San Francisco's accomplishment for other communities by arguing that such a result could only be achieved in an urban community, not a rural one, because of poverty and antiquated views of animals. When No-Kill was achieved in the rural Tompkins County, New York animal control shelter, it was argued by some that it could not be done in the South. When it was achieved at an animal control shelter in Charlottesville, Virginia, these same groups claimed it could not be similarly achieved in developing communities that are seeing tremendous population growth and urban sprawl, because the influx of new people and animals would overwhelm the infrastructure of animal control, forcing shelters to kill. The 90 percent rates of lifesaving in the communities in and around Reno, Nevada, a more than 50 percent drop in killing and doubling of the adoption rate in less than one year, despite rapid population growth and approximately 16,000 dogs and cats entering the system annually, disproves that, too (Brown, 2008).

These and other cities have either achieved No-Kill, are close to doing so, or have begun moving aggressively in that direction by implementing the programs and services of the No-Kill Equation. Building the capacity to save lives after years of failing to do so may take time, but that does not obviate the fact that shelter killing is a result of shelter practices and not pet overpopulation. Furthermore, no-kill shelter advocates say, the argument that success in less affluent, more rural, or Southern areas is precluded by some peculiarity of lack of caring is not only wrong, elitist, and mean-spirited; it is simply another example of making excuses. It ignores the success in rural Tompkins County. It ignores the tremendous success in Charlottesville, Virginia. It goes against a study by a South Mississippi humane society that found 69 percent of people with unsterilized dogs or cats would get them spayed/neutered if it was free, a fact which is not surprising for a state with some of the lowest per capita income levels in the United States (Winograd, 2007).

Fourth, no-kill shelter advocates note that these arguments mimic the claims in other areas of animal rights that reject practical or utilitarian considerations over ethical or rights-oriented ones. Just as the animal rights movement rejects other ideas that violate the rights of animals even in the face of some human benefit or other practical consideration, it too should reject the idea that killing them is acceptable because of the claim, even if one were to accept it as fact, that there are too many for the arguably too few homes which are available.

What the Future Holds

Since No-Kill is a nascent movement, it is still undergoing a turbulent period prior to acceptance and sustainability. It is also highly dependent on the actions and success of committed individual leaders. For No-Kill to succeed in the long term, advocates believe that shelters must build a culture of accountability and lifesaving that allows agencies to continue on their path to No-Kill even when their visionary leaders move on to other pursuits.

To do that, shelters need to create a no-kill-oriented board of directors, staff, and volunteer corps, and share their success publicly until the community accepts it. Shelter reform legislation, which lays out the roles and responsibilities of shelters, must also be codified and enforced. This will provide a defense against backslid-

ing later, by creating the expectation of lifesaving among a shelter's board, volunteers, and the community at large.

The more successful this effort is, the more No-Kill will shift from being personality based, a result of the efforts of individual leaders, to becoming institutionalized as the doctrine of the shelter and the expectation of the community. Given the increasing acceptance of broader animal rights issues, even when people do not have a personal connection or relationship to the animals involved, the long-term prognosis for the success of the No-Kill paradigm is good. Underpinning the philosophy is the building of a new consensus, which rejects killing as a method for achieving results. But even within the philosophy, there are some contradictions and challenges which need to be resolved and which will increasingly rise to the forefront.

Animal activists are not suggesting that hopelessly ill or injured sheltered animals be put up for adoption, and few, if any, are calling for truly vicious dogs to be adopted into homes in the community. Under the prevailing No-Kill philosophy, these animals would not be counted under killing for purposes of population control (Keith, 2007).

While more than 90 percent of dogs and cats entering shelters are neither hopelessly suffering nor vicious and would fall outside this limited range of exceptions, however, it does not follow that the remainder should be killed. The reality is that, while fewer than 10 percent of shelter animals are ineligible for adoption, the vast majority are not suffering and as long as they are not suffering, their killing raises a host of ethical issues. In fact, not only are some unadoptable animals living without pain, they enjoy a good quality of life and can continue to do so, at least for a time. These include, for example, cats diagnosed with feline leukemia, animals in the early stages of renal disease, and aggressive dogs.

The fact that shelters cannot and should not adopt out vicious dogs, for example, does not mean that killing them isn't ethically problematic. Today, the great challenge in sheltering is between No-Kill advocates working to ensure that healthy animals, animals with treatable medical conditions, and feral animals are no longer killed in shelters versus the voices of tradition, which argue that killing under the guise of euthanasia is both necessary and proper. As the No-Kill paradigm's hegemony becomes more established, however, the humane movement will have to confront other ethical quandaries within even this philosophy.

These ethical quandaries, for example, killing dogs that are aggressive but can lead happy lives in sanctuaries, giving hopelessly ill animals hospice care as opposed to what is considered mercy killings or true euthanasia in shelters, will become paramount. In fact, even today the very idea of killing is being challenged by a small but growing movement of sanctuaries and hospice care groups. They argue for a third door between adoption and killing (Johnson, 2008). That these issues have not been rigorously debated as a movement does not mean animal advocates must wait to demand that these animals be saved as well. From the animal rights perspective, compassion must be embraced whenever it presents itself, especially when it gives meaning to an animal's right to live.

The right to life is universally acknowledged as a basic or fundamental right, because the enjoyment of the right to life is a necessary condition of the enjoyment of all other rights. A movement cannot be rights-oriented and ignore the fundamental right to life. If an animal is dead, the animal's rights become irrelevant.

In addition, it is the relationship between Americans and their animal companions that can open a door to larger animal rights issues. In their daily interactions with their dogs, cats, and other animal companions, people experience an animal's personality, emotions, and capacity for both great joy and great suffering. They learn empathy for animals. It is not a stretch to say that someone who is compassionate and passionate about their animal companions would, over time and with the right information, be sympathetic to animal suffering on farms, in circuses, in research facilities, and elsewhere. The No-Kill philosophy which seeks to save companion animals in shelters can provide the bridge.

Moreover, given the public's progressive attitudes regarding companion animals, and the ability to end the population control killing of these animals in shelters, to achieve that goal would set a powerful precedent for the rights of other animals, and should therefore be a goal which the animal rights movement should seek and support.

Further Reading

Bradley, Janis. 2005. *Dogs bite: But balloons and slippers are more dangerous*. Berkeley: James & Kenneth Publishers.

Brown, Bonney. 2008. How we did it. Nevada Humane Society, February 15, 2008, available at http://www.nevadahumanesociety. org/pdf/HowWeDidIt2-15-08.pdf.

Clifton, Merritt. 1993. Can we outlaw pet overpopulation? *Animal People*, May.

Clifton, Merritt, ed. 2007. U.S. shelter killing toll drops to 3.7 million dogs & cats. *Animal People*, July-August: p. 1.

Duvin, Ed. 1989. Unfinished business. *Animals Voice* magazine.

Johnson, Annysa. 2008. Unwanted no more: Sanctuary saves unadoptable dogs, cats from euthanasia. *Milwaukee Journal Sentinel*, June 15, 2008, available at http://www.json line.com/story/index.aspx?id=762372.

Keith, Christie. 2007. Interview with Richard Avanzino of Maddie's Fund, November 24, 2007, available at http://www.petconnec tion.com/blog/ck-ratranscript/.

Keith, Christie. 2007. What's in a name? Playing the Orwellian card, November 20, 2007, available at http://www.petconnection.com/blog/2007/11/20/whats-in-a-name-playing-the-orwellian-card/.

No Kill Advocacy Center, The 90 percent rule. *The No Kill Advocate*, Issue #3 2008, at pp. 20–21.

Winograd, Nathan J. 2007. *Redemption: The myth of pet overpopulation & the no kill revolution in America*. Los Angeles: Almaden Books.

Nathan J. Winograd

SIGNALS AND RITUALS OF HUMANS AND ANIMALS

Ritual is a universal feature of human behavior. While rituals differ from culture to culture, the defining features that distinguish them from ordinary behaviors are surprisingly consistent across all human societies. Rituals tend to be formal, stereotyped, repetitive, and traditional. They are therefore easily distinguished from other behaviors. Rituals help pattern and predict social interactions. For example, when two people meet, they have expectations about how the social interaction will proceed. In Western societies, meetings commence with a handshake and a simultaneous "How are you?" or some similar formality. While none of us invented the handshake, we all recognize it as a greeting ritual.

Religious rituals are particularly easy to detect, as they tend to be more elaborate than other rituals. They also generally include music, chanting, or dance, which further distinguishes them from other behaviors. Masks, icons, special settings, extraordinary garments, and even distinctive languages may be used to further demarcate religious ritual from the ordinary. While religious rituals frequently appear to be shrouded in mystery, their formality and elaborateness make it clear to participants and observers alike that they are rituals. Nobody mistakes Sunday morning church for the Sunday afternoon football game.

The same underlying features of ritual that allow us to recognize it across widely diverse human cultures also enable us to recognize ritual in nonhuman species. Wild dogs, wolves, and chimpanzees all perform highly ritualized greeting ceremonies that include muzzle-to-muzzle contact, hugging, and choral vocalizations whenever the members of a social group meet. Wolf spiders, salamanders, and sandhill cranes all perform intricate dances as part of their courtship. Parrots and Pacific humpback whales engage in improvisational, synchronized singing during mating and group rituals. Chimpanzees have been observed to engage in occasional drumming of tree trunks and sporadic group rain dances. Ritual conveys significant social information about participants in both human and animal groups. It permits and promotes social interaction by creating frameworks of expectancy that lay the foundation for the prediction of behavior by others. But to fully appreciate the similarities between human and animal rituals, and to understand why they are similar, we first need to understand ritual's less complicated parent, signals.

Signals as Cooperative Communication

We all use signals in everyday life. Colored lights that regulate traffic flow and sirens that alert us to possible danger are examples of common human visual and auditory signals. Animals, too, use many different kinds of signals to communicate with other members of their groups. The scent marking of dogs, the alarm calls of monkeys, and the changing colors of Siamese fighting fish are all signals that convey information about the state, condition, or intent of the sender. Some signals, such as the croak pitch of male frogs, directly convey the physical and physiological characteristics of the sender. Since croak pitch is a function of body size, larger males produce deeper croaks. This direct relationship between body size and sound pitch makes it possible for both females and competitor males to estimate the size of unseen males based solely on their croaking. Such indexical signals convey reliable information about a signal sender because they are directly linked to attributes that cannot be concealed or manipulated by the sender.

Most signals used in human and animal communication are not indexical, but still provide reliable information about the sender. They have evolved over time because they benefit both the sender and the receiver. Numerous conventional signals, such as the pecking response of herring-gull chicks to red dots, are the result of genetically programmed fixed-action patterns. Such signals automatically elicit or release evolved preprogrammed behaviors in signal receivers. In the case of the herring chicks, pecking at the red dots on the mother's bill provides the chick with food. Grouper fish exhibit innate responses to the dance performed by sucker fish. Even when reared in isolation, groupers exposed to the sucker fish dance lie down on the sand, spread their fins, and allow the sucker fish to clean the algae from their scales.

Such genetically encoded fixed-action pattern signaling systems have evolved because the benefits they provide for both the sender and receiver outweigh the costs involved in signaling.

Although it was once thought that all animal signals result from these genetically programmed fixed-action patterns, ethologists have since found that many animal signals are much more complex, incorporating both genetic and learned components. The alarm calls of vervet monkeys provide a good example of such complex signals. Vervets inhabit woodland areas in eastern Africa, and use alarm calls to alert other members of the social troop to the presence of predators. Vervets emit a bark in the presence of a jaguar, a cough in the presence of an eagle, and a chutter in the presence of a snake. Young vervets have an innate tendency to respond to calls and to make different calls in response to different stimuli. However, young monkeys are not born with preprogrammed knowledge of jaguar, eagle, and snake calls. They

A sage grouse fans his tail and puffs up his chest to attract females. (Photos.com)

must learn the specific call to emit for each particular predator. While vervets are preprogrammed to learn these calls, young monkeys must hear the different calls used within the appropriate context in order to learn the correct call for each particular predator.

This innate capacity to learn species-specific signals during a particular developmental period is seen in many other species. The courtship songs of many birds involve both genetic programming and developmental learning; male birds are genetically primed to learn their species' song, but must be exposed to it during a specific developmental window in order for learning to occur. In humans, both music and language learning integrate this same combination of genetic predispositions and culturally-transmitted learning during specific developmental periods.

Signals as Deception

Sometimes signaling contexts involve senders and receivers who have conflicting interests. Under such circumstances, there is great incentive for signalers to use deception in order to influence receiver responses. Camouflage, mimicry, and deception are widespread throughout the animal kingdom. Many species have evolved color patterns and special structures to deceive potential predators and prey. Viceroy butterflies fool potential predators through their mimicry of the unappetizing Monarch. Angler fish lure unsuspecting prey with a specially-evolved mouth appendage. Females of the predatory firefly genus *Photuris* mimic the mating flashes of the related genus *Photinus* in order to lure *Photinus* males close enough to attack and consume them. Humans bluff, cheat, and lie in cards, war, and love.

Honest Signals

Signal receivers clearly have an incentive to detect dishonest signals and prevent such manipulation. Receivers should seek out signals that provide honest information. In many species this has resulted in the evolution of quality signals that provide receivers with reliable information about the general condition of the sender. In birds, the intensity of plumage color is negatively correlated with parasite load; the brighter the plumage, the healthier the bird. Females seek out males with the most brilliant plumage. As a result, the color brilliance of males has evolved to be a quality signal for females. In humans, a similar quality signal is provided by facial symmetry, which is positively correlated with health. Numerous studies have shown that males and females worldwide find symmetrical faces more attractive. In various songbird species, male song repertoire size is an important quality signal for females. Males with large song repertoires are less likely to be infected by malarial parasites and more likely to bring larger caches of food for their offspring.

Quality signals that benefit the receiver frequently incur costs for the sender. Male peacocks with the longest, brightest tails and male songbirds with the largest repertoires not only expend more energy on the development and maintenance of these traits, they also attract more predators than less showy individuals. Biologist Amotz Zahavi has proposed that such high cost signals are adaptive for signalers precisely because they handicap the sender. Since only those peacocks and songbirds with sufficient resources are able to successfully produce and maintain the longest, showiest tails and the largest and most captivating song repertoires, it would be impossible for less fit competitors to fake these signals. Likewise, Mazeratis and mega-mansions constitute culturally-constructed quality signals in contemporary human societies, since only the wealthiest can afford the direct, maintenance, and opportunity costs of such luxuries.

Ritual as a Signal

Rituals are the costliest of signals. The four basic elements of ritual, formality, pattern, sequence, and repetition incur high time, energy, and resource costs for ritual performers. Yet these four features make up the structure of ritual in species as diverse as horned toads, chickens, and humans. Laboratory experiments have shown that these elements of ritual are optimally effective in engaging and focusing attention, heightening discrimination, enhancing multidimensional generalization, and improving associative learning. The formality of ritual captures the attention of the audience and focuses it on the signal elements most likely to evoke receiver response. Ordinary traits and behaviors may be exaggerated in order to make them extraordinary. The eyes of a peacock's long, iridescent tail prominently displayed during his ritual dance, the changing body colors of male squid as they gently jet water over a potential mate, and the ornate garments worn by human brides all represent formal elements of ritual that engage and focus the attention of ritual participants.

By exaggerating and elaborating ordinary features, the formality of ritual alerts brain structures such as the reticular formation, the basal ganglia, and the amygdala which function to prime emotions and prepare the body to react. Once attention is focused, the sequence, pattern, and repetition of ritual optimize the processing time critical for memory and learning.

Ritual has other impacts on neuroendocrine function as well. Changes in the levels of neurotransmitters, neuromodulators, and hormones of both the sender and the receiver occur during ritual, resulting in changes in the physiological, immunological, and behavioral responses of ritual participants. Biologist Russell Fernald's studies of cichlid fish from Lake Tanganyika in Africa dramatically illustrate ritual's effects on physiology. He found that antagonistic displays between cichlid males induce major changes in the hormones, external appearance, brain neuron sizes, and even the gene expression of winners and losers. Fernald observed aggressive and brilliantly colored black, yellow, blue, and red males almost instantly morph into much less aggressive drab brown satellite fish when ousted from their territories by rivals. If the satellite later acquired a new territory, his color, hormones, hypothalamic neuron sizes, and gene expression changed again. Similar neuroendocrine changes have been recorded in songbird responses to ritual as well. The ritualized vocalizations of male songbirds impact female sexual receptivity by inducing hormonal changes in the female, but they also impact the brain neurons and song-related genes of the signaler. In wolves and nonhuman primates, ritualized dominance and submission behaviors can alter participants' cortisol, dopamine, and testosterone levels. Across animal species, the ability of ritual to alter individual neurophysiology and behavior is critical to its adaptive value.

The Relationship of Human and Animal Signaling Systems

Many human signaling systems share deep phylogenetic roots with our closest primate kin. Like bonobos and chimpanzees, humans everywhere use similar facial expressions to identify and convey basic emotional states. Likewise, laughter, body language, and shouts of alarm are universally understood within both chimp and human societies. Yet, the most elaborate and distinctive human rituals, including synchronized chanting, music, and dance, are notably rare in other primates. While our distant cousins, the pair-bonded gibbons, do engage in male-female singing duets, the ritualized use of collective song and dance is conspicuously absent among all of our closest kin, including gorillas, bonobos, and chimpanzees.

Collective song and dance are, however, found in many other animal species. Wolves and wild dogs engage in choral howling, humpback whales sing synchronized group songs, and a multitude of bird species chorus, sing, and dance. Across human and nonhuman species alike, such ritualized behaviors evoke emotional and physiological responses that impact individual health and behavior while defining, facilitating and patterning social interaction. Understanding the nature and function of animal ritual not only broadens our understanding of nonhuman species, it also deepens our understanding of ourselves.

See also Music, Dance, and Theater—Music and Animals;

Music, Dance, and Theater—Music as a Shared Trait among Humans and Animals.

Further Reading
Alcock, J. (2005). *Animal behavior: An evolutionary approach* (8th ed.). Sunderland, MA: Sinauer Associates, Inc.

Alcorta, C. (2008). Music and the miraculous: The neurophysiology of music's emotive meaning. In J. Harold Ellens (Ed.), *Miracles, God, science, and psychology in the para-normal, Vol. 3 Parapsychological perspectives.* Westport, CT: Greenwood Publishers.

Alcorta, C., & Sosis, R. (2005). Ritual, emotion, and sacred symbols: The evolution of religion as an adaptive complex. *Human Nature, 16*(4), 323–59.

d'Aquili, E., Laughlin, Jr., C. D., & McManus, J. (1979). *Spectrum of ritual.* New York: Columbia University Press.

Fernald, R. (2002). Social regulation of the brain: Status, sex and size. In D. Pfaff, A. Arnold, A. Etgen, S. Fahrback, & R. Rubin (Eds.), *Hormones, brain and behavior* (pp. 435–44). New York: Academic Press.

Hauser, M. D., & Konishi, M. (Eds.). (1999). *The design of animal communication.* Cambridge, MA: The MIT Press.

Searcy, W. A., & Nowicki, S. (2005). *The evolution of animal communication: reliability and deception in signaling systems.* Princeton, NJ: Princeton University Press.

Smith, W. J. (1979). Ritual and the ethology of communicating. In E. G. d'Aquili, C. D. Laughlin, Jr., & J. McManus (Eds.), *The spectrum of ritual* (pp. 51–79). New York: Columbia University Press.

Sosis, R. (2004). The adaptive value of religious ritual. *American Scientist, 92,* 166–72.

Zahavi, A., & Zahavi, A. (1997). *The handicap principle: A missing piece of Darwin's puzzle.* Oxford: Oxford University Press.

Candace S. Alcorta and Richard Sosis

THE SILVER SPRING MONKEYS

In 1981 the Institute for Behavioral Research (IBR) in Silver Springs, Maryland was raided by police as a result of accusations of cruelty to animals. It was the first time in American history that a scientific research laboratory had been raided by police as a result of alleged cruelty to animals, and it quickly became a landmark case that set legal and political precedents across the United States.

The research at the IBR, led by Dr. Edward Taub, was funded by the National

Institutes of Health, and focused on somatosensory deafferentation research in primates, in which one or both forelimbs had all sensation surgically removed. The extent to which the animals then used or could use their limbs was evaluated. It was believed that voluntary movement was impossible in the absence of sensory feedback, a conclusion disproved by the research at IBR.

In the earl summer of 1981, an animal activist named Alex Pacheco asked Taub for a job at IBR. Taub told Pacheco that there was no paying job available at the Institute, but that he was welcome to work at the laboratory on a volunteer basis. Taub was not aware that Pacheco was one of the founding members of People for the Ethical Treatment of Animals (PETA). During his five months at IBR, Pacheco took photographs of the conditions in the facility. In addition, while Taub was away on vacation, he brought five scientists, two zoo veterinarians and three animal activists, two of whom were primatologists, into the facility to witness the conditions in the laboratory.

On September 22, 1981, in response to the affidavits of the five scientists alleging grossly unsanitary conditions and inadequate care, and the photographs provided by Pacheco, the Montgomery County police raided IBR, confiscating the primates, and seizing laboratory records. Taub was subsequently charged with cruelty to animals. In November 1981, Taub was found guilty of providing inadequate veterinary care to six of the seventeen primates. The other 113 charges were dismissed. Taub appealed the conviction, demanding a second trial before a jury, and was found guilty on a single count of inadequate veterinary care. He appealed to the Maryland Supreme Court, which dismissed the case because, it argued, the Maryland anti-cruelty statute did not apply to federally-funded research. The NIH subsequently determined that the IBR facilities and program violated several aspects of NIH animal research policies, and first suspended and then terminated Taub's funding.

The case has had a tremendous impact on the animal research debate and on resulting public policy. At the time of the police raid, Congress had scheduled hearings on several animal research bills. The news coverage of the raid and the publicity generated by Pacheco's photographs refocused the Congressional hearings, which spent one of the two scheduled days grilling representatives of the various federal agencies about their oversight of research. NIH also found its own policies too vague to deal adequately with the events, and initiated a major revision of its animal research policies. The research community, particularly the American Psychological Association and the Society for Neuroscience, were very concerned about the case, and rallied behind Taub to defend him from his critics. In contrast, two laboratory animal veterinarians testified for the prosecution that the conditions pictured at IBR were grossly substandard for the care of primates. Subsequently, PETA and NIH fought over the fate of the Silver Spring monkeys, especially the deafferented animals whose nerves were destroyed; they ultimately ended up at the Delta Primate Research Center. The monkeys continued to be the focus of court battles well into the 1990s until the last animal was euthanized because of failing health.

Andrew N Rowan

SIZEISM

Sizeism is a form of speciesism that specifically relates to a failure to empathize with or give small animals the same consideration that would be given to large animals. This is manifest when people carry out invasive procedures in small animals that they would not do if the animal was larger or older. Moreover, if those procedures were carried out on large and adult animals they would normally be given an anesthetic. The reasons for doing such mutilations relate to research, agricultural practices, and cosmetic procedures in companion animals. Furthermore, there are good physiological reasons to suspect that young animals feel pain just as much as large or adult animals. In fact there is growing evidence that young animals may feel more pain then when they are adults. Examples of the surgical procedures performed include amputation of digits in rats, mice, and puppies compared with horses and cows; docking of tails in lambs, piglets puppies, kittens, and calves compared with adult animals; castration in lambs, piglets, puppies, kittens, and calves compared with adult animals, particularly dogs and cats, and cardiac puncture and intracerebral injections in mice and rats compared with cardiac puncture in horses, dogs, sheep, and cows. Intracerebral injections are normally carried out only in mice.

David B. Morton

SOCIOLOGY OF THE ANIMAL RIGHTS MOVEMENT

Behavioral scientists have used several approaches to understanding the sociology and psychology of those who oppose the use of animals. Some researchers have distributed surveys; other investigators have collected qualitative data based on extended interviews with animal activists. All of these studies show that animal activists are a diverse group with varying philosophies and approaches toward the treatment of animals, but who often share some common characteristics.

Virtually all research indicates that women are much more likely than men to be involved in animal protection. This was also true of the Victorian Era anti-vivisection movement. The reasons for the current and historical predominance of women among rank and file activists are unclear. Animal activists tend to come from middle- and upper class backgrounds and are much more likely than the average American to be Caucasian. As a group, their median income is higher than average and they tend to be well educated; a 2006 study reported that 79 percent of activists attending a national animal rights meeting had undergraduate or graduate degrees. Relatively few activists live in rural areas or small towns. Most studies indicate that two-thirds of animal activists consider themselves vegetarians or vegans, and the overwhelming majority share their homes with companion animals.

Activists tend to identify with the moderate to left side of the political spectrum. Most activists indicate that they also support the goals of other social movements. Among these are the environmental, women's, and gay rights movements. Several studies have reported that the majority of activists are not affiliated with mainstream organized religions, and a substantial proportion report being atheists or agnostics. However, several researchers have pointed out that the animal rights movement itself has quasi-religious characteristics.

Public attitudes toward the animal rights movement are mixed. Several surveys have reported that a majority of Americans have positive attitudes toward the animal rights movement. For example, a 1994 public opinion poll reported that most respondents had either a very favorable or a mostly favorable view of the animal rights movement. On the other hand, fewer than 10 percent say that they agreed with both the agenda of the animal rights movement and its strategies. A 2003 Gallup Poll found lukewarm support for animal rights. The poll found that 96 percent of Americans believed animals deserve some protection, and 25 percent of those believed that animals should have the same rights as humans. While 62 percent of Americans believed that there should be strict laws governing the treatment of farm animals, only 38 percent believed that there should be no laboratory testing of animals, and only 35 percent believed that there should be no medical research on laboratory animals.

See also Gender Gap and Policies toward Animals; Student Attitudes toward Animals.

Further Reading

Gallup. 2003. Public lukewarm on animal rights. http://www.gallup.com/poll/8461/Public-Lukewarm-Animal-Rights.aspx.

Groves, J. 1997. *Hearts and minds: The controversy over laboratory animals.* Philadelphia: Temple University Press.

Herzog, H. A. 1993. The movement is my life: The psychology of animal rights activism." *Journal of Social Issues* 49: 103–119.

Jamison, W., and Lunch, W. 1992. Rights of animals, perceptions of science, and political activism: Profile of animal rights activists. *Science, Technology, and Human Values* 17: 438–458.

Jasper, J. M., and Nelkin, D. 1992. *The animal rights crusade: The growth of a moral protest.* New York: Free Press, 1992.

Lowe, B. M. 2006. *Emerging moral vocabularies.* New York: Lexington Books.

Harold A. Herzog

SPECIES ESSENTIALISM

Essentialism is the claim that every member of a real kind shares some one quality with all and only others of that kind. What is now in doubt is that such kinds, which may well exist, can ever be identified with biological species. One can question whether, to be a dog, it is necessary to share some quality with all and only dogs, and whether it is necessary to suppose that there are pure dogs, having no other qualities than dogs require. Biologists typically blame Aristotle and his followers for species essentialism, for supposing that there are real, discrete biological kinds such that there are perfect specimens of each such kind. The truth is that Aristotle insisted that there were no absolute divisions in nature; we could conveniently classify living things, but would always find that there were hybrids and intermediates in any system.

Aristotle was correct: the existence of cross-species hybrids and the supposed existence of ancestral species from which several modern species have evolved show that nature is a continuum. A species is a set of interbreeding populations, not a natural kind. There need be no one quality that every member of a species shares with all and only others. Not all members even resemble all their conspecifics (members of the same species) more than they resemble creatures of other species. Nor is there any perfect specimen of a given species; any member of a species, however unusual, is equally and perfectly a member.

Nothing says that any individual can have fertile intercourse with any conspecific of the other sex, nor that every individual of that species shares any one particular character with every other, nor that its failure to have some feature shared by most is any real defect. Realizing this may have moral as well as scientific benefits; we no longer need to think that unusual specimens are defective. On the contrary, diversity within a population is an evolutionary advantage. Some groups, closed off from others, will be highly uniform, others will not, yet the differences do not grow into true species differences unless the group happens to split up. Sometimes one species will turn into two only because some crucial, intermediate population has perished, without any change in any other population. It is not even entirely true that genetic information cannot pass between real species. Occasional hybrids aside, viral infection transfers genetic material. New species are also formed by symbiosis, collaboration between organisms originally of very different species.

With respect to the human species, it turns out not to be a natural kind; it is just the relevant set of interbreeding populations. There may have been and there may yet be more than one such human species. What the individuals concerned were or will be like, and what our duties might be toward them, cannot be settled by deciding on their species.

See also Evolutionary Continuity.

Further Reading

Clark, S.R.L. 1994. Is humanity a natural kind? In T. Ingold, ed., *What is an animal?* London: Routledge,.

Douglas, Mary. 1973. *Natural symbols*. Harmondsworth: Penguin.

Ellis, Brian *Scientific essentialism* (Cambridge: Cambridge University Press 2007)

Gotthelf, A., and Lennox, J. G., eds. 1987. *Philosophical issues in Aristotle's biology*. Cambridge: Cambridge University Press.

Margulis, Lynn. 2002. *Acquiring genomes: A theory of the origins of species*. New York: Perseus Books.

Mayr, Ernst. 1963. *Animal species and evolution*. Cambridge, MA: Belknap Press of Harvard University Press.

Oderberg, David. 2007. *Real essentialism*. London: Routledge.

Sober, Elliott. 1994. *From a biological point of view*. Cambridge: Cambridge University Press.

Stephen R. L. Clark

SPECIESISM

The term speciesism was first coined by Richard Ryder in 1970. In 1985, the *Oxford English Dictionary* defined speciesism as "discrimination against or exploitation of certain animal species by human beings, based on an assumption of mankind's superiority." This definition marked the official acceptance of speciesism into the language. Peter Singer did much to establish its use.

Speciesism became a useful campaigning term from 1970 onwards. Ryder was a member of the Oxford Group of anti-speciesist thinkers of the period, which included Ros and Stanley Godlovitch, John Harris, Andrew Linzey and, a little later, Peter Singer and Stephen Clark. The term first appeared in Ryder's leaflets and then in Godlovitch and Harris's seminal *Animals, Men and Morals* (1971), to which Ryder contributed a chapter. Ryder turned down Singer's invitation to coauthor *Animal Liberation*, which emerged in 1975, but the term was employed here by Singer. Ryder helped popularize the term on British radio and television, arguing that treating the suffering of different species equally follows logically from Darwinism. Richard Dawkins, too, in his classic *The Selfish Gene* (1976), used the term *speciesism* in supporting those who campaign for animals. The RSPCA's *Declaration Against Speciesism* was signed by 150 delegates at the world's first animal liberation conference held at Trinity College, Cambridge, in 1979, and was extensively used by Ryder when he was RSPCA Chairman, and as Director of the Political Animal Lobby, in the successful campaigns to put animals into British and European politics.

By drawing the parallel between speciesism, sexism, and racism, campaigners have been able to attract the attention, and often the support, of liberals, democrats, and others who might otherwise have remained indifferent to the interests of nonhumans. Thus, although the concept has proved useful on the philosophical level, for example, as a means to address the subject without any commitment to the idea of rights, it has had value on the psychological and political levels also.

Further Reading

Ryder, Richard D. 1971. Experiments on animals. In Stanley Godlovitch, Roslind Godlovitch, and John Harris, eds., *Animals, men and morals: An enquiry into the maltreatment of nonhumans*. London: Victor Gollancz.

Ryder, Richard D. 1983. *Victims of science: The use of animals in research*. London: Davis-Poynter; Centaur Press.

Ryder, Richard D. 1989, 2000. *Animal revolution: Changing attitudes towards speciesism*. Oxford: Basil Blackwell Ltd.; Berg.

Ryder, Richard D., ed. 1992. *Animal welfare and the environment*. London: Gerald Duckworth.

Ryder, Richard D. 1998. *The political animal: The conquest of speciesism*. Jefferson, NC: McFarland.

Singer, Peter. 2001. *Animal liberation*, 2nd ed. New York: Harper.

Richard D. Ryder

SPECIESISM: BIOLOGICAL CLASSIFICATION

Speciesism is the attribution of weight given to species when evaluating the ethical treatment of individuals. When we say that all, and only, human life is sacred, we are embodying speciesism in a basic moral principle. When we treat nonhuman animals as mere means to our ends, while condemning the same attitude in the case of human beings, we are incorporating speciesism into our practices. The question is whether we are justified in drawing such a line, that is, in granting humans a different, and higher, status.

A deep-seated tradition tends to give differential treatment of members of other species an air of respectability. Recently, however, speciesism has been equated with racism and sexism as a form of arbitrary discrimination. Some philosophers have pointed out that if we reflect on the human rights theory, we realize that we have already settled similar questions of relevance. People generally believe that race and sex should play no role in our morality. To be consistent, the same judgment should be made in the case of species membership. In this view, the very idea of human equality tells us that speciesism is ethically objectionable.

This, however, does not solve the problem, for one should explain what is wrong with racism and sexism. The answer seems evident. Races and sex are biological classifications. As such, they are concerned with purely physical characteristics such as skin color and reproductive role, rather than with psychological properties such as the capacity for being harmed or benefited. Since ethics is an autonomous theoretical subject, endowed with its own standards of justification; criteria coming from different disciplines have no bearing on it. Against this, it can be said that there is a correspondence between race or sex and the possession, or lack, of some characteristics that are morally relevant, so that group membership may be appealed to as a mark of this difference. This can be called the correspondence approach. Thus, for example, racists often claim that members of other races are less intelligent than members of their own race. However, even if the claim were true, this approach would not work. First, if the underlying reference is to other characteristics, drawing a line through race membership is uselessly confusing. Second, what we shall find will be overlap, not mutual exclusion, between races, and to treat individuals, not on the basis of their own qualities, but on the basis of what is allegedly normal for their group, would be irrational.

Thus it seems that racism and sexism are in fact arbitrary discriminations. But can we really say the same for speciesism? Many have disputed this. Since it is undeniable that species is a biological characteristic just as race and sex are, the objections to the parallel have focused on the correspondence approach. While seen as unacceptable in the case of humans, this approach has claimed to be sensible in the case of other animals, because the gulf between us and them is allegedly so large as to prevent overlap.

However, since the work of Charles Darwin, we have abandoned the idea of a gulf between us and the other animals; we see the animal world as composed of a multitude of organisms that resemble one another in some ways, but differ in others, and we hold that differences among species should be viewed as differences in degree rather than in kind. Moreover, if some people want to stick an arrangement of beings in a linear, ascending scale, they still have to be concerned with the presence within our species of disabled, disturbed, or brain-damaged individuals. All in all, it seems that racism, sexism, and speciesism are arbitrary discriminations.

If this conclusion is sound, we can only preserve our valued belief that there are no morally relevant barriers within our species at the price of abandoning the belief that there is a morally relevant barrier around our species.

Further Reading

Cavalieri, Paola. 2001. *The animal question.* New York: Oxford University Press.

Johnson, Edward. 1976, 1977. *Species and morality.* Ph.D. dissertation, Princeton University, July 1976; Ann Arbor, MI: University Microfilms International,1977.

Pluhar, Evelyn. 1988. Speciesism: A form of bigotry or a justified View? *Between the Species* 4(2): 83–96.

Rachels, James. *Created from animals: The moral implications of Darwinism.* Oxford: Oxford University Press.

Singer, Peter. 1993. *Practical ethics,* 2nd ed. Cambridge: Cambridge University Press.

Tooley, Michael. 1983. *Abortion and infanticide.* Oxford: Oxford University Press.

Paola Cavalieri

SPECIESISM: ETHICS, LAW, AND POLICY

The term speciesism, coined in 1970 by Richard Ryder, has become a widely used tool for describing how some humans have thought of and treated other living beings. It is useful to review how speciesism works as a concept, and how that concept can be used to gain an understanding of the nature and history of ideas about other animals that still dominate many people's thinking.

To understand why speciesism troubles some people, consider how this notion works as a concept. The term works well to describe a long-prevailing attitude that has two basic features. On the positive or inclusive side, speciesist attitudes expressly include any and all humans in a moral circle, such that all humans are then understood to be so important as to merit moral and legal protections. Coupled with this important inclusive move honoring any and all humans is an equally decisive negative move, namely, exclusion of all other living beings from basic moral or legal protections whenever such exclusion benefits any humans in any way. Speciesism is thus a two-step process of inclusion and exclusion.

Speciesism as an idea is rooted in a biological category, namely, membership in the human species. Attitudes grounded in speciesism are by no means the only biologically-based attitudes that have played an important historical role in humans' views and actions. Racism and sexism, for example, are also biologically based views. Exclusions on the basis of racist or sexist views, it is now well known, have some peculiarly unfair features. It is not the qualities of the individual that determine how that individual is treated, but instead mere membership in a particular biological class. Favoring members of one race over another, or privileging members of one sex over members of the opposite sex, ignores the fact that members of the disfavored group can, on their own terms, be truly deserving of protections.

The exclusion of all other animals' interests simply because they are not members of the human species has these peculiar features as well. It ignores the fact that some nonhuman animals can be quite deserving of humans' moral concern and legal protections even though they are not members of the human species.

Speciesism has become a valuable tool for describing moral views, legal protections, existing policies, and the reasoning of some people who insist that only humans merit moral and legal

protections. As such, the word and the underlying concept are now widely used when humans discuss the moral status of other animals. In particular, the word has helped many focus on the structure of the species-based thinking that underlies the ways many people justify the current status quo under which some human societies and governments deny nonhuman animals basic protections, whether moral or legal. In modern, industrialized societies, such justifications are not only common, but also systematic and persistent. They are the bedrock of many modern legal systems under which all humans are deemed legal persons while all nonhuman animals are relegated to the unprotected category of legal things. Modern property notions in some legal systems, such as that of the United States, are speciesist in nature. Humans cannot be owned by another human because humans are, by definition, legal persons. But legal persons can, of course, own any nonhuman, for all living beings outside the human species are in the category of legal things, along with chairs, automobiles, and other inanimate objects. This kind of arrangement is a paradigmatic example of speciesism in action. But this need not be so. As is well known, property notions in the law have been malleable over time. At one point, it was legal to own other humans, but such ownership is not legal in most legal systems today. Some legal systems already outlaw the ownership of some nonhumans (chimpanzees, for example), and it is thus possible that in many legal systems the idea of legal property will be severely qualified or even eliminated altogether with regard to certain nonhuman individuals and species.

Despite the prevalence of speciesist reasoning and justifications, these ideas nonetheless can take altogether peculiar forms. For example, speciesist attitudes promote justifications that even the minor interests of humans, such as cosmetic appearance, recreation, or convenience, outweigh the major interests of other animals, such as keeping their lives, and remaining free from captivity and experimentation. This is one reason that even though many people feel that it is immoral for humans to be cruel to other animals, severe cruelties and deprivations are nonetheless tolerated in many industrial practices, such as factory farming, slaughterhouses, circuses, and even zoos. Captivity and its deprivations and cruelties are often rationalized as acceptable because some humans own and generate income from harsh uses of particular nonhuman animals. or because our society as a whole still goes forward on the belief that public exhibition of captive animals is educational in some way or another. When such reasoning prevails, the minor interests of the humans involved, namely making money, or enjoying the limited educational benefits that captivity might offer, prevail over the major interests of the captive nonhumans.

Overriding the interests of other animals has traditionally been supported by assertions that other animals exist for humans. Aristotle made such a claim (*Politics,* Book I, Section 8) in the fourth century BCE, and three centuries later Cicero made similar human-centered or anthropocentric claims (*De Natura Deorum,* II, p. 14). Such claims are still made in great earnestness not only by the food production and entertainment industries, but also by some religious institutions. For example, the revised 1994 *Catholic Catechism* claimed, "Animals, like plants and inanimate things, are by nature destined for the common good of past, present and future humanity." (Paragraph 2415).

There are, however, many in religious circles who, following the lead of exemplary figures like Francis of Assisi or Albert Schweitzer, dispute claims that other animals are by nature destined for our use. Asserting that humans have a moral obligation to other living beings, many have challenged justifications that invoke speciesist reasoning, that is, that humans deserve the privilege of using other animals merely by virtue of being members of the human species. Such challenges question justifications which assume that there is no moral problem when basic moral or legal protections, such as the protection of life, liberty, and freedom from intentional infliction of avoidable harm, are denied to any and all nonhumans.

The exclusions that speciesist claims require are sometimes framed as morally justified because the focus is solely on the human side of the issue. Inclusion of all humans is, of course, a highly respected position today, especially because everyone is painfully aware that there have been long stretches of human history in which exclusion of many humans was not only tolerated, but even promoted as morally acceptable. But speciesist attitudes entail more than an affirmation of all humans because, by definition, they also require the all-important exclusion of any and all nonhumans from moral and legal protections whenever doing so benefits humans in even a minor way. Exclusion is as fully a core feature of speciesism as is the inclusivist feature that honors humans.

It is precisely this exclusion, not the inclusion of all humans, which is the target of anti-speciesism advocates. Such challenges focus on ways in which it is unfair to exclude all other animals solely because those other beings are not members of the human species. Challenges to speciesist attitudes, claims, and practices have often taken their cue from the fact that many cultures and religions have not promoted speciesist views. In fact, the speciesist claim that there is such a profound divide between humans and all other animals that it is moral to deprive all other animals of protections has by no means been universal. For example, many indigenous peoples have viewed other living beings as morally important, as have many religious traditions.

Speciesist views have lately, however, taken some particularly virulent forms, as evident in scientific research, commercial ventures, and environmental destruction. A feature of many contemporary speciesist claims is that their proponents often treat exclusive focus on humans in the same manner that fundamentalists in religions treat the belief that they alone have revealed truth. In other words, some proponents of speciesism will accept no possible challenge to their exclusion of all other animals. Those who challenge speciesist views appeal to open-minded and closely-reasoned inquiries arising out of a passionate commitment to accurate description of the surrounding world and the actual realities of various nonhuman animals.

Another feature of speciesism is that it is part of institutions that may be religious or secular, philosophical or non-philosophical, or science-driven or nonscientific altogether. As such, species-based exclusions are widespread and continue to lead to important, visible consequences which can be measured by the historian, theologian, philosopher, natural or social scientist, or public policy analyst, although doing so can lead to extreme disfavor in academic, government, or business circles.

One can, it should be noted, exclude from the list of speciesist acts those actions by individual humans chosen in order to ensure one's own or one's family's immediate survival. What stimulates and continues to drive the charge of speciesism is the crass justification of many avoidable, nonessential human activities. Instrumental use of other animals, sport hunting, factory farming, testing of cosmetics, biomedical experiments that can be conducted without animals, roadside animal shows and recreational animal parks involve intentional, avoidable damage to other animals' interests, are paradigmatic examples of speciesism.

Historically, the first proponents of speciesism often compared the exclusion of all nonhuman animals to the exclusions of racism and sexism. Despite the emotional value of drawing analogies between speciesist practices and the historically pervasive, biologically-based means of exclusion along the lines of race and sex, there are several basic problems in doing this. Analogies of speciesism to human-on-human forms of discrimination such as racism, sexism, or slavery are at best partial comparisons, and they are often so emotionally charged that others object to what seems to them a comparison of humans to other animals. Importantly, discriminations against other humans are, in at least one important respect, unlike discriminations across the species line, for discrimination against other humans for whatever reason is subject to the obvious challenge of inconsistency. The theologian James Cone once said of any Christian minister who backs racism, "He is an animal . . . We need men who refuse to be animals and are resolved to pay the price, so that all men can be something more than animals." (Cone, 1969, p. 80). Here a speciesist devaluation is invoked in order to challenge racist exclusions, and this suggests important differences between speciesism and racism. The species line is, in biological terms, qualitatively more significant than are race-based divisions, the latter often being culturally-influenced and easily manipulated. Racial divisions, then, are subject to a particularly forceful objection based on a lack of consistency because not all humans are being treated the same.

For many reasons, reliance on analogies to humans discrimination against other humans is not likely to be totally illuminating with regard to the reasoning or practices of speciesism, even if such comparisons are useful in some respects. For example, such analogies can have limited value, for they can help some people discern that the exclusivist attitudes many have toward other animals are, like the prejudices people have with regard to members of other races or toward the opposite sex, supported by flimsy reasoning, bias, and ignorance which can blind us to our own exploitation and oppression of other living beings.

Unfortunately for members of all species other than humans, perpetuation of speciesist views remains a central feature of the most influential secular and religious institutions in Western culture, thereby anchoring the human-centeredness or anthropocentrism of traditional ethics. Some who challenge speciesist forms of ethics, namely, those forms of ethics that favor only members of the human species, reason that fundamental changes can be achieved only incrementally, that is, only by breaching the species line at

a particular point. One well known attempt to breach the species barrier is The Great Ape Project (Cavalieri and Singer, 1994), an effort based on the notion that it is in the interest of many other animals if some nonhumans are now brought into the protected circle. The Great Ape Project has thus focused on humans' closest genetic cousins, reasoning that a first step in dismantling the traditional prejudices which draw their life from speciesist beliefs and practices confining fundamental protections to members of the human species can be taken now on behalf chimpanzees, bonobos, gorillas, and orangutans, because so much good science clearly shows that these nonhumans are deserving of fundamental protections. It remains to be seen if creating protections for our closest evolutionary cousins will reduce barriers to granting important moral and legal protections to other animals generally.

Further Reading

Cavalieri, Paola, and Singer, Peter, eds. 1993, 1994. *The Great Ape Project: Equality beyond humanity* London: Fourth Estate; New York, St Martins Press.

Catechism of the Catholic Church. 1994. London: Gregory Chapman.

Cone, James H. 1969. *Black theology and Black Power.* New York: Seabury Press.

Dunayer, Joan. 2004. *Speciesism.* Derwood, MD: Ryce.

Midgley, Mary. 1984. *Animals and why they matter.* Athens: University of Georgia Press.

Ryder, Richard. 1975. *Victims of science: The use of animals in research.* London: Davis-Poynter.

Ryder, Richard. 1989. *Animal revolution: Changing attitudes towards speciesism.* Oxford: Basil Blackwell.

Singer, Peter. 2001. *Animal liberation*, 2nd ed. New York: Harper.

Waldau, Paul. 2001. *The specter of speciesism: Buddhist and Christian views of animals.* New York: Oxford University Press.

Waldau, Paul, and Patton, Kimberley. 2006. *A communion of subjects: Animals in religion, science and ethics.* New York: Columbia University Press.

Paul Waldau

SPORTS AND ANIMALS

Sports may be best understood as a quest for entertainment and excitement, whether experienced personally as an active participant, or vicariously as a spectator. Sports are about competition, tension, and resolution, and serve as outlets for some socially acceptable forms of violence. The greatest satisfaction is achieved when one side wins and the other loses.

How do animals figure into sports? Animals may be used as a source of entertainment themselves, or as a tool to increase people's excitement during sporting competitions. The incorporation of animals and sport can take any of three different forms: animals competing against other animals for the entertainment of humans, humans competing against animals, and humans competing with animals against some other arbitrary measure, such as time. In some cases, more than one of these may apply. For example, sport hunting with dogs involves dogs co-operating with humans against other animals. Additionally, animal sports such as hunting, bull-fighting, pit sports can be considered consumptive if a participant dies, thereby removing it from the gene pool/ecosystem. Non-consumptive sports, where everyone usually lives to compete again, include racing, catch-and-release fishing, rodeo, and many dog sports.

The pursuit of excitement that plays such a large part in sport applies in especially large measure to those events considered blood sports, where the object of the game is the death of a participant. These activities surface in man's earliest writings, and feature animals on sometimes grand scales.

Hunting, as one of the earliest blood sports, did not originate as a sport at all, but instead as a means of procuring necessary food. Hunting wild animals for entertainment appeared in ancient art and literature after the advent of agriculture, and described the beginnings of a perceived gap between man and the natural world. Ancient Greek writings portrayed hunting as a confrontation between culture and nature, a war by humanity against the wilderness. Hunting was seen as cultivating manly virtues necessary in combat, and since the lives of prey animals were assumed to have no intrinsic worth, and therefore warranted no concern, no thought was given to their suffering. The kill was the end goal, and the anticipation and effort required to arrive at this result were merely added benefits.

Hunting inspired great excitement and satisfaction, not only for participants, but also for spectators. The public delighted in bloody staged hunts (*venationes*) conducted by the ancient Romans for the entertainment of the assembled masses. Gladiatorial battles fought on colossal scales featured scores of animals pitted against human soldiers, slaves, or other animals. Even as some classical writers lamented the gruesome spectacles as the source of man's inhumanity to man, wealthy rulers continued to keep, exhibit, and hunt exotic animals for pleasure (Cartmill, 1993, p. 44). Objections centered not on the brutality of the sport toward the creatures involved, but on how the violent behavior demonstrated by human sponsors and participants could extend to further violence against people.

As the medieval period gave way to the Middle Ages, a civilizing process changed social perceptions regarding behavior and the appropriate balance between pleasure and restraint (Elias & Dunning, 1986). With regard to sport, the pleasure manifest in the excitement of the hunt's kill became tempered with restraint in the form of increasingly complex rules. Putting stipulations and codes of behavior on sport hunting was designed to prolong the action, increasing anticipation and pleasurable tension before the kill. During this time, too, quiet murmurs of discontent were occasionally heard as dismay over cruel treatment of animals portended an emerging welfare attitude. Cruelty, however, would not go away quickly, as people found new and increasingly grisly ways to amuse themselves using animals.

If fox hunting was the domain of the landed gentry, pit sports found favor with all social classes. As the civilizing process deemed it increasingly inappropriate to settle disputes using physical violence against people, pit sports provided an acceptable outlet for daily frustrations, as well as spirited entertainment for visiting royalty, foreign ambassadors, and the unwashed masses. Bull-baiting, bear-baiting, cockfighting, and dog fighting pitted animals against other animals in mortal combat. All occurred with regularity across England and Europe, and later North America. Baiting involved tethering an animal to a stake, then allowing one or several dogs to attack the restrained creature. All manner of beasts were baited, including bears, bulls, badgers, apes, mules, and occasionally horses. For some time it was believed that bull-baiting was necessary to ensure flavorful, tender meat, and was compulsory before a bull could be butchered (Thomas, 1983, p. 93). Baiting sports became illegal in England in 1835.

As opposed to animals used for baiting, pit fighting contestants were often highly valued and vigorously prepared for battle. Value was a subjective term, and referred not to the animal's intrinsic worth, but instead to its value in terms of the fight. Animals that exhibited prowess in the ring were granted better treatment and valued higher than their less capable brethren. Roosters used for cockfighting were brought up on special diets and exercised to increase their stamina. Wattles and combs were cut off to reduce targets for opponents (behavior not considered cruel to the animal), and artificial spurs were strapped to their feet to increase the lethality of attacks. Spirited wagering accompanied matches, and flaring emotions occasionally spurred spectator riots. Gambling also accompanied organized dog fights. Breeds such as the Staffordshire bull terrier and, later, the American Pit Bull Terrier, their aggressive tendencies rerouted from the baiting arena to the dog-fighting pit, were trained by sacrificing smaller, docile dogs for practice. Although largely outlawed today, organized underground networks continue to engage in cockfighting and dog fighting, and large sums of money are often recovered from police raids, along with dead, injured, and mutilated animals. Disgraced former Atlanta Falcons football quarterback Michael Vick made headlines in 2007 when his Bad Newz Kennels dog-fighting operation was uncovered, revealing over 70 fighting dogs in various states of health, with the result that Vick and three others were convicted and sentenced on federal charges.

The ritualized Spanish sport of bullfighting developed out of Roman gladiatorial battles, combined with the rites of Mithraism, an early religion of the Roman Empire. Some consider the pageantry of bull-fighting to be an art form. It remains a popular spectator sport in Spain and Mexico, though interest is waning in Spain, invariably ending with the death of the bull. Matador gorings are not uncommon, and occasionally deadly.

The Romantic period's changing ethic and outcry against cruelty, including the use of animals in some sports, might have been lauded as a sign of an enlightened public concern for animal pain and suffering. However, early changes in doctrine regarding proper treatment of animals arrived out of concern not for the animals themselves, but that mistreatment of animals would lead to depravity against men. This anthropocentric view was combined with various interpretations of the Old Testament, which ensured human dominion over all animals, but cautioned against unnecessary suffering. The resulting combination of religious piety and bourgeois sensibilities led to the banning and social stigmatization of many previously popular animal sports. Hunting, however, remained largely outside these restrictions, with the separation between legitimate meat acquisition, predator/vermin eradication, and pleasurable entertainment used as justification.

The American West had its own evolution of thought regarding animal sports. Early settlers were amazed at the number and diversity of game animals, and greed brought on by the fur trade and apparent inexhaustibility of animal resources led to some mind-boggling excesses. The demise of the passenger pigeon and great buffalo herds are two of the most impressive examples of human myopia, much of it carried out in the name of sport. Theodore Roosevelt was an enthusiastic sportsman, killing thousands of animals on hunting trips across North America and Africa. He believed that shooting game was necessary "to cultivate that vigorous manliness" that comes from close contact with nature (Mighetto, 1991, p. 34). He was also concerned that continued exploitation of resources could lead to wildlife shortages, a sentiment shared by a small but influential group of men of his time. He is credited with cofounding the Boone and Crockett Club in 1888, one of the first conservation organizations. The members of the club were not so much concerned with animal welfare as worried that overhunting would leave future generations without adequate stocks of game species. Land for habitat was set aside in establishment of the first wildlife preserves to help ensure future sport hunting opportunities.

During this time, too, the sport of rodeo emerged as a series of competitive activities associated with the cattle drives of the American South and Midwest. Cowboys, bored by endless days in the saddle, amused themselves by holding impromptu contests of skill and bravery, including bull and bronc riding, calf roping, and steer wrestling, among others. When the great cattle drives ended, rodeo continued on as an organized sport, and still enjoys a wide audience across the United States and Canada.

Today, moral concerns related to animals in sport lie with the welfare and capabilities of animals themselves, a shift from the anthropocentric and largely utilitarian mindset. This change in climate around some controversial aspects of sport such as rodeo calf-roping and bronc riding, open field coursing, canned hunts, and rattlesnake roundups, for example, has lent encouragement to nonconsumptive sports such as racing, catch-and-release fishing, and flyball,

Bull riding competition at the National Finals Rodeo in Las Vegas, Nevada. (AP Photo/Joe Cavaretta)

and cooperative sports such as dog agility, rally obedience, and freestyle, among others. Even among supposedly nonconsumptive sports, well-publicized events have altered public sentiment regarding the human desire for competition and success. The catastrophic breakdown and subsequent euthanization of the thoroughbred filly Eight Belles after the 2008 Kentucky Derby focused a harsh spotlight on the entire thoroughbred racing industry, and led to the questioning of competitive breeding practices that may emphasize early brilliance and speed at the expense of durability and soundness. Questions have been raised concerning the morality of requiring animals to perform physically and psychologically demanding activities for human benefit. These concerns are countered by advocates who believe it may be more inhumane to deny an animal the opportunity to perform an activity it has been expressly bred to do and actively enjoys. Others whether the sport of dog breeding/exhibition, which contributes to a decline in health and welfare of some breeds due to increasingly exaggerated physical constructs, is perpetuated in pursuit of some unattainable standard. Does the life of the bull up until the fight, the vast majority spent at pasture, living a good life, compared with the relatively short, but ultimately fatal time in the ring as a performer, justify the sacrifice? Research supporting the idea that animals are capable of complex emotions and thought is rapidly causing us to reinterpret our treatment of animals during sporting events. Sport hunting has become increasingly polarized, with a huge base of popular support, countered by a vocal opposition. People will always enjoy the competition, excitement, and culminating resolution of sport. It remains to be seen how the changing mores of society will ultimately influence man's inclusion of animals as part of sporting entertainment.

See also Blood Sports; Entertainment and Amusement: Circuses, Rodeos, and Zoos; Fishing as Sport; Hunting, History of Ideas.

Further Reading

Atyeo, D. 1979. *Blood and guts: Violence in sports.* New York: Paddington Press Ltd.

Cartmill, M. 1993. *A view to a death in the morning: Hunting and nature through history.* Cambridge: Harvard University Press.

Elias, N., & Dunning, E. 1986. *Quest for excitement: Sport and leisure in the civilizing process.* Oxford: Basil Blackwell Ltd.

Franklin, A. 1999. *Animals and modern cultures: A sociology of human-animal relations in modernity.* London: SAGE Publications.

Fulton, J. 1971. *Bullfighting.* New York: The Dial Press.

Hummel, R. 1994. *Hunting and fishing for sport: Commerce, controversy, popular culture.* Bowling Green: Bowling Green State University Popular Press.

Mighetto, L. 1991. *Wild animals and American environmental ethics.* Tucson: University of Arizona Press.

Thomas, K. 1983. *Man and the natural world: Changing attitudes in England 1500–1800.* London: Allen Lane.

Toynbee, J.M.C. 1973. *Animals in Roman life and art.* Baltimore: The Johns Hopkins University Press.

Cindy McFadden

STEREOTYPIES IN ANIMALS

A stereotypy is a repeated, relatively invariant sequence of movements that has no obvious function. It is the repetition of the same behavior pattern which makes the stereotypy so obvious to an observer, and the abnormality is also indicated by the distinction from useful repetitive behaviors such as breathing, walking, or flying. Among the most striking abnormal behaviors shown by some animals in zoos and in confined conditions on farms are stereotypies such as route-tracing, bar-biting, tongue-rolling, or sham-chewing. As an example, a female mink, in a cage 75 x 37.5 cm and 30 cm high on a mink farm, would repeatedly rear up, cling to the cage ceiling with her forepaws, and then crash down on her back.

Stereotypies can be shown by humans with neurological disorders, by those with some degree of mental illness, and by those in situations where they have little or no control over aspects of their interaction with their environment. People with no illness may show stereotypies when confined in a small cell in prison, or when exposed to situations like waiting for an important interview, or for their wife to give birth.

The causes of stereotypies in nonhuman animals seem to be very similar to those in humans. Frustrated individuals, especially those unable to control their environment for a long period, are the most likely to show the behavior. Individuals treated with particular drugs, especially psychostimulants such as amphetamines and apomorphine, may show stereotypies, but it is not clear what this tells us about the causation of stereotypies. Animals with irritant disease conditions such as sheep scab show rubbing and oral stereotypies. Many stereotypies seem to be related to oral movement or to locomotion, so the control systems for such movements are clearly susceptible to being taken over by whatever causes repetition. The age of the individual and the amount of time in the housing condition can affect the stereotypies shown, for example, horses changing from crib-biting to wind-sucking, or from side-to-side pacing to head-weaving, and confined sows changing from bar-biting to sham-chewing. Movements can also become more complex with age.

In most cases we do not know whether a stereotypy is helping the individual to cope with the conditions, has helped in the past but is no longer doing so, or has never helped and has always been just a behavioral abnormality. None of the studies that demonstrate a relationship between the extent of occurrence of stereotypies and opioid receptor blocking or opioid receptor density measurement tell us with certainty whether or not stereotypies have any analgesic or calming function, but in all cases the stereotypy indicates that the individual has some difficulty in coping with the conditions, so it is an indicator of poor welfare. Animals that have larger home ranges in natural conditions have been found to be more likely to show stereotypies in zoos. Some stereotypies must indicate worse welfare than others, but any individual showing them has a problem.

Stereotypies are sometimes ignored by those who keep animals, and may be taken to be normal behavior by those people if they see only disturbed animals. For example, zookeepers may see route-tracing by cats or bears, laboratory staff may see twirling around drinkers by rodents, and farmers may see bar-biting or sham-chewing by stall-housed sows without realizing that these indicate that the welfare of the animals is poor. A greater awareness of the importance of stereotypies as indicators of poor welfare is resulting in changes in animal housing. More complex environments that give the individual more control, and hence result in the occurrence of fewer stereotypies, are now being provided in good animal accommodations. These environments also give opportunities for a larger proportion of the full behavioral repertoire to be expressed, and for the patterns of movements in the repertoire to be varied. The consequent reduction in frustration and increase in the proportion of an individual animal's interactions with an environment that is under its control can improve its welfare.

Further Reading

Broom, D. M. 1983. Stereotypies as Animal Welfare Indicators. In D. Smidt, ed., *Indicators relevant to farm animal welfare: Current topics in veterinary medicine and animal science* 23: 81–87.

Broom, D. M., and Fraser, A. F. 2007. *Domestic animal behaviour and welfare,* 4th ed. Wallingford: CABI.

Broom, D. M., and K. G. Johnson. 2000. *Stress and animal welfare.* Dordrecht: Kluwer.

Lawrence, A. B., and Rushen, J. eds. 1993. *Stereotypic animal behaviour: Fundamentals and applications to welfare.* Wallingford: CAB International.

Mason, G. J. 1991. Stereotypies: A critical review. *Animal Behaviour* 41: 1015–1037.

Mason, G.J., Clubb, R., Latham, N. and Vickery, S. 2007. Why and how should we use environmental enrichment to tackle stereotypic behaviour? *Applied Animal Behaviour Science,* 102: 163–188.

Donald M. Broom

STRESS AND LABORATORY ROUTINES

For almost as long as animals have been used in tests and experiments that harm them, the practice has drawn criticism concerning animal suffering. The focus of that criticism is usually on how the experiments may cause pain, distress, and death to the animal subjects. It doesn't require much imagination to conclude that having a household product dripped repeatedly onto the eye causes pain and stress for a rabbit (Draize test), or that being given cancer is nasty for a mouse.

Often overlooked in the vivisection debate is the animals' experience outside the experiments themselves. When data are not being collected, most of an animal's time in the lab is spent in a cage. Thus, day-to-day routines of laboratory life play a big role in the animals' overall welfare. What are these routines, and how may they affect the animals' welfare?

Animals' Sensitivity

To appreciate an animal's potential vulnerability to stressful events or stressors in a laboratory, it helps to have some understanding of an animal's sensory world. The species most commonly used in research are the house mouse and the Norway rat. American laboratory veterinarian Larry Carbone has estimated that there are now upwards of 100 million of these rodents used each year in American labs.

Although there are some important differences between rats and mice, there are many similarities in how they sense their worlds. Both species rely heavily on smells and sounds to communicate. Mice leave tiny droplets of urine wherever they go, and other mice can read a lot of information from these cues, including age, sex, individual identity, social and reproductive status, and even parasite load. Both species use many sounds in the ultrasonic spectrum, that is, above human hearing. Both are mostly nocturnal, preferring to explore and forage in low light.

Other species commonly used in laboratories include rabbits, hamsters, rhesus monkeys, dogs, and cats. Each of these species has a suite of senses adapted to the natural habitats in which they evolved. Be it through good vision, hearing, touch, and/or smell, each is acutely tuned in to their surroundings.

Laboratory Environments

Laboratory settings are profoundly different from natural settings. Animals are kept in small, usually metal cages whose area is thousands of times smaller than the smallest home ranges of their wild counterparts. Cages tend toward barrenness, with little opportunity to engage in such natural behaviors as burrowing, climbing, exploring, foraging, and choosing social partners. Many animals are housed alone, and many have no shelter to hide in. Studies in which animals are given a preference, for example, between a cage with or without company, or a shelter, indicate that meeting these needs is very important to these animals. Lack of tactile contact with other rats, for instance, is believed to underlie the self-biting and tail manipulation observed in isolated rats.

Many seemingly silent laboratory activities produce intense noise in the ultrasonic spectrum to which rodents are sensitive. These include computing equipment, cage washers, hoses, running taps, squeaky chairs, some fluorescent lighting, and husbandry procedures such as emptying food pellets into hoppers. Loud noises can trigger seizures, reduce fertility, and cause various metabolic changes. Chemical solvents used to clean cages, and detergents, perfumes, and companion animal scents on human handlers are known to be aversive to rodents. Feeding regimens, consisting of cakes of processed chow, tend to be monotonous.

In sum, laboratory environments are highly restrictive, and they afford the animals little opportunity to perform highly motivated natural behaviors, or to control their situation. It is sometimes claimed that laboratory-bred animals are unlike their wild ancestors, and that they don't have the same needs. But animals bred in cages for hundreds of generations still retain their ancestral behaviors, as shown by studies in which they are released into natural habitats.

Routine Sources of Stress

Animals in laboratories are commonly subjected to a variety of husbandry, monitoring, and sampling procedures. These include cage cleaning and moving, picking the animal up, and restraining the animal for other routine procedures such as weighing, injections, and gavage or force-feeding. Blood is often required to monitor animals' progress before, during, and/or after a study, and various methods are used to bleed the animal. In rodents, blood may be drawn by needle from the tail vein, from the tail by cutting off the tip, or from just behind the eye with a very thin broken glass tube called a micro-pipette, a technique called post-orbital puncture. In rabbits, blood is often drawn from the ear, where veins are easily seen. Monkeys can be trained to offer a forearm to have blood taken in return for a treat, but routinely they are restrained instead.

It may be asked why procedures that don't cause physical pain, such as handling and cage maintenance, can be stressful to animals in labs. Context helps provide an answer. Life in the wild, while not always safe or easy, nevertheless provides a starkly different living situation for an animal. In the wild, animals have responsibilities. They exercise much more control over their lives. They decide when and where to forage or explore, what to eat, who to associate with and who to avoid, how much light they are exposed to. If they get too hot, they seek a cooler place, and vice versa. If they tire

of eating a certain food, they can choose to go in search of another.

In the laboratory, these decisions are not theirs. The loss of autonomy and opportunity to engage in meaningful behaviors can cause stunted brain growth and the development of abnormal behaviors such as stereotypies, self-mutilation, excessive aggression, etc. It is well established that these circumstances, coupled with unpleasant or painful stimuli, can produce harmful levels of stress. In the laboratory, the appearance of someone wearing a white laboratory coat often precedes a painful or otherwise unpleasant event, so we may expect animals to become stressed.

Why Study Stress?

There has been a large amount of research on stress, partly because prolonged stress can make individuals more vulnerable to illness. In turn, stress may be an important confound, that is, an undesirable factor that makes it harder to interpret the meaning of experimental results. Many scientists, interested in the possible effects that day-to-day laboratory routines might be having on animals' stress levels in the laboratory, have set out to measure stress responses to these procedures. Most of these studies have involved caged mice or rats, but there are also data on stress responses in monkeys, hamsters, rabbits, and several wild species.

How to Measure Stress?

There are various ways to measure stress in animals that can't be asked how they are feeling. One of the most common is to measure blood components, especially the stress hormone corticosterone. To do this, blood is typically collected and analyzed before and after the stressful event. Because blood collection is itself stressful, it is important that the blood be collected quickly to minimize this confound. One method is to have a tube permanently inserted into a vein so that blood can be automatically drawn at any time without causing any additional stress to the animal.

Other blood measures associated with stress include growth hormone, glucose, insulin, epinephrine, and prolactin. Glucose and epinephrine (more commonly known as adrenaline), for example, are released into the bloodstream as a preparation for a fight-or-flight response to a perceived threat. Other, less invasive measures include blood pressure, heart rate, and body temperature, all of which tend to rise with stress. Behavioral measures of stress include freezing in place, that is, remaining completely still, a classic fear response in small mammals, moving and rearing less and, in rodents, defecating more. However, some non-stress situations might also cause increased activity, so one has to be careful in interpreting the meaning of results.

Stress Responses to Laboratory Routines

Jonathan Balcombe has reviewed eighty previously published studies documenting the potential stress associated with three routine laboratory procedures commonly performed on animals: handling, blood collection, and gavage. Handling was defined as any noninvasive manipulation occurring as part of routine husbandry, such as picking the animal up, moving the cage, and/or cleaning the cage. Most of these studies were performed on mice and rats.

In rodents, handling procedures generated average maximal increases in blood corticosterone levels ranging from 63 to 338 percent above a baseline measured before the stressor. Heart rates rose by between 20 and 46 percent, and blood pressure by between 15 and 34 percent. Blood collection caused comparable increases in corticosterone, though one study documented an increase of 595 percent among ten male mice bled from the tail tip. In three studies of rabbits bled from the ear vein, blood glucose levels rose between 24 and 120 percent. Six studies of monkeys bled from a leg vein documented cortisol increases from 40 to 66 percent. Force-feeding of rats generates a suite of short- and long-term stress responses, including corticosterone increases up to 596 percent, weight loss, death of liver cells, and death.

These responses were not fleeting. Typically, they lasted from 30 to 60 minutes, and sometimes longer. This is consistent with a lasting painful and/or emotional response rather than a brief feeling of excitement or anticipation. Overall, these results indicate that despite their routine use in laboratory studies, these procedures are acutely stressful for animals. As one of the scientists who conducted some of these studies noted: "Care should be exercised in dismissing a procedure as non-stressful just because it is simple or routine."

Several studies have also addressed the possibility that animals might be stressed by witnessing another animal in pain or distress. Being in a room where other rats were subjected to routine cage changes, handling and weighing, blood collection, or killing (by beheading) caused significant increases in various stress measures in rats. Mice and monkeys have shown similar witnessing effects. In a study conducted at McGill University in 2005, mice were injected with painful irritants into the stomach or paws. This caused these mice to writhe in pain. When a writhing mouse could be observed by another mouse, the witnessing mouse became more sensitive to pain, but only if the writhing mouse was a familiar individual. This result suggests that mice can show empathy for another whom they know, such as a mate, a social companion, or a sibling.

In summary, routine procedures commonly performed in laboratories are stressful to the animals being used. It may be concluded that significant fear and stress are predictable consequences of routine laboratory procedures. Animals can remember past events that were painful or unpleasant, and they can anticipate and fear a repeat performance. As science reveals more about the sensitivities and emotions of animals once dismissed as unfeeling things, whether or not humans should be deliberately harming animals in laboratories is likely to come under closer scrutiny.

Further Reading

Balcombe, J. P., Barnard, N., Sandusky, C. 2004. Laboratory routines cause animal stress. *Contemporary Topics in Laboratory Animal Science* 43: 42–51.

Balcombe, J. P. 2006. Laboratory Environments and Rodents' Behavioural Needs: A Review. *Laboratory Animals* 40: 217–235.

Berdoy M. 2002. The laboratory rat: A natural history. Film. 27 minutes: www.ratlife.org.

Carbone L. 2004. *What animals want: Expertise and advocacy in laboratory animal welfare Policy.* Oxford: Oxford University Press.

Jennings M., Batchelor G. R., Brain, P. F., Dick, A., Elliott, H., Francis, R. J., et al. 1998. Refining rodent husbandry: the mouse. *Laboratory Animals* 32: 233–259.

Langford, D. J., Crager, S. E., Shehzad, Z., Smith, S. B., Sotocinal, S. G., Levenstadt, J. S., et al. 2006. Social modulation of pain as evidence for empathy in mice. *Science* 312: 1967–1970.

Jonathan Balcombe

STRESS ASSESSMENT, REDUCTION, AND SCIENCE

What Is Stress?

Evidence is gradually accumulating that the majority of mammalian research animals, particularly rodents, are mentally stressed by their living conditions. Stress in rodents will be specifically addressed because they account for about 90 percent of all research animals. Stress is generally defined as a state in which an individual perceives that the needs for adaptation to a new or excessive demand or to a different environment exceed the personal resources that they have available. Thus psychological as well physical components play a role in the stress response, at least in the more complex animals. The physical aspect of the stress response is fuelled by stress hormones that flow through the body, altering every organ and biochemical function, with wide-ranging effects on metabolism, growth, and reproduction. These changes may not necessarily result in a reduction in physical fitness, at least initially. In addition, the animals' physiological systems will be affected to a varying degree according to the threshold of the stress response for each individual animal.

Why Care If Laboratory Animals Are Stressed?

If animals are under stress, they can have permanently raised concentrations of stress hormones. In the case of rodents, those that cope by increasing their physical activity show high stimulation of the sympathetic autonomic nervous system and consequent release of epinephrine and norepinephrine, whereas in those who cope passively, the parasympathetic nervous system is activated. In both cases, the hypothalamus/adrenal/pituitary axis is stimulated, leading to greater release of corticosterone. Other stress-induced biochemical changes can include reduced concentrations of sex hormones and compromised immune systems. Although it is true that stress does not always compromise health and welfare, and that the stress response is necessary for survival in the wild, stress, particularly when elicited repeatedly, disturbs the body's homeostasis and imposes a cost on the body. This cost arises if stress-induced mediators such as adrenal hormones, neurotransmitters, and cytokines are released too often. Another problem, more pertinent to animal research than to animal welfare, is that the degree to which a given stressor elicits these responses will vary between animals, even between those of a similar species and strain. These uncontrolled variables make such animals unsuitable subjects for scientific studies.

Researchers often dismiss questions concerning environmental influences on their experimental data by claiming that such effects cancel out because all of their control animals are housed under the same conditions. But the conclusions drawn from such experiments are specific to the stressed animals and cannot necessarily be extrapolated to healthy animals. The increasing use of genetically modified mice since their advent twenty years ago amplifies this problem. Genetically modified mice either lack a specific gene or gene-pair (knock-out mice) or carry a piece of foreign DNA integrated into their own chromosomes (transgenic mice), and are used to deduce the functions of particular genes. Studies are beginning to show that an animal's environmental conditions can completely change the results of genetic studies. As stated by Poole (1997): "It is essential that the stress status of laboratory animals is monitored and controlled because stress may alter the experimental data obtained from those animals."

How Do We Know If Laboratory Animals Are Stressed?

The most practical method for determining whether laboratory animals are stressed is by observing their appearance and wakeful behavior. Of course, taking blood samples and measuring concentrations of stress hormones would appear to be a more accurate way to evaluate stress levels, but this process alone can actually cause stress. Symptoms of stress in rodents which are easily observable include redness around the eyes and on the scruff of the neck, reflecting inflammation. These symptoms mainly arise from excessive grooming of themselves and their cage-mate(s), an activity that is seen in a range of stressful situations. Other behavioral responses to stress include increases in total activity and rearing onto the hind legs. However, individual rodents, even those from the same strain, cope with stress in different ways. Coping mechanisms may vary even within the same rat. Active copers show more active behavior, driven by the fight or flight response, whereas passive copers respond by hiding or, if that is not possible, by freezing. Although, with practice, it is fairly easy to determine whether or not rodents are stressed just by observing them, many laboratory rodents continue to be housed in a stressful environment. This is partly because rodents are nocturnal and most of the observations of their behavior are made during the day when researchers are performing experiments. Unless the rodents' light/dark cycle is reversed, this means that most observations will be made when the animals are asleep, and so any behavioral indicators of stress will not be manifested.

What Makes Laboratory Animals Stressed?

Although laboratory animals do not lack essential physical needs such as food and water, there are many major and minor environmental perturbations encountered in animal facilities that can significantly contribute to a stress response in the animals. For example, when an animal is moved to a new cage, increases in blood pressure, heart rate and locomotor behavior occur, which are indicative of a stress response. With regard to the environment in the rooms inhabited by laboratory animals, the lighting, temperature and humidity are usually well controlled. However, there are many uncontrolled sources of noise in animal facilities, most of which derive from human activities. These include high-pressure hoses, cage cleaners, and air-conditioners or heaters, squeaking doors, carts, and movable chairs, and jangling keys. Rodents, in particular, are sensitive to these noises, and studies show that this sensitivity does not diminish with time, as is commonly assumed. These noises can alter rodents' behavior and even adversely affect their health. Yet, surprisingly, many scientists are unaware that loud noises in their animal facilities can affect research outcomes and compromise their data.

The noise and vibration of building construction have caused major problems with rat behavioral studies and experiments requiring unstressed control rats. One study in rats (Shepherd et al., 2004) even showed that building-induced stress rapidly inhibited glucose absorption by the intestinal transporter, GLUT2. Several studies have shown that noise in animal care facilities can reach as high as 90–100 dB. Such levels of noise can induce physiological and behavioral responses in laboratory rodents such as increased plasma corticosterone levels, reduction in body weight, decrease in gastric secretion, changes in immune response and tumor resistance, and a decrease in reproductive function. Much of the noise in institutional animal facilities is caused by personnel activity, because measurements have shown that environmental noise levels decrease dramatically at night and during the weekends.

Apart from noise problems, research animals are often housed in small cages with no source of enrichment, such as wheels, shelves, or tubes. Such devices enable animals to exert some control over their environment, such as escaping an attack from a cage-mate by moving to another level in the cage or hiding. Often, researchers are unwilling to include such items in their animals' cages because other researchers do not. However, rigorous standardization of the environment, particularly if it leads to barren surroundings, increases the risk of obtaining results that, because they are specific to a narrow set of conditions, cannot be compared with other researchers' results. The word boredom is used to describe the experience of animals who spend their lives in highly monotonous environments. Sometimes the animals fill the time with abnormal behaviors including excessive grooming of self and cage-mate(s), and repetitive patterns of movement known as stereotypies. The excessive grooming can cause regions of inflammation, especially on the neck area. These behaviors disappear when the animals are provided with a chance to make choices, such as to enter a tunnel which provides them with low light conditions and the security of a confined space, or to move to a different level of the cage to avoid the aggressive overtures of a cage-mate.

How Can We Reduce the Stress?

Although the effective control of all environmental variables all the time is impossible, reasonable attempts could be made to control those variables most likely to interfere with the work. In order to ensure the validity and usefulness of animal experimentation, it is necessary to provide conditions that minimize stress-related activities and that allow the animals to perform the behaviors normal for their species. Probably the problems that require the most immediate attention are the absence of adequate species-specific enrichment items and the lack of adequate noise control. To address the need for appropriate items of enrichment, normal and aberrant behaviors for each species could be agreed upon institutionally and a list made available to all investigators. Designing cage environments to suit animals' psychological and physiological needs would be far preferable to the minimalism, otherwise known as standardization, that is currently employed. The exact conditions used to achieve these goals would probably vary between laboratories, but the end result would be similar. Both the welfare of research animals and the quality of the science would be markedly improved, leading to data that could be meaningfully applied to our quest for medical knowledge.

With regard to the noise issue, the latest edition of the *Guide for the Care and Use of Laboratory Animals* recommends that noise control should be considered in facility design and operation. To the greatest extent possible, activities that might be noisy should be conducted in rooms or areas separate from those used for housing animals, especially rodents. In addition, the guide suggests that excessive and intermittent noise can be minimized by training personnel in alternatives to practices that produce noise, and by the use of cushioned castors and bumpers on carts, trucks, and racks. However, it is difficult to estimate the degree to which these particular recommendations are currently being followed. A recent survey (Baldwin et al., 2007) indicates that such precautions to minimize noise are still often ignored in animal facilities in the United States. In addition, the guide's recommendations for noise levels may not be adequate to protect research animals. While the guide specifies a value of 85 dB SPL as the maximum allowable noise level, studies in the literature have shown that noise intensities as low as 73 dB SPL can significantly increase the concentration of stress hormones in the plasma of rodents.

There are several relatively simple and inexpensive solutions to the noise problem. For example, noise levels in an animal facility have been reduced by as much as 15 dB using readily available industrial and architectural acoustical panels. Electronic noise-canceling equipment is now available, and the cost of this technology is becoming more reasonable. Animal research facilities are a prime site for justifying installation of such systems. Principal investigators with research animal programs can be provided with data on environmental stressors, particularly noise, that is recorded in the rooms in which their animals are housed. With minimal additional effort, continuous tracking of significant changes in noise, temperature, air flow, and light intensities could be added to facility monitoring, enabling remedial actions to be effected in hours or days, rather than weeks, months, or even years.

Husbandry and laboratory technicians should be made aware that typical, apparently minor, noise sources that they either encounter or produce may affect their animals' level of stress, and that this may confound research outcomes. They must also appreciate the importance of performing their duties quietly, and of reporting noisy incidents, either acute or chronic, to their supervisors and to the principal investigators. Even the jangling of keys can disturb rodents and produce variable alterations in their physiology. Riley (1981), who demonstrated that mice in conventional animal facilities had plasma corticosterone values more than ten times greater than mice from special low stress housing, stated that "few technicians or research scientists are good judges of moderate stress." At present little formal training is required for animal caretakers and animal technicians in universities in the United States. Although the American Association for Laboratory Animal Science operates animal technician education programs, little emphasis, if any, is placed on the deleterious effects of noise on the validity of data obtained from experimental animals. Such information should be a required component of institutionally conducted training courses required prior to working with animals.

The United Kingdom Farm Animal Welfare Council sets forth the following basic requirements for farm animal in its Welfare Code in terms of freedoms. These are: Freedom to express normal behavior by providing sufficient space, proper facilities, and company of the animal's own kind, and freedom from fear and distress by ensuring conditions and treatments that avoid mental suffering. These freedoms are described as being ideals that anyone with responsibility for animals should aim to provide. Many researchers believe that we are just as surely required to ensure that these freedoms are also provided to laboratory animals.

Further Reading

Dallman, M. F., Akana, S. F., Bellman, M. E., Bhatnager, J., Choi, S., Chu, A. et al. 1999. Warning! Nearby construction can profoundly affect your experiments. *Endocrine* 11:111–113.

Fraser, A. F., and Broom, D. M. 1990. *Farm animal behaviour and welfare,* 3rd ed. London, Bailliere Tindall; New York: Saunders.

Milligan S. R., Sales, G. D., and Khirnykh, K. 1993. Sound levels in rooms housing laboratory animals: An uncontrolled daily variable. *Physiology and Behavior* 53:1067–1076.

National Research Council. 1996. *Guide for the Care and Use of Laboratory Animals.* Washington (DC): National Academic Press.

Poole, T. 1980. Happy animals make good science. *Laboratory Animal Science* 30: 422–439.

Riley, V. 1981. Psychoneuroendocrine influences on immunocompetence and neoplasia. *Science* 212:1100–1109.

Shepherd, E. G., Halliwell, P. A., Mace, O. J., Morgan, E. L., Patel, N., and Kellet, G. L. Stress and glucocorticoid inhibit apical GLUT 2-trafficking and intestinal glucose absorption in rat small intestine. *Journal of Physiology* 560:281–290.

Ann Baldwin

STUDENT ATTITUDES TOWARD ANIMALS

Since the publication of Singer's *Animal Liberation* in 1975, print and electronic news media, movies and television sitcoms, and textbooks and popular books have increasingly concerned themselves with issues relating to the treatment of animals. As a result, students have been exposed to and have formed opinions about issues ranging from hunting and trapping to the use of animals in research, product testing, and the classroom. The diversity of their views is indicated by a study which distinguished 10 different attitudes toward animals found in the American public. Some of these are ecologistic, humanistic, moralistic, dominionistic, aesthetic, utilitarian, and negativistic. While there is a considerable diversity of attitudes, individuals hold hard attitudes. This means that at an early age individuals form strong views toward animals, and these particular views are enduring.

Numerous studies have established that gender is the most powerful predictor of an individual's general attitude toward animals. For example, one investigator found that, in 10 of 15 countries studied, with a trend in the same direction in the remaining five countries, women opposed animal research significantly more than men (Pizer, Shimutzu, and Pifer, 1994). The reasons for this gender gap are not fully understood, but implicate differences in parental views of girls and boys, such as the importance given in the socialization of girls to developing caring and nurturing relationships.

Age is also an important variable, with younger people being more concerned with animal welfare. It is not known whether the link to age is a generational one, that is, younger people like animals more, or whether it is true of this cohort, in other words, that these people will retain these views as they get older. Studies on vegetarianism showed that a greater number of young people are vegetarian than had been the case in the previous decade. Another group of investigators suggests that "decline in [laboratory] work with animals stems largely from changing student attitudes" and that these attitudes "... are in tune with current widely shared concerns for the natural environment and animal welfare" (Driscoll, 1992). These studies suggest that the concern with animal welfare is changing and is not just a youthful phase.

The correlation between attitudes toward animals and amount of education, specifically science education, is also unclear. One study found no significant relation between degree of scientific knowledge and attitude, while a second found that more scientifically knowledgeable young adults were less likely to oppose animal research.

Attitudes toward animals are also related to political and ideological positions. Liberalism, as compared to conservatism, is associated with more pro-animal views. As compared to a group of college students, animal rights activists attending a large national protest are more likely to have a high degree of confidence that moral behavior will really produce positive results, and to have a moral philosophy that rejects relativism and relies on universal principles. Further, those who take up the cause of animals are also more likely to be concerned about discrimination against certain classes of people. Support for animal rights is associated with more tolerance of human diversity, specifically, acceptance of rights for women, homosexuals, and ethnic minorities. Concern for the welfare of human and nonhuman animals is typically held by the same individual. One final variable is personality type. People who rely more on intuition and feeling and are more focused on relationships are more likely than thinking types to oppose animal research.

In terms of actual positions on the issues, there is as indicated a diversity of views. Taking attitudes toward animal research as an example, evidence as to the general level of opposition to the use of animals in research is mixed. Although a number of studies found that on average individuals espouse a middle position, the extensive study of individuals in 15 countries discussed earlier found a high level of opposition.

See also Dissection, Student Attitudes to; Sociology of the Animal Rights Movement; Student Rights and the First Amendment.

Further Reading

Driscoll, J. (1992). Attitudes towards animal use. *Anthrozoös, 5,* 32–39.

Galvin, S., & Herzog, H. (1992). Ethical ideology, animal rights activism, and attitudes toward the treatment of animals. *Ethics and Behavior, 2,* 141–149.

Kellert, S. (1989). Perceptions of animals in America. In R. Hoage (Ed.)., *Perceptions of Animals in American Culture* (pp. 5–24). Washington DC: Smithsonian Institution.

Pifer, L, Shimuzu, K, & Pifer, R. (1994). Public attitudes toward animal research: Some international comparisons. *Society and Animals, 2,* 2, 95–113.

Kenneth J. Shapiro

STUDENT OBJECTIONS TO DISSECTION

Increasingly, students have been objecting to dissection in the classroom on ethical grounds and demanding the student rights option, a policy that guarantees the right of a student to an alternative educational exercise. As a legal issue, their objections pit the rights of students to freedom of religion or, more broadly, of conscience, under the First Amendment of the federal Constitution against teachers' rights to academic freedom. The claim against dissection is based on the civil liberties of a human animal, the student, and only indirectly implies a claim to rights for animals. To date, in several cases, the courts appear to be sympathetic to student claims.

A second issue raised by dissection in the classroom is whether using animals in laboratory exercises is an effective way of teaching anatomy, medicine, or behavior. Based on over 30 published studies, it is clear that the use of alternatives such as computer software, models, and transparencies are at least as likely to achieve the intended instructional goals. Increased technological advances, such as imaging that allows the student to view, for example, the nervous system at any level, to rotate the image, to make certain layers opaque and others transparent, to cut away certain layers, and to repeat these operations in reverse, add an overwhelming advantage to alternatives.

Supporters of dissection frequently argue that hands-on experience is essential to the student's education. There is no evidence supporting this claim. Further, the term must be redefined to reflect current practices. Increasingly, as techniques of observation and intervention become more sophisticated, both for scientist and surgeon, hands-on is coming to refer more to the microscope, computer, and television monitor than to direct observation and manipulation of organs and body parts.

A number of studies have explored the impact of the experience of dissection on student attitudes and psychology (Balcombe, 2000; Shapiro, 1991; Hepner, 1994). There is evidence that it generally decreases sensitivity and empathy. In a study of adults formerly involved in classes involving dissection, it was found that most people remember their first lab dissection vividly, with strong associated feeling, and many consider it an important experience of their childhood or adolescence. For a minority of these, the memory has some features of a traumatic event; it is easily remembered and negatively emotionally loaded. Interviews with these adults and with students currently involved in classroom dissection suggest several reasons why this experience is emotionally loaded for most individuals, and negatively so for a minority: (1) Unresolved issues around the early exploration of death by young people in this culture are part of what gives emotional loading to the experience of dissection. Whereas children are exposed to death and violence graphically through television and other media every day, often they are shielded from direct exposure to serious illness, dying, and death when it strikes loved ones. For this reason, the killing, dying, and death of a frog or rat in the classroom tends to assume significant psychological importance. (2) Dissection teaches lessons that are, for some, strikingly at odds with the constructive adolescent self-discovery process. Instead of being associated with individuality, integrity, and privacy, the body is objectified, reduced to internal workings, and publicly displayed. (3) In dissection, there is public encouragement and sanction of the otherwise censured impulse to kill and/or mutilate. This likely arouses a developmentally early form of evil called defilement, a common childhood experience exemplified by pulling the wings off a butterfly or tormenting other small animals. The impulse to defile is a mixture of disgust and fascination at the suffering of another individual.

See also Alternatives to Animal Experiments in the Life Sciences; Alternatives to Animal Experiments: Reduction, Refinement, and Replacement; Student Attitudes toward Animals; Student Rights and the First Amendment.

Further Reading

Balcombe, J. 2000. *The use of animals in higher education: Problems, alternatives, and recommendations.* Washington, DC: Humane Society Press

Francione, G., & Charlton, A. 1992. *Vivisection and dissection in the classroom: A guide to conscientious objection.* Jenkintown, PA: American Anti-Vivisection Society.

Jukes, N., and Chiuia, M. 2003. *From guinea pig to computer mouse*, 2nd ed. Leicester, UK: Interniche.

Hepner, L. 1994. *Animals in education: The facts, issues, and implications.* Albuquerque, NM: Richmond.

Shapiro, K. 1991. "The psychology of dissection." *The Animals' Agenda, 11,* 9, 20–21.

Kenneth J. Shapiro

STUDENT RIGHTS AND THE FIRST AMENDMENT

The free exercise clause of the First Amendment to the U.S. Constitution provides that "Congress shall make no law. . . prohibiting the free exercise" of religion. Although the U.S. Supreme Court has not yet had an opportunity to interpret this First Amendment guarantee in the precise context of a student objection to dissection and vivisection in the classroom, the Court has guaranteed First Amendment protection in cases that are relevant to the issue.

The Supreme Court has long drawn a distinction between belief and conduct in the context of interpreting the constitutional guarantee of freedom of religion. In *Cantwell v. Connecticut* (1940), the Court held that the free exercise clause "embraces two concepts—freedom to believe and freedom to act. The first is absolute but, in the nature of things, the second cannot be. Conduct remains subject to regulation for the protection of society." That is, government cannot regulate religious belief and can only regulate religious conduct, a notion that was upheld in *Thomas v. Review Board* (1981).

The legal framework established by the Court and Congress involves six elements for evaluating the suitability of the regulation of conduct that is claimed to be protected by the free exercise clause of the First Amendment. First, the regulation must constitute state action. The reason for this requirement is that, with certain exceptions not relevant here, the U.S. Constitution protects us only from the action of some branch of government. Although there may be other federal and state laws that apply to the actions of private institutions, a claim under the First Amendment requires that the student show that there is a legally relevant relationship between either federal, state, or local government and the challenged regulation, so that the regulation may be treated as an act of the state itself. For example, a requirement to vivisect or dissect imposed by a state university would constitute state action. The same requirement imposed by a private school, even one that receives state money, may not qualify as a state action, depending on the relationship of the private institution to the government.

Second, the First Amendment's guarantee of freedom of religion protects only religious or spiritual beliefs and does not protect bare ethical beliefs. It is important to understand, however, that the Supreme Court has held quite clearly that the religious belief need not be theistic or based on faith in a God or Supreme Being, and that the claimant need not be a member of an organized religion. So, for example, a

person who accepts reverence for life as a spiritual belief, but who does not believe in God per se, would qualify for First Amendment protection. Finally, it is not necessary that the belief be recognized as legitimate by others who claim to be adherents of a religious or spiritual doctrine. So, for example, it is not relevant to a claim that the killing of animals is contrary to Christian belief that others who identify themselves as Christians feel that animals have no rights and should not be the subject of moral concern.

Third, the student who asserts a First Amendment right must be sincere. If, for example, a student objects to vivisection on the ground that it violates the student's belief in the sanctity of all life, the fact that the student eats meat, wears leather, and trains fighting dogs for a hobby may indicate that the student's asserted concern for the sanctity of all life is insincere and should not be protected.

Fourth, the state action must actually burden the religious belief. This requirement is not usually a problem in the context of student rights to oppose animal exploitation, because in most cases the state is conditioning the receipt of a benefit, that is, an education, on the performance of an act, that is, vivisection or dissection, that is proscribed by the student's religious belief system.

Fifth, once it is determined that the state is placing a burden on a sincerely-held religious or spiritual belief, then the state may have the burden to prove that the regulation serves a compelling state interest. That is, the state must prove that there is a very important reason for the regulation. In the last decade, the U.S. Supreme Court has stated that if a law is neutral and of general applicability, the state does not have to justify it by a compelling state interest even if the law has the incidental effect of burdening a religious practice. Normally, schools argue that the state has a compelling interest in establishing educational standards. That may very well be true, but if the school has exempted other students from having to vivisect or dissect because, for instance, they happened to be ill on the day of the lab, then the objecting student can show that the requirement of dissection or vivisection is not being neutrally applied, and the claim that the state has a compelling interest in particular educational standards has less force.

Sixth, the state must show that the requirement is the least restrictive means of satisfying any state interests. For example, if there are educationally sound non-animal alternatives to the vivisection/dissection requirement, then the state must allow such alternatives. The quality and availability of educational materials that do not use animals has improved significantly in recent years.

In addition to the protection afforded the free exercise of broadly defined religious and spiritual beliefs protected by the First Amendment, there may be other federal and state laws that are relevant to the student's claim, depending on the particular case. Other relevant federal laws concern freedom of speech and association, due process and equal protection, procedural due process, and civil rights. Other relevant state laws include state, as opposed to federal, constitutional guarantees, as well as laws concerning contract, tort, and discrimination within educational institutions.

Several states (California, Florida, Illinois, New Jersey, New York, Oregon, Pennsylvania, Rhode Island, and Virginia) have provided for a limited statutory right to object to vivisection and dissection. These laws usually apply to students in kindergarten through high school, and provide the student with the right to choose a non-animal alternative without being penalized. Other states have or are developing educational policies approving of alternatives.

See also Dissection, Student Objections.

Further Reading
Cantwell v. Connecticut, 310 U.S. 296 (1940)
Thomas v. Review Board, 450 U.S. 707 (1981)
Church of the Lukumi Babalu Aye, Inc. v. City of Hialeah, 508 U.S. 520 (1993)
Francione, Gary L., and Charlton, Anna E. 1992. *Vivisection and dissection in the classroom: A guide to conscientious objection.* Jenkintown, PA: American Anti-Vivisection Society.

Anna E. Charlton

T

TELEOLOGY AND TELOS

Following the Scientific Revolution, epitomized in Newtonian physics, the fundamental metaphor encapsulating society's conceptual characterization of nature was the machine. As articulated most clearly in Descartes, even biology came to be seen as best expressed in terms of physics and chemistry, culminating in the ascendance of molecular biology, and reductionism as the aim of science in the 20th century.

It is thus important to recall that historically the longest reigning approach to understanding nature was the teleological, functional worldview of Aristotle, which held sway from 300 BC until the Newtonian revolution. In Aristotle's conceptual scheme, emerging from his orientation as a biologist and as a philosopher of ordinary experience, teleology meant that the world was an assemblage of functions defining the natures or essences of natural kinds of things. A thing was what it did; its essence was its final cause. Contrary to some of his predecessors such as the atomists Democritus and Leucippus, there was no ultimate science of all things for Aristotle. Explanation, for Aristotle, was optimally done by reference to the laws and regularities specific to the sort of thing being explained, not by invoking general laws that apply to everything. If any science was the master science, it was biology, because the functional categories natural to explaining living things were the model for all explanation. Contrary to Descartes, physics became the biology of dead matter. Thus for Aristotle, rocks fell when dropped because their natural place was the center of the Earth, which was also the center of the universe.

Since the Aristotelian worldview became, in the hands of Thomas Aquinas, the worldview of medieval Catholicism, teleological explanations acquired a patina of conscious design by God never envisioned by Aristotle, and thus became seen by scientific revolutionaries as inherently equated with religion and superstition. Spinoza's blistering and unfair attack on references to final causes became emblematic of how scientists dismissed teleology.

It is essential to recall that teleological explanations do not entail either conscious divine design or consciousness on the part of the entity being explained teleologically. To say, for example, that the adrenal gland secretes adrenalin to prepare the body for fight or flight does not entail that it was consciously designed to do so or that it consciously strives to do so. In the same vein, saying that the thermostat regulates the room's temperature does not entail reference to awareness on the part of the thermostat. Darwin's account of natural selection is thus benignly teleological, and clearly not incompatible with a mechanistic account of a genetic basis for selection. Similarly, it would be difficult to teach physiology without reference to functions and bodily purposes, yet such talk is not at loggerheads with the reduction of physiology to molecular mechanisms. Again, ecological explanations are inherently functional and teleological in explaining ecosystemic interactions. Obviously, explanations of human behavior by reference to conscious intentions are a form of teleological explanation, but not every teleological explanation involves reference to conscious intentions.

The notion of an animal's telos—the pigness of a pig, the cowness of a cow, its ultimate end—was rescued from the cemetery of dead ideas by Bernard Rollin's *Animal Rights and Human Morality* (1981) with the emergence of animal ethics in the 1970s and 1980s. The first contemporary book on animal ethics was Peter Singer's pioneering *Animal Liberation* (1975), based philosophically on the ethical theory known as utilitarianism, which defines morally good action as that which maximizes the pleasure of sentient beings in a given situation, or minimizes their pain. Since at least higher animals are sentient and capable of feeling pleasure and pain, they are to be included in the scope of moral concern, and our treatment of them is to be judged morally in terms of pleasure and pain. In Rollin's view, pleasure and pain are inadequate tools for analyzing human obligations either to animals or people. For Rollin, not all harm done to animals can be rationally encompassed under the rubric of pain. Causing fear in animals, or boredom, or immobilization, or separation from others they naturally interact with, as in pack or herd animals; in short many of the consequences of how we keep animals in agriculture or zoos, or use them in experimentation, are not naturally or reasonably characterized as pain, though certainly many such uses do involve pain. Indeed, it is demonstrable that animals will endure physical pain to escape from traps or to avoid a highly confined environment such as that of hens in battery cages.

In Rollin's construction of animal ethics, drawn from logically extending the basic moral principles we use in our societal ethic to evaluate treatment of people, it is patent that what we do to animals matters to them, but there are more sorts of mattering than what we call pain. According to Rollin, what we owe animals morally can best be captured by reference to the fundamental needs and interests embodied in their biological and psychological nature or telos. As the song goes, "fish gotta swim, birds gotta fly."

Rollin then argues that the societal ethic of Western democratic societies such as the United States protects key features of human nature such as speech, property ownership, assembly, and belief from oppression for the sake of society as a whole, by building protective fences around them known as rights. The Bill of Rights is a paradigm example. Violation of these elements of human nature matters to people. By the same logic, the concept of rights determined by telos should be extended to animals, and encoded in law, as restrictions on how animals are used. Rollin also argues that much of modern industrialized agricultural and research uses of animals significantly violate the interests dictated by their telos, for example, keeping a breeding sow in a 2' x 3' x 7' cage or crate for her entire productive life, or a nocturnal burrowing animal in a polycarbonate cage under illumination.

If one examines emerging laws for animals, and non-legislated changes in animal use across Western societies, one can indeed find evidence that the public is greatly concerned that animal telos be respected. This is particularly manifest in Europe, where many agricultural systems violative of animals' basic interests have been abolished. In the same way, zoos that fail to respect animal telos, the state of the art fifty years ago, have largely been eliminated. Austere and impoverished environments for laboratory animals are being modified in favor of environmental enrichment, mandated by U.S. laboratory animal laws, as is control of pain and distress.

As far as providing a guidepost for fair treatment of the animals we utilize for human benefit, the notion of respect for telos provides a commonsensical, intuitively plausible consensus template for actualizing our moral obligations to other creatures.

Further Reading

Aristotle, *De Anima*.

Aristotle, *Metaphysics*.

Aristotle, *Physics*.

Rollin, Bernard E. 1995. *The Frankenstein syndrome: Ethical and social issues in the genetic engineering of animals*. New York: Cambridge University Press.

Rollin, Bernard E. 2005. Telos, value and genetic engineering. In Harold Baillie and Timothy Casey, eds., *Is human nature obsolete? Genetic engineering and the future of the human condition*. Cambridge: MIT Press.

Rollin, Bernard E. 2006. *Animal rights and human morality*, 3rd ed. Buffalo: Prometheus Books.

Bernard E. Rollin

TOXICITY TESTING AND ANIMALS

There is a movement to refine, replace, and reduce the number of animals used in toxicity experiments in scientific research. The term "the 3Rs" was generated by Russell and Birch in their 1959 book *The Principles of Humane Experimental Technique*. The 3Rs refer to reduction, replacement, and refinement of whole animal use in scientific research. The concept of alternatives does not necessarily refer to a complete eradication of animals from the research arena, but to an attempt to decrease the suffering of laboratory animals by reduction, replacement, and refinement.

A reduction alternative substantially decreases the number of whole animals necessary to perform a test. A number of *in vitro* assays are now being used as screening tests for the Draize test, a test for ocular irritancy, reducing the number of animals required to fully evaluate the potential irritancy of a chemical. A reduction in the numbers of animals could be possible if testing techniques are refined and made more sensitive in screening processes. A replacement alternative is one that entirely eliminates the need for whole-animal testing. For example, a replacement could be the use of an invertebrate instead of a vertebrate animal. Refinement alternatives are those that improve the design and/or efficiency of the test, therefore lessening the distress or discomfort experienced by laboratory animals. An example of a refinement to the Draize test is that the test is no longer performed using substances that are known to be severely irritating to the eye.

History of the Movement to Refine, Reduce, and Replace

By the 18th and 19th centuries, animal research had become commonplace. According to Andrew Rowan, author of the book *Of Mice, Models and Men: A Critical Evaluation of Animal Research*, several medical advances influenced public perception concerning the benefits of animal research, including the development of ester anesthesia (1846), antiseptic surgical practices (1860s), and the identification of many disease-causing bacteria. Not everyone was in favor of using animals in experimentation, and the formation of organizations that opposed animal cruelty also started to appear at this time, particularly in Europe. The surge in animal protection was influenced by Darwin's theory of evolution, Jeremy Bentham's utilitarian argument against using animals ("The question is not can they reason, nor can they talk, but can they suffer"), and the prevailing Victorian sentiment of the time, that preventing cruelty to animals was seen as an extension to preventing cruelty to human beings.

The National Anti-Vivisection Society (NAVS) was the first organization formed to oppose animal experiments. Formed in 1875, with the help of Miss Frances Power Cobbe, the society helped motivate England's Parliament to pass the first national antivivisection law, the Cruelty to Animals Act, in 1876. The law required all experimenters to have permits, and it established guidelines for researchers. However, the Cruelty to Animals Act did not ban animal research entirely, and was therefore opposed by some antivivisection societies.

The first American antivivisection organization was formed in 1883, followed by the formation of the New England Anti-Vivisection (NEAVS) in 1895. The results obtained by American antivivisection, however, were far less impressive than those in England. The U.S. scientific community resisted most attempts to regulate the use of animals in research. Although U.S. antivivisection bills were frequently introduced in Congress beginning in the 1890s, none passed.

The antivivisection movement lost momentum after World War I, when the focus for animal humane societies shifted to establishing humane education programs and enforcing animal cruelty laws. During the 1950s, many humane societies established programs concerned with the humane care of laboratory animals, but such organizations were not publicly opposed to animal experimentation, and many humane societies compromised with medical establishments. For example, in 1952, the Metcalf-Hatch bill was passed in New York, which mandated pound seizure. Pound seizure required shelters to sell their cats and dogs to institutions that performed animal research. Andrew Rowan attributes the marginalization of the antivivisection movement to the success of well-organized, powerful lobbying efforts of medical researchers opposing antivivisection efforts, the lack of antivivisection resources, and the lack of credibility of antivivisection societies, which supported unorthodox medical theories such as repudiation of germ theory and vaccinations.

In the 1960s, the first efforts to raise funds for the development of alternative tests were successful, and groups formed with the specific aim of incorporating alternatives in accordance with the 3Rs philosophy. In 1962, the first European trust for moneys dedicated for the search of alternatives to animal testing appeared in the name of the Lawson Tait Trust. This British group formed the trust with the goal of working with medical researchers. Now known as the Human Research Trust, the fund continues to play an important role in funding the development of alternatives. The U.S. group, United Action for Animals, formed in 1967 and campaigned specifically for replacement alternatives. In 1969 the UK group, Fund to Replace Animals in Medical Research (FRAME), raised funds and launched a quarterly publication called *Alternatives to Laboratories Animals* (ATLA) to disseminate pertinent information to scientists concerning the search for alternatives. In 1973, the Lord Dowding Fund, established by the National Antivivisection Society, began dispensing research grants to individuals *not* holding licenses under the UK Cruelty to Animals Act for research projects that did *not* involve the use of live animals for experiments likely to lead to the alleviation of human or animal suffering.

Private companies also donated to the search for alternatives to animal testing. A report that the Draize eye irritancy test seemed amenable to alternative options spurred animal activist Henry Spira to begin his campaign against the cosmetic industry to deter use of the test. The Draize test was a good test to challenge, because the suffering rabbits endured from the test could be made visible to the public. Spira pursued a very public campaign by exposing people to graphic images of suffering animals and asking society "if another shampoo was worth blinding rabbits." The illustration in Figure 1 was run as a full-page ad in the *New York Times* in 1980. Cosmetics companies were vulnerable because they promoted beauty, which contrasted starkly with appalling images of rabbits undergoing a Draize test. Cosmetic corporations are also reliant upon the public's financial support, in the form of purchasing products, for their longevity. Spira did not insist that the cosmetics industry end all animal testing immediately. Instead he asked cosmetics companies such as Revlon to donate money (one-hundredth of one percent of Revlon's net profit = $170,000) to search for alternatives and, in the meantime,

to reduce the number of animals being used in Draize testing. Other companies followed suit. Avon and Bristol-Myers Squibb allocated one million dollars to John Hopkins University to establish the Center for Alternatives to Animal Testing (CAAT) in 1981, and in 1982 Colgate Palmolive provided $300,000 for the investigation of the chorio-allantonic membrane system (see CAAT web site: http://http://caat.jhsph.edu/).

Legislation and the 3Rs

European legislative initiatives to aid in the search for alternatives to animal testing appeared well before U.S. attempts. In 1971, Council of Europe Resolution 621 suggested that an alternatives database be established. The first grant to be given to a group to aid in the search for alternatives to animals in testing was given to FRAME in 1984.

In the United States, by 1981, there were seven bills on alternatives and/or regulation of animal experimentation pending in the Subcommittee on Science, Research, and Technology of the House Committee on Science and Technology. In a subcommittee hearing, a draft of the bill that incorporated aspects of the seven prior bills was generated and referred to the House Committee on Energy and Commerce. Eventually, in 1985, the provisions of this bill were incorporated into the Health Research Extension Act, requiring those awarded research monies by the National Institute of Health to consider the use of alternatives.

Validation of Alternative Tests

The major obstacle to reduction, refinement, and replacement of animals in toxicity testing is validation of alternative tests by appropriate regulatory agencies. Validation of *in vitro* tests required the establishment of new agencies to coordinate the processes of development, acceptance, and dissemination of information between scientists, regulatory agencies, and the public. Directive 86/609/EEC regulates the use of animals for experimental and other scientific purposes in the European Union. A Communication from the Commission to the Council and the Parliament in October 1991, pointing to a requirement in Directive 86/609/EEC for the protection of animals used for experimental and other scientific purposes, led to the establishment of the European Centre for the Validation of Alternative Methods (ECVAM). The U.S. National Institutes of Health Revitalization Act of 1993 established criteria for the validation and regulatory acceptance of alternative testing, and recommended the creation of a process to scientifically validate alternative methods so that they can be accepted for regulatory use. This Act prompted the creation of the Interagency Coordinating Committee on the Validation of Alternative Methods (ICCVAM) and its support center, The National Toxicology Program Interagency Center for the Evaluation of Alternative Toxicological Methods (NICEATM). ICCVAM uses information from toxicological test methods to support human health or environmental risk assessments, and represents 14 different U.S. regulatory agencies. The recommendations regarding the usefulness of test methods provided by ICCVAM allows regulatory agencies to assess the risks of various test methods and make regulatory decisions.

Validation of alternative tests has proved to be an obstacle in the search for alternatives. Animal testing has long been the standard method companies utilize to test the safety of their products, and regulatory agencies have accepted animal tests as valid. Regulatory agencies have to be convinced to accept the validity of newer, alternative methods. There are many obstacles to convincing regulatory agencies, such as the Consumer Product Safety Commission, the Environmental Protection Agency, the Occupational Health and Safety Association, and the U.S. Department of Agriculture, to accept new methods of testing. Some obstacles include tradition, prior regulations, lack of validated *in vitro* methods, lack of a process to determine validity, and resistance from the biomedical community. Understanding the mechanisms by which *in vitro* tests work takes effort and training by those evaluating them. There is a certain comfort level in familiar tests that have always satisfied regulations.

There are a few alternative tests that are validated by the relatively newly formed ICCVAM, and therefore are accepted by regulatory agencies as qualified substitutions for traditional testing, including the Local Lymph Node Assay and Corrositex. The process of validation by ICCVAM is seen as a major success for the alternatives movement, because validating alternatives has been a complicated process. Validation of tests for toxicity will aid in the European drive to ban animal testing for cosmetics. There is currently a ban currently on finished cosmetic products tested on animals in the European Union. A future aim is to ban animal testing of cosmetic product ingredients, effective September 2009.

Validation of alternative methods will need to extend globally as the United States develops, imports, and exports more products. ICCVAM has a firm relationship with ECVAM, and both groups will be instrumental in the validation of alternative processes. In the last few weeks of 2000, the ICCVAM Authorization Act was passed, which firmly established the organization's role in validating alternative methods in the future. Another success for the alternatives movement was the acceptance of an animal-friendly approach, Test Smart, toward the numerous studies that will be conducted in the future by various agencies to determine the hazardous potential of 2,200 U.S.-produced chemicals, namely, the High Production Volume Chemical Challenge. A future endeavor of ICCVAM will be to pursue alternatives to animal tests used to assess the toxins contained in popular anti-wrinkle treatments such as botox.

Replacement, refinement, and reduction will continue in the United States and Europe, decreasing pain and distress for laboratory animals. Total replacement of animal testing, however, will not take place in the near future. Some animal testing seems necessary at this point in time in order for regulatory agencies to fulfill their responsibility to the public to provide safe consumer products.

See also Alternatives to Animal Experiments: Reduction, Refinement, and Replacement

Further Reading

Altweb, the Alternatives to Animal Testing Web site. http://altweb.jhsph.edu/about.htm.

Frazier, J. M. (1992). *In vitro toxicity testing: Applications to safety testing.* New York: Marcel Dekker.

John Hopkins University Center for Alternatives to Animal Testing Web site: http://caat.jhsph.edu/.

Rowan, A. N. (1984). *Of mice, models and men: A critical evaluation of animal research.* Albany: State University of New York Press.

Rowan, A. N., & Loew, F. M., with Weer, J. (1995). *The animal research controversy: Protest, process and public policy. An analysis of strategic issues.* N. Grafton, MA: Tufts Center for Animals and Public Policy, School of Veterinary Medicine.

Rudacille, D. (2000). *The scalpel and the butterfly: The war between animal research and animal protection.* New York: Farrar, Straus and Giroux.

Singer, P. (1998). *Ethics into action: Henry Spira and the animal rights movement.* Lanham, Boulder, New York, Oxford: Bowman and Littlefield.

Nicole Cottam

TRAPPING, BEHAVIOR, AND WELFARE

For an activity that affects millions of wild animals each year, little is known about the full impact of trapping on individual animals, wildlife populations, and ecosystem health; reviews and extensive references can be found in Papouchis, 2004 and Fox, 2004a,b. Political forces and lobbies have greatly influenced trapping research, especially in the United States, where commercial fur trapping and predator control trapping are considered sacred cows. Many scientists working with animals believe that trap researchers and wildlife management agencies should establish research protocols that ensure that behavioral and welfare parameters are included in trap research, and standards should be developed that adequately measure all trapping related impacts. Traps that fail to meet these standards should be immediately prohibited. By resisting and undermining efforts to reduce adverse effects of trapping, wildlife management agencies and trap researchers open themselves to public and scientific criticism, and face increasing pressure to address these issues. Animal rights advocates believe that, ultimately, as society places greater value on wildlife and the humane treatment of all animals, use of traps and other management methods known to harm individual animals, wildlife populations, and ecosystem health will no longer be condoned.

More animals are trapped in the United States than in any other nation. Roughly three to five million animals are trapped and killed in the United States annually by commercial and recreational trappers (Fox, 2004a). Millions more are trapped and killed by wildlife damage and predator control trappers, researchers, and wildlife managers. Notably, there are few comprehensive assessments of the effects of trapping on animal welfare, behavior and physiology, and wildlife population dynamics.

The paucity of research on the effects of trapping on animal behavior and welfare reflects fundamental flaws and political biases in current trap-testing programs and the development of national and international mammal trap standards (Fox, 2004b). For example, the U.S. government is currently conducting a national Best Management Practices (BMP) trap-testing program to test leghold traps and other restraining traps on animals commonly trapped by recreational and commercial fur trappers. The BMP program was implemented as a result of pressure from the European Union to prohibit use of leghold traps in those countries that still allow their use. Instead of banning leghold traps, however, the U.S. government threatened trade reprisals if the fur ban was implemented, and instead agreed to conduct a national trap-testing program of traditional restraining traps. According to Tom Krause, editor of *American Trapper* magazine, one of the stated goals of the BMP program is to "maintain public acceptance" of trapping (Fox, 2004a). However, while injury rates, capture efficiency, and selectivity are part of the testing protocols, behavioral and overall physiological analyses are not. Previous studies that have considered the behavioral and physiological responses of animals caught in traps have shown that the trauma of being caught in a trap can alter the behavior of released animals, reduce survival rates, and disrupt the social dynamics of territorial species. That behavioral and physiological assessments are not part of the BMP trap testing protocols suggests that trapping proponents are unwilling to conduct comprehensive evaluations of traditional trapping devices for fear that such information could challenge the status quo and require that wildlife management agencies question the appropriateness of certain trap types and trapping practices.

Assessing the Impacts of Trapping

Research on trapping in the United States has focused primarily on trap injury rates, selectivity, and efficiency. In an effort to standardize the assessment of the injuries caused by body-gripping traps, several injury or trauma scales have been developed to quantify trap-induced injuries in restraining traps, and time to unconsciousness and death in killing-traps (Colleran et al., 2004). Physical trauma, however, is not the only measurement of trap impact. Psychological distress, fear, physiological stress, and pain can also be observed and assessed in trapped animals through behavioral analyses and stress-related hormonal and blood-cell measurements. However, at present no scoring system for restraining traps integrates physical injuries with behavioral and physiological responses (Proulx, 1999). Without such analyses, no comprehensive evaluation of the full impact of trapping on animals or the dynamics of wildlife populations can be made.

By intentionally underestimating the adverse effects of traps on animals, use of inhumane traps can be more easily justified by those with a vested interest in ensuring their continued use. The steel-jaw leghold trap, a device condemned as inhumane by the American Veterinary Medical Association, the American Animal Hospital Association, the World Veterinary Association, and the National Animal Control Association, is still one of the most widely used traps in the United States today. Leghold traps can cause severe swelling, lacerations, joint dislocations, fractures, damage to teeth and gums, limb amputation, and death (Papouchis, 2004). Many injuries result from the animal's struggle to escape, while others are incurred from the clamping force of the trap's metal jaws on the animal's limb. Unpadded steel leghold traps have been shown to cause significant injuries to a number of commonly trapped species, and generally fail to meet basic trap standards with regard to injuries.

These traps are widely used by the U.S. government in its federal predator control programs. While more than 80 countries have banned the controversial device, leghold traps remain legal in most U.S. states and public land systems. In 1995, the European Union banned the use of leghold traps in member states and sought to bar the import of furs from countries still using these traps. The United States, one of the world's largest fur producing and consuming nations, continues to defend commercial fur trapping and the use of the leghold trap, and even threatened the EU with a trade war if it prohibited the importation of fur from countries

allowing the use of leghold traps (Fox, 2004b).

In addition to leghold devices, kill-traps are commonly used by wildlife managers and commercial fur trappers throughout North America. Kill-traps, also called rotating-jaw traps, have been shown to cause extreme trauma, pain, and stress to trapped animals. Conibear traps and other common models of kill-traps may not cause instant death, because of the numerous variables needed to produce a killing blow to the neck or head, that is, a correct-sized animal entering the trap at the correct angle and speed.

While few studies of kill-traps conducted in the United States have included comprehensive trap impact assessments, several studies conducted outside the United States have analyzed the behavioral, physiological, pathological, and/or clinical responses of trapped semi-aquatic mammals in drowning sets. Most of these studies have been conducted in Canada and other countries, where trap researchers are often more independent and less influenced by political lobbies than in the United States. Killing traps employed underwater reduces their efficiency, so that when the strike is of insufficient strength or improperly placed to kill the animal, they act as restraint devices, and death is caused by drowning. Leghold and submarine traps act by restraining animals underwater until they drown. Most semi-aquatic animals, including mink, muskrat, and beaver, are adapted to diving by means of special oxygen-conservation mechanisms. The experience of drowning in a trap must be extremely terrifying; animals have displayed intense and violent struggling, and death was found to take up to four minutes for mink, nine minutes for muskrat, and 10 to 13 minutes for beaver (Gilbert and Gofton,

1982). Mink have been shown to struggle frantically prior to loss of consciousness, an indication of extreme trauma. Because most animals trapped in aquatic sets struggle for more than three minutes before losing consciousness, Proulx (1999) concluded that they did not meet basic trap standards and could therefore not be considered humane.

Capture Myopathy and Post-Release Survival

The survivability and fitness of trap-injured wildlife remains largely unknown because of the lack of research in this area. However, several published reports document long-term adverse effects of capture and handling in carnivore species including black bears, grizzly bears, and otters, as well as reduced post-release survival as a result of trapping related injuries (Hartup et al., 1999; Powell and Proulx, 2003, Papouchis. 2004, Lossa et al. 2007; Cattet et al. 2008). Restraint in a trap can cause psychological stress or fear for an animal, as well as physical and physiological damage, including various forms of capture myopathy, a stress-induced condition in wild animals that frequently occurs following prolonged capture or chase also called capture myopathy (Hartup et al., 1999; Cattet et al., 2008) and can disrupt behavior and the social dynamics of territorial species (Banci and Proulx, 1999). Hornocker and Hash (1981) suggest that intensive trapping contributes to behavioral instability and home range overlap among resident adults.

Carnivores released with internal trap injuries to feet, limbs, teeth, or other body parts may be so severely injured that they are unable to survive in the wild due to their physiology and the methods by which they obtain food (Van Ballenberghe, 1984, p.1428). Tooth damage from biting on the trap and claw loss may also affect a carnivore's ability to catch wild prey (Lossa et al., 2007). Restricted blood flow to the limbs caused by leghold traps can lead to gangrene and subsequent reduced survival if a trapped animal is released without an examination of internal injuries and subsequent rehabilitation.

A recent study on the long-term effects of trapping and capture on bear found that "[s]ignificant capture-related effects might go undetected, providing a false sense of the welfare of released animals" (Cattet et al., 2008: p. 973). In measuring blood serum levels to assess muscle injury in association with different methods of capture, the authors found that both grizzly and black bear suffered significant physiological damage, including capture myopathy. The proximate cause for capture myopathy is likely a combination of fear and anxiety accompanied by muscle exertion. Fear is the single most important factor in capture myopathy. Wild animals frequently die of capture myopathy, but death may occur hours, days, or weeks after capture and release. Cattet et al. (2008) showed that injuries were most severe as a result of being captured in leghold restraining devices. The authors emphasized that such injuries may not be detectable to the naked eye, and cautioned all wildlife researchers who capture wildlife with traps to seriously consider the long-term effects of trapping wildlife, regardless of species.

It seems plausible that different species also will respond similarly when faced with similar stressors. This possibility should at the very least challenge persons capturing wild animals to evaluate their capture procedures and research results very carefully. (Cattet et al., 2008: pp. 986–987).

Minimizing Impacts of Trapping in Wildlife Research

Whether trapping animals for scientific research, relocation, or reintroduction programs, wildlife researchers and managers require state-of-the-art, humane live traps. They need to know, for example, if a particular trap type may negatively alter an animal's behavior after it is released. Powell and Proulx (2003) argue that researchers should choose traps that minimize pain, stress, and discomfort, if for no other reason than to minimize the effect on the behavior and survival of animals, which ultimately affects research results.

Non-target animals trapped in leghold traps and then released may be so severely injured that they are unable to survive in the wild. Redig (1981) reported that 21 percent of the bald eagles admitted to the University of Minnesota Raptor Research and Rehabilitation Program over an eight-year period had been caught in leghold traps. Of these, 64 percent had sustained injuries that proved fatal. Oftentimes, trap-related injuries may be internal and therefore less readily apparent. Furthermore, the somatic and psychological stress to wild animals that can result from trapping can suppress their immune systems and significantly compromise their post release recovery (Jordan, 2001).

As animal ethologists and ethicists continue to demonstrate the cognitive, emotional, and behavioral similarities between humans and other animals (Fox, 2001), it will become increasingly difficult to justify continued testing and use of traps known to inflict fear, pain, and

suffering on wildlife. Ideally, in the field of wildlife research, trapping will be replaced with less invasive methods that preclude the need for trapping. Track plates, hair traps, remotely triggered cameras, and DNA hair testing offer non-invasive alternatives to trapping. When trapping is necessary, researchers should ensure that traps minimize physical injury as well as behavioral and physiological stress. Researchers must also be aware that when they conduct what appears to be benign, least-invasive research that involves trapping, there may be post-release impacts that affect individual animal(s), and ultimately their research results (Powell and Proulx, 2003; Cattet et al., 2008).

See also Predator Control and Ethics; Wildlife Abuse; Wildlife Services

Further Reading

Banci, V. and Proulx, G. 1999. Resiliency of furbearers to trapping in Canada. In G. Proulx, ed. *Mammal trapping*, 1–46. Sherwood Park, Alberta: Alpha Wildlife Research and Management Ltd.

Cattet, M., Boulanger, J., Stenhouse, G., Powell, R. A., and Reynolds-Hogland, M. J. 2008. An evaluation of long-term capture effects in Ursids: Implications for wildlife welfare and research. *Journal of Mammalogy* 89:973–990.

Colleran, E., Papouchis, C., Hofve, J., and Fox, C. 2004. The use of injury scales in the assessment of trap-related injuries. Chapter 5 in C. H. Fox and C.M. Papouchis, eds., *Cull of the wild: A contemporary analysis of wildlife trapping in the United States.* Sacramento, CA: Animal Protection Institute.

Fox, C. H. 2004a. The status of fur trapping: An historical overview. Chapter 1 in C.H. Fox, and C.M. Papouchis, eds., *Cull of the wild: A contemporary analysis of wildlife trapping in the United States.* Sacramento, CA: Animal Protection Institute.

Fox, C. H. 2004b. The development of international trapping standards. Chapter 4 in C.H. Fox, and C.M. Papouchis, eds. *Cull of the wild: A contemporary analysis of wildlife trapping in the United States.* Sacramento, CA: Animal Protection Institute.

Fox, M. W. 2001. *Bringing life to ethics: Global bioethics for a humane society.* Albany: New York State University Press.

Gilbert, F. F., and Gofton, N. 1982. Terminal dives in mink, muskrat and beaver. *Physiology & Behavior* 28:835–840.

Hartup, B. K., Kollias, G. V. et al. 1999. Capture myopathy in translocated river otters from New York. *Journal of Wildlife Diseases* 35:542–547.

Hornocker, M.G., and Hash, H.S. 1981. Ecology of the wolverine in northwestern Montana. *Canadian Journal of Zoology* 59:1286–1301.

Lossa, G., Soulsbury, C. D., Harris, S. 2007. Mammal trapping: A review of animal welfare standards of killing and restraining traps. *Animal Welfare* 16:335–352.

Papouchis, C.M. 2004. Trapping: A review of the scientific literature. Chapter 6 in C.H. Fox, and C.M. Papouchis, eds., *Cull of the wild: A contemporary analysis of wildlife trapping in the United States.* Sacramento, CA: Animal Protection Institute.

Powell, R. A., and Proulx, G. 2003. Trapping and marking terrestrial mammals for research: Integrating ethics, performance criteria, techniques, and common Sense. *ILAR Journal* 44:259–276.

Proulx, G. 1999. Review of current mammal trap technology in North America. In G. Proulx, ed. *Mammal trapping*, 1–46. Sherwood Park, Alberta: Alpha Wildlife Research and Management Ltd.

Redig, P. 1981. Significance of trap-induced injuries to bald eagles. In *Eagle Valley environmental technical report* BED 8145–53. St. Paul: University of Minnesota.

Van Ballenberghe, V. 1984. Injuries to wolves sustained during live-capture. *Journal of Wildlife Management* 48:1425–1429.

Camilla H. Fox

U

URBAN WILDLIFE

The 21st century continues to bear witness to the relentless growth of human populations, along with the cities that have become our principal habitation. In 2008, an unheralded boundary was crossed when more humans globally came to live in cities than outside them. The transition from humans living in small social groups to a massive, urban, cosmopolitan populace has taken place in less than one percent of the time we have been identifiable as a species. We are, it seems, villagers confronting the challenges of big city life, and seemingly poorly equipped to deal with problems ranging from obvious social discord to our near-suicidal mistreatment of the natural world. Proponents of concepts such as biophilia and nature deficit disorder tell us that one of the more important consequences of urban life is that we are also becoming increasingly alienated from the natural world, in ways that can produce a lack of empathy, concern, and connection to other living things, humans included.

Abetting a moral and personal alienation from nature is the ever-growing burden of the urban ecological footprint. Cities not only dominate, directly and indirectly, the global ecology, they are also important ecosystems in their own right (Hadidian & Smith, 2001). The urban environment is characterized by both a high degree of landscape heterogeneity and a rapid change of landforms, primarily as a result of constant development activities. Wild animals that have long been urban residents, for example, squirrels, must cope with these, and species that are colonizing urban habitats, for example coyotes, must adapt. An ever growing and expanding zone of human-animal contacts characterizes city and suburb, wherein the majority of interactions are undoubtedly positive, while the more noticed, discussed, and attended to are undoubtedly negative. Any wild animal living in the urban environment can be, and certainly at one time or more has been, labeled a pest. Historically, wildlife authorities in North America have conducted pest control by using traditional approaches—hunting, trapping, and poisoning being preferred. Derived largely from an agricultural context, such practices have been deemed necessary as economic measures, but are harshly criticized for their moral presuppositions (Fox and Papouchis, 2004; Robinson, 2005).

Controversy and polarization arise from differing ethics of how we ought to relate to and live with nonhuman animals. Both specific practices, as well as the principles underlying the treatment of wild animals in the urban context, are rightly being questioned. Traditional wildlife control practices, such as the drowning of nuisance animals that have been caught in traps, deserve obvious criticism, because science informs us that the method itself is inhumane (Ludders et al., 1999). Less obviously, but equally important, ethics tells us that the principle of conducting lethal control should be criticized as inhumane when it fails to establish a justifiable rationale for removing an animal in the first place.

The variety of wildlife conflicts within the urban or humanized environment is extremely diverse, the context always challenging. Beaver build dams in the floodplains from which they were long ago trapped out, but in which they are naturally appropriate occupants, while humans are not. Coyotes, at home in a variety of landscapes, including highly populated urban centers, prey on cats and dogs, bringing urbanites face to face with the realities of living with predators. Deer, after decades of propagation and habitat management to increase their numbers, become so abundant that it is claimed they are destroying entire forest ecosystems, not to mention Mrs. Smith's prized roses. Human culpability, in creating the fragmented landscapes of suburbia with edge habitat that promotes high deer and coyote densities, or occupying floodplains better left alone, typically goes unmentioned as either a causative or correctable factor when discussing how to address wildlife problems. Humans seem to always be the last to assign their own responsibility for conflicts with wild animals, while being the first to impose solutions on them that disregard the natural processes and balances that will provide the most lasting, environmentally responsible, and humane alternatives available. The ethics of such situations are being raised, whether invited and recognized, or not (Lynn, 2005b).

In many cities around the world numerous different species of urban wildlife can be seen. Here, a squirrel perches on top of a trash can looking for food. (Photos.com)

Contemporary human-wildlife conflicts have scientific, political, and moral dimensions that are not well addressed by traditional approaches in wildlife management. There is a critical need for a dialogue about these shortcomings to coincide with a growing recognition of the need for a dialogue on ethics in all fields of wildlife, as well as biodiversity management (Eggleston et al., 2003; Minter & Collins, 2005). Addressing this need through urban wildlife can provide a bridge to the social and biological dimensions of wildlife issues that addresses the real and practical concerns and needs of urbanites who, as the demographic dominating the political environment, are the decision-making majority. The superstructure of this bridge is ethics, to help guide and inform the wildlife profession and the policies by which it operates. A practical ethic guiding our response to human-animal conflicts is both warranted and necessary.

Practical ethics is a very old paradigm of ethics that focuses on the full range of moral values that inform our lives, such as what is right, good, just, and caring. It looks to moral theories and concepts, as well as individual cases and their empirical realities, for insight in making ethical decisions. By honoring the insights coming from many moral theories, practical ethics creates a reservoir of concepts to triangulate on the best understanding of a moral problem. Because it is especially attentive to concrete and specific cases, practical ethics provides more fitting and contextual guidance for our thoughts and actions. Altogether, we term this a situated moral understanding (Lynn, 2005a). This is the approach in ethics that should inform urban wildlife policy and management, as well as articulate a vision of our place in a mixed community of people and animals.

It is fairly straightforward, not to mention scientifically defensible, that such principles as justification, proof of benefits, necessity, feasibility, minimization of harm, and humaneness all be included in an ethic for urban wildlife. But what about questions regarding intrinsic value? Here, the idea is that animals have value in and of themselves, independent of their extrinsic value, that is, what uses someone might have for them. The extrinsic value of animals is dominant in public discourse, but it may not be dominant in the public mind. Consider for example the many wildlife managers who treat wildlife as fungible units of ecosystems, but hold entirely different feelings of love and duty to the individual dog or cat they have brought into their lives. Some animals are pets, and some are pests, it seems, without much thought being given to what all animals are in the first place, which is individuals with their own unique life histories, personalities, interests, and accomplishments. Attending to questions of intrinsic value can help us sort out these tacit understandings. We cannot make wise decisions about urban wildlife management without acknowledging the intrinsic and extrinsic values at play.

Ethics, as it relates to urban wildlife, offers hope that a reconnection with nature can be made in a way that helps revitalize what is best about our relationship with the natural world. Where studies have been conducted to examine the attitudes and beliefs of urbanites toward wildlife a great deal of empathy and concern has been found (Kellert, 1997). These qualities, apparently fundamental to our species, seem to be amplified by the urban experience, perhaps because urbanites no longer rely directly on wild species for subsistence needs, or because urban living affords us the opportunity to view animals as subjects and members of a mixed community, not simply as objects for human use. We are becoming increasingly aware that our urban environments are not simply environmental wastelands, but are in their own way thriving and complex ecological communities. That is to say, they are at least potentially so, assuming we take prudent care of our urban spaces and all their inhabitants. Acknowledging that urbanites are a part of, and not apart from, the natural world will go a long way toward a needed rebalancing of values. Such acknowledgement also allows us to view and experience our relationship with urban wildlife in new and different ways. For example, instead of seeing the coyote family inhabiting the green spaces in our community as frightening and alien predators, we can see them for what they truly are: natural and vital components of a dynamic and vibrant ecosystem deserving of respect and understanding. Such changes of mind would imply not only recognition of an obligation to work with natural processes, but to give urban communities, human and nonhuman both, the moral consideration to which they are due.

Further Reading

Eggleston, J. E., Rixecker, S. S., and Hickling, G.J. 2003. The role of ethics in the management of New Zealand's wild mammals. *New Zealand Journal of Zoology* 30: 361–376.

Fox, C.H., and Papouchis, C.M., eds. 2004. *Cull of the wild: A contemporary analysis of wildlife trapping in the United States.* Sacramento, CA: Animal Protection Institute.

Hadidian, J., and Smith, S. 2001. Urban Wildlife. The State of the Animals 2001. In D. J. Salem and A. N. Rowan, eds., 165–182. Washington, DC: Humane Society Press.

Kellert, S. R. 1997. *Kinship to mastery.* Washington, DC: Island Press.

Ludders, John W., Schmidt, R. H., Dein, F. J., and Klein, P. N. 1999. Drowning is not euthanasia. *Wildlife Society Bulletin* 27: 666–670.

Lynn, W. S. 2005a. Between science and ethics: What science and the scientific method can and cannot contribute to conservation and sustainability. In D. Lavigne, ed., *Gaining ground: In pursuit of ecological sustainability*, Limerick, Ireland: International Fund for Animal Welfare and the University of Limerick.

Lynn, W. S. 2005b. Finding common ground in a landscape of deer and people. *Chicago Wilderness Magazine* 8: 12–15.

Minter, B. A., and Collins, J. P. 2005. Ecological ethics: Building a new tool kits for ecologists and biodiversity managers. *Conservation Biology*19(6): 1803–1812.

Robinson, M. J. 2005. *Predatory bureaucracy.* Denver: University Press of Colorado.

John Hadidian
Camilla H. Fox
William S. Lynn

UTILITARIANISM

The term utilitarianism is often used to describe any ethical stance that judges whether an action is right or wrong by considering whether the consequences of the action are good or bad. In this broad sense of the term, utilitarianism is equivalent to what is sometimes called consequentialism. It is opposed to rule-based ethical systems, according to which an action is right if it is in conformity with moral rules and wrong if it is in violation of these rules, irrespective of its consequences.

An example may help to make this more concrete. Is it wrong to break a promise? Those who base ethics on a set of moral rules and include keeping promises among these rules would say that it is. On the other hand, a utilitarian would ask: What are the consequences of keeping the promise, and what are the consequences of breaking it? In some situations the good consequences achieved by breaking the promise would clearly outweigh the consequences of keeping it.

This gives rise to a further question: What kind of consequences are relevant? According to the classic version of utilitarianism, first put forward in a systematic form by the English philosopher and reformer Jeremy Bentham, what ultimately matters is pleasure or pain. Thus classic utilitarians judge acts right if they lead to a greater surplus of pleasure over pain than any other act that the agent could have done. Bentham included in his calculations the pleasures and pains of all sentient beings. In rejecting attempts to exclude animals from moral consideration, as virtually everyone did in his day, Bentham wrote: "The question is not, Can they reason? nor Can they talk? but, Can they suffer?"

Nowadays there are many who continue to call themselves utilitarians who, while still holding that the rightness of an act depends on its consequences, think that the idea that pleasure and pain are the only consequences that should count is too narrow. They argue that some people may prefer other goals. For example, a writer might be able to achieve a life of luxury by working for an advertising agency, but may prefer the long and lonely work of writing a serious novel. Bentham could claim that she thinks that she will get more lasting pleasure from writing the novel, but it is also possible that she simply considers writing something of lasting literary value to be more worthwhile, irrespective of how much pleasure it is likely to add to her life and the lives of others, than writing advertising copy. Considering such cases has led to the development of a form of utilitarianism known as preference utilitarianism. Preference utilitarians judge acts to be right or wrong by attempting to weigh up whether the act is likely to satisfy more preferences than it frustrates, taking into account the intensity of the various preferences affected. In this view, too, animals will count as long as they are capable of having preferences, and an animal who can feel pain or distress can be presumed to have a preference to escape that feeling.

Utilitarianism has great appeal because of its simplicity, and because it avoids many of the problems of other approaches to ethics, which can require you to obey a rule or follow a principle, even though to do so will have worse consequences than breaking the rule or not following the principle. On the other hand, this very flexibility may also mean that the utilitarian reaches conclusions that are at odds with conventional moral beliefs. Hence one of the most popular ways of attempting to refute utilitarianism is to show that it can, in appropriate circumstances, real or imaginary, lead to the conclusion that it is right to break promises, tell lies, betray one's friends, and even kill dear old Aunt Bertha in order to give her money to a worthy cause. To this some utilitarians respond by retreating to some form of a two-level view of morality, based on the idea that at the level of everyday morality we should obey some relatively simple rules that will lead us to do what has the best consequences in most cases, while in some special circumstances, and when assessing the rules themselves, we should think more critically about what will lead to the best consequences. Other, more tough-minded utilitarians say that if our common moral intuitions clash with our carefully checked calculations of what will bring about the best consequences, then so much the worse for our common moral intuitions.

Further Reading

Bentham, Jeremy. 1789, 1948. *An introduction to the principles of morals and legislation.* New York: Hafner.

Hare, R. M. 1981. *Moral thinking.* Oxford: Clarendon Press.

Mill, John Stuart. 1863, 1960. *Utilitarianism.* London: Dent.

Sidgwick, Henry. 1907. *The methods of ethics,* 7th ed. London: Macmillan.

Sinnott-Armstrong, Walter. 2007. Consequentialism. *The Stanford encyclopedia of philosophy (Spring 2007 Edition),* Edward N. Zalta, ed.: http://plato.stanford.edu/archives/spr2007/entries/consequentialism/.

Smart, J.J.C., and Williams, Bernard. 1973. *Utilitarianism, for and against.* Cambridge: Cambridge University Press.

Peter Singer

UTILITARIANISM AND ASSESSMENT OF ANIMAL EXPERIMENTATION

Many defenders of animal experimentation maintain that the practice is justified because of its enormous benefits to human beings. While it is true that animals die and suffer, the defenders say that is morally insignificant when compared with experimentation's benefits. It is important to notice that this utilitarian defense of the practice assumes that animals have moral worth. Unless animals had moral worth, it would make no sense to say that we must include their deaths and suffering on the scales. If they are without value, or their value were morally negligible, the impact of experimentation on them would never enter the moral equation.

Utilitarians can judge conflicts between members of different species by saying that the moral worth of an action would be the product of the moral worth of the creature that suffers, the seriousness of the wrong it suffers, and the number of such creatures that suffer.

Many defenders of research often speak as if utilitarian cost-benefit calculation is easy. Frequently they cast the public debate as if the choice to pursue or forbid animal experimentation were the choice between your baby or your dog. However, this way of framing the question can be grossly misleading. The choice has not been, nor will it ever be, between your baby and your dog. Single experiments on single animals don't confirm biomedical hypotheses. Only a series of related experiments can confirm such hypotheses. Animal experiments are part of a scientific framework. Thus, we must change the moral question to: Is the practice of animal experimentation sufficiently beneficial to justify its costs?

Whatever the precise details of this utilitarian analysis, animal experimentation clashes with the moral codes against doing evil to promote some good, and inflicting suffering on one creature of moral value to benefit some other creature of moral value. That is, we do an evil to animals to provide good for humans. Moreover, the evil we do, inflicting suffering on animals, is definite, while the good we promote, preventing the suffering of humans, is only possible. Additionally, the creatures that suffer will not be the ones that benefit from that suffering. Dogs pay the cost of experimentation; humans reap the benefits.

The force of these codes of conduct is deep in our ordinary morality. Although undergoing a painful bone marrow transplant to save the life of a stranger is noble, we think that requiring a person to undergo that procedure would be wrong. Even if we think people should be required to make some sacrifices for other members of their species, most of us think that requiring the ultimate sacrifice would be inappropriate. For instance, even if we assume that nonhuman animals have less moral worth than humans, most people think there are some sacrifices animals should not have to make.

Abandoning these codes of conduct, though, would mean that nonconsensual moral experiments on humans could be justified if the benefits to humans were substantial enough. It would also require abandoning the idea of the moral separateness of creatures, a view central to all Western concepts of morality. For instance, virtually everyone would be opposed to requiring people to give up one of their kidneys to save someone else's life. Thus, even if we assume that animals have less value than humans, this latter imbalance means that researchers must show staggering benefits of experimentation to justify the practice morally.

Moreover, when determining the gains relative to the cost of animal experimentation, we must include not only the costs to animals, which are direct and substantial, but also the costs to humans and animals of misleading experiments. For instance, we know that animal experiments misled us about the dangers of smoking. By the early 1960s, researchers found a strong correlation between lung cancer and smoking. However, since efforts to induce lung cancer in nonhuman animal models had failed, the government delayed acting.

Furthermore, since we should include possible benefits on the scales, we must also include possible costs. For example, some researchers have speculated that AIDS was transferred to the human population through an inadequately screened polio vaccine given to 250,000 Africans in the late 1950s. Although the hypothesis is likely false, something like it might be true. We know, for instance, that one simian virus (SV_{40}) entered the human population through inadequately screened vaccine. In fact, several hundred thousand people have been exposed to SV_{40} through vaccines and, in *in vitro* tests, the virus causes normal human cells to mutate into cancerous cells. Therefore it is difficult to know how researchers could possibly claim that there would be no substantial ill-effects of future animal experimentation. These possible ill-effects must be counted.

Finally, and perhaps most important, the moral calculation cannot simply look at the benefits of animal experimentation. It must look instead at the benefits that only animal research could produce. To determine this utility, the role that medical intervention played in lengthening life and improving health, the contribution of animal experimentation to medical intervention, and the benefits of animal experimentation relative to those of nonanimal research programs have to be ascertained. Since even the American Medical Association recognizes the value of non-animal research programs, then what goes on the moral scales should not be all the supposed benefits of animal experimentation, but only the increase in benefits compared with alternative programs. Since we do not know what these other programs would have yielded, determining the increase in benefits would be impossible to establish.

Further Reading

Bailar, III, J., and Smith, E. 1986. Progress against cancer? *New England Journal of Medicine,* 314, 1226–31.

Brinkley, J. 1993. Animal tests as risk clues: The best data may fall short. *New York Times National,* (23 March) C1, C20–1.

Cohen, Carl. 1990. Animal experimentation defended. In S. Garattini and D.W. van Bekkum, eds., *The importance of animal experimentation for safety and biomedical research.* Dordrecht: Kluwer Academic Publishers

Elswood, B. F., and Stricker, R. B. 1993. Polio vaccines and the origin of AIDS (letter to the editor). *Research in Virology,* 144, 175–177.

LaFollette, H., and Shanks, N. 1996. *Brute science: The dilemmas of animal experimentation.* London: Routledge.

McKinlay, J. B., and McKinlay, S. 1977. The questionable contribution of medical measures to the decline of mortality in the United States in the twentieth century. *Health and Society,* 55, 405–28.

Hugh LaFollette and
Niall Shanks

V

VEGANISM

Vegans (pronounced VEE-guns) are people who choose not to eat any animal products, including meat, eggs, dairy, honey, and gelatin. Vegans do not wear fur, leather, wool, down, or silk, or use cosmetics or household products that were tested on animals or contain ingredients that were derived from animals. Most vegans also do not support industries that feature captive and/or performing animals, including circuses, zoos, and aquaria.

The American Vegan Society (2006) defines veganism as "an advanced way of living in accordance with Reverence for Life, recognizing the rights of all living creatures, and extending to them the compassion, kindness, and justice exemplified in the Golden Rule."

The word vegan was derived from the word vegetarian in 1944 by Elsie Shrigley and Donald Watson, the founders of the UK Vegan Society. Shrigley and Watson were disillusioned that vegetarianism included the consumption of dairy products and eggs. They saw vegan as "the beginning and end of vegetarian," and used the first three and last two letters of vegetarian to coin the new term.

An April 2008 "Vegetarianism in America" survey conducted by Harris Interactive Service Bureau indicated that 7.3 million American adults are vegetarians; approximately one million of these are vegans. The poll also indicated that 11.9 million people are "definitely interested" in following a vegetarian-based diet in the future: http://www.prnewswire.com/cgi-bin/stories.pl?ACCT=104&STORY=/www/story/04-15-2008/0004792955&EDATE=.

A Mintel survey showed that U.S. sales of vegetarian and vegan food increased by 64 percent from 2000 to 2005, and that the vegetarian food market was forecast to grow to over $1.7 billion in sales by 2010. The increase is attributed to concerns about animal welfare, personal health, and/or the environment.

Ethical Reasons for Veganism

Approximately 27 billion cows, pigs, chickens, turkeys, and other animals are killed for food each year in the United States ("Chew on This," People for the Ethical Treatment of Animals. Retrieved March 13, 2007, http://www.goveg.com/feat/chewonthis/). Our modern factory farming system strives to produce the most meat, milk, and eggs as quickly and cheaply as possible and in the smallest amount of space possible.

Some people, such as Jewish Nobel Prize-winning author Isaac Bashevis Singer, have equated the treatment of animals in slaughterhouses with the treatment of humans during the Holocaust. Having fled Nazi Europe in 1935, Singer took a room above a slaughterhouse and watched as cows were prodded, kicked, and sworn at as they were herded down a ramp to their deaths. He proclaimed that "as long as human beings go on shedding the blood of animals, there will never be any peace" (Dujack, 2003).

There is evidence of cows, chickens, pigs, and other meat animals being raised in poor conditions, where they may be fed high-bulk food, such as grains, or substandard or inappropriate food. They are sometimes kept in very small spaces in order to raise as many animals as possible. Most disturbing, there is evidence that, at slaughterhouses, animals are not always humanely killed, as, for example, when stun-guns do not work. U.S. Department of Agriculture inspection records documented 14 humane slaughter violations at one processing plant, including finding hogs that "were walking and squealing after being stunned [with a stun gun] as many as four times" (Warrick, 2001). During slaughter, animals are hung upside-down and their throats are slit, sometimes while they're still conscious. Many are still alive while they're skinned, dismembered, or scalded in defeathering tanks.

A survey conducted in 2004 by the Social Responsibility Initiative at Ohio State University suggested that the majority of people, even those who are not vegetarians or vegans, are concerned about farmed animal welfare. Surveys were sent to 3,500 randomly selected Ohioans, and 56 percent responded. Of the respondents, 92 percent agreed or strongly agreed that it is important for farm animals to be well cared for, 85 percent indicated that the quality of animal lives is important, even though some animals are used for meat production, and 81 percent said that the wellbeing of farm animals is just as important as the well-being of companion animals: http://www.smallfarms.cornell.edu/pages/quarterly/archive/fall06/Fall_2006_Page_20.pdf.

According to Mintel's *Eggs and Egg Substitutes—U.S., June 2004,* consumers who are concerned about animal welfare choose to buy eggs from hens that are not raised in cages. Vegans also believe that it is wrong to use animals for their milk or eggs. It is estimated that each vegetarian saves more than 100 animals every year.

Health Reasons for Veganism

Animal products, particularly meat, eggs, and dairy, are generally high in saturated fat, cholesterol, and concentrated protein. Numerous studies have linked the consumption of certain animal products to serious illnesses, such as heart disease, strokes, diabetes, and breast, colon, prostate, stomach, esophageal, and pancreatic cancer.

Unlike animal products, plant-based foods are cholesterol free, generally low in fat, and high in fiber, complex carbohydrates, and other vital nutrients. Researchers from the University of Toronto have found that a plant-based diet rich in soy and soluble fiber can reduce cholesterol levels by as much as one-third (Fauber, 2003). According to David Jenkins, professor of nutrition and metabolism at the University of Toronto, "the evidence is pretty strong that vegans, who eat no animal products, have the best cardiovascular health profile and the lowest cholesterol levels" (Callahan, 2003).

Studies have shown that, on average, vegetarians and vegans are at least 10 percent leaner and live six to ten years longer than meat eaters. The ADA has reported that "vegetarians, especially vegans, often have weights that are closer to desirable weights than do non-vegetarians" (American Dietetic Association, 1993).

In *Dr. Spock's Baby and Child Care,* the late Dr. Benjamin Spock, an authority on child care, wrote, "Children who grow up getting their nutrition from plant foods rather than meats have a tremendous health advantage. They are less likely to develop weight problems, diabetes, high blood pressure, and some forms of cancer" (Spock, 1998).

According to the ADA and Dietitians of Canada, "well-planned vegan and other types of vegetarian diets are appropriate for all stages of the life cycle, including during pregnancy, lactation, infancy, childhood, and adolescence" (ADA Web site).

It is possible to get most vital nutrients from a vegan diet; however, because vitamin B-12 is primarily found in animal sources, vegans need to take a multivitamin or B-12 supplement to get ample B-12. Vitamin B-12 is also found in nutritional yeast and many fortified cereals and soy milks.

Environmental Reasons for Veganism

The process of turning cows, pigs, chickens, and turkeys into meat, pork, and poultry takes a toll on the environment. According to *E: The Environmental Magazine,* almost every aspect of animal agriculture, from grazing-related loss of cropland and open space, to the inefficiencies of feeding vast quantities of water and grain to cattle in a hungry world, to pollution from factory farms, can cause an environmental disaster with wide and sometimes catastrophic consequences (Motavalli, 2002).

The November, 2006, United Nations report *Livestock's Long Shadow* indicated that raising animals for food generates more greenhouse-gas emissions than all the cars, trucks, trains, ships, and planes in the world combined. The report concluded that the livestock sector is "one of the top two or three most significant contributors to the most serious environmental problems, at every scale from local to global," and that the livestock industry should be "a major policy focus when dealing with problems of land degradation, climate change and air pollution, water shortage and water pollution, and loss of biodiversity": http://www.nhbs.com/livestocks_long_shadow_tefno_150529.html.

When Gidon Eshel and Pamela Martin of the University of Chicago compared the amount of fossil fuel needed to cultivate and process various foods, including running machinery, providing food for animals, and irrigating crops, they found that the typical U.S. diet generates nearly 1.5 tons more carbon dioxide per person per year than a vegan diet with an equal number of calories (*New Scientist*, 2005).

The Environmental Protection Agency has reported that factory farms pollute our waterways extensively. Animals raised for food produce approximately 130 times as much excrement as the entire human population, 87,000 pounds per second (PETA Vegetarian Starter Kit).

Livestock waste emits ammonia, nitrous oxide, carbon dioxide, and other toxic chemicals into the atmosphere. A study by Duke University Medical Center showed that people living downwind of pig farms are more likely to suffer from tension, depression, fatigue, nausea, vomiting, headaches, shallow breathing, coughing, sleep disturbances, and loss of appetite than the general population (Schiffman et al., 1995).

Raising animals for food also requires massive amounts of water and land. It takes 2,500 gallons of water to produce a pound of meat, but only 60 gallons of water to produce a pound of wheat, and a meat-based diet requires more than 4,000 gallons of water per day, whereas a vegan diet requires only 300 gallons of water a day (Robbins, 1987). Approximately 3 1/4 acres of land are needed to produce food for a meat-eater; food for a vegan can be produced on just 1/6 of an acre of land: http://www.goveg.com/world Hunger-animalAgriculture.asp.

In the United States, animals are fed more than 70 percent of the corn, wheat, and other grains we grow (PETA). The world's cattle consume a quantity of food approximately equal to the caloric needs of 8.7 billion people; around 1.4 billion people could be fed with the grain and soybeans fed to U.S. cattle alone.

See also Religion—Veganism and the Bible

Further Reading

American Dietetic Association. (1993). Position of the American Dietetic Association: Vegetarian diets. Retrieved March 13, 2003 from http://www.fatfree.com/FAQ/ada-paper.

American Dietetic Association. *Vegetarian diets.* Retrieved March 23, 2005 from: http://www.eatright.org/cps/rde/xchg/ada/hs.xsl/advocacy_933_ENU_HTML.htm.

American Vegan Society. *What is Vegan?* Retrieved March 27, 2006 from http://www.americanvegan.org/vegan.htm.

ARAMARK, 2005. Vegan options more popular than ever on college campuses: ARAMARK focuses on meeting consumer needs in honor of Vegan World Day, June 21. http://www.aramark.com/PressReleaseDetailTemplate.aspx?PostingID=552&ChannelID=210.

Callahan, M. (2003). Inside veggie burgers. *Cooking Light,* June 2003, 74.

"Chew on this," People for the Ethical Treatment of Animals. Retrieved March 13, 2007, http://www.goveg.com/feat/chewonthis/.

Corliss, R. (2002). Should we all be vegetarians? Do you consider yourself a vegetarian? (July 15), *Time.* 160, (3), 48.

Dujack, S. R. (2003, April 21). Animals suffer a perpetual "Holocaust." *Los Angeles Times.*

EG Smith Collective. (2004). *Animal ingredients A to Z.* Oakland: AK Press.

Fauber, J. (2003, July 22). Ape diet shown to lower cholesterol. *Milwaukee Journal Sentinel,* A01.

Marcus, E. (2000). *Vegan: The new ethics of eating.* Ithaca: McBooks Press.

Motavalli, J. (2002). The case against meat. *E: The Environmental Magazine,* Vol. 13, 1, 26.

New Scientist. (2005, December 17), It's better to green your diet than your car, issue 2530, 19.

People for the Ethical Treatment of Animals. PETA Media Center-Vegetarian Fact sheets. Retrieved from http://www.peta.org/mc/factsheet_vegetarianism.asp. March 27, 2006

People for the Ethical Treatment of Animals. *Vegetarian Starter Kit* Retrieved from http://www.petaliterature.com/VEG297.pdf. March 27, 2006

Robbins, J. (1987). *Diet for a new America.* Walpole: Stillpoint Publishing.

Robbins, J. (2001). *The food revolution.* Berkeley: Conari Press.

Schiffman, S., Saitely Miller, E., Suggs, M., & Graham, B. (1995). The effect of environmental odors emanating from commercial swine operations on the mood of nearby residents. *Brain Research Bulletin,* *37*(4), 360–375.

Scully, M. (2002). *Dominion: The power of man, the suffering of animals.* New York: St. Martin's Press.

Singer, P. (1975). *Animal liberation.* New York: Avon Books.

Spock, B. (1998). *Dr. Spock's baby and child care,* 7th ed. New York: Simon & Schuster, Inc.

Stepaniak, J. (1998). *The vegan sourcebook.* Lincolnwood: Lowell House.

Stepaniak, J. (2000). *Being vegan.* Lincolnwood: Lowell House.

USA Today (2004, February 23). Choosing a meat-free option: http://www.usatoday.com/educate/et/ET04.06.2004.pdf

Warrick, J. (2001, April 10). They die piece by piece. In Overtaxed plants, humane treatment of cattle is often a battle lost. *The Washington Post*, A01I.

Heather Moore

VEGETARIANISM

Vegetarians are of two main types: those who include some animal products in their diet and those who do not. The former are usually called vegetarians and the latter vegans. Vegetarianism refers to these dietary regimes, but more importantly, to the belief system that supports vegetarian practice. Paradoxically, not all vegetarians subscribe to such a belief system. They may, for example, just not like the taste of meat, But most, especially vegans, do have an outlook that proscribes eating animals. Many people today, whether or not they are vegetarians, recognize that livestock production, especially by means of large-scale intensive or factory farming, causes the worst abuses of animals and is an extremely wasteful way of securing food.

There have been vegetarians in all eras of recorded history. Often their dietary choices have been regarded as either subversive or eccentric, but their voices, although in the minority, seldom go unheard by people of conscience. In what follows, a number of arguments that can and often do contribute to a vegetarian stance will be summarized.

1. *Health.* Whether a vegetarian diet is as healthy as or healthier than one including meat is a subject of much controversy. It may seem that good health is simply a matter of one's long-term self-interest, but some philosophers, notably Immanuel Kant, have argued that

A piglet declares, "No, I don't have any spare ribs!" in an advertisement supporting vegetarianism from People for the Ethical Treatment of Animals (PETA). PETA also advocates a vegan lifestyle—using no animal products, including milk and eggs. (PETA)

we have duties to ourselves, and others such as Aristotle have argued that we must always strive to attain a virtuous or morally praiseworthy kind of life. In both of these views, health, and thus a sound diet, would be a precondition of our being able to carry out any moral obligations, including duties to ourselves and acting virtuously, and is therefore itself a matter of moral concern. Persons to whom we have responsibilities likewise have a stake in our health, as does society, which has an interest in our being productive, non-burdensome members. If a vegetarian diet were healthier overall, as vegetarians contend, then it would be the one we should choose for both our own self-regard and our concern for others.

2. *Animal suffering and death.* There is no method of rearing food animals without pain and suffering, and how ever this is done, death is the animals' fate. Confinement, transportation, and slaughtering are the main sources of abuse in the process of extracting consumer products from animals. Factory farming, a widespread phenomenon of our time, maximizes the problems, and its cruelties are well documented. Utilitarians are typically concerned with promoting pleasure and other interests of sentient beings, and with reducing or eliminating pain, suffering, and other conditions that frustrate such beings' welfare. Vegetarian diets are conducive to realizing these desirable outcomes. Animal rights theorists hold that many animals are irreplaceable individuals who have morally significant interests and hence rights, including the right to live and not be caused gratuitous pain and suffering. From this view, even totally painless meat production that gave great pleasure to human eaters, and that might therefore satisfy utilitarian ethical demands, would still be unacceptable, because death is an ultimate harm to rights-bearers.

3. *Impartiality and moral wellbeing.* An impartial person who is well informed about animals understands that they have morally significant interests, such as staying alive and having a certain wellbeing, health, and contentment which can only be respected if we refrain from eating them. Using animals instrumentally for food violates the condition of impartiality and demonstrates speciesism.

4. *Ecological concerns.* Large-scale meat production by agribusiness causes great environmental depletion and degradation, including huge demands on water and fossil fuel supplies, deforestation, desertification, and loss of wild animal habitat, an infusion of greenhouse gases, and excrement influx into waterways. A worldwide shift to vegetarian diets is seen as a way to lessen or eliminate these impacts, and as a necessity in view of the unsustainability of the meat economy.

5. *World hunger and social justice.* Food animal production that relies on feeding animals in feedlots rather than letting them naturally forage is extremely wasteful, yielding far less protein output than the protein input required to make it work. Vegetarian diets would aid in freeing up resources to feed the world's hungry by undermining the artificially created economy of scarcity.

6. *Interconnected forms of oppression.* Some ecofeminists have argued that various forms of domination, oppression, and exploitation are causally and conceptually intertwined. Those who are more powerful than others tend to exercise control over them, see them as inferior, and treat them as merely serving their own interests. A vegetarian way of life can contribute to breaking out of this traditional pattern by changing the dynamics

of food production, distribution, and consumption.

7. *Universal compassion and kinship.* Evolutionary considerations of biological kinship reinforce the idea that humans should exercise compassion toward other animals. Vegetarianism accords with a compassionate approach to life.

8. *Universal non-violence (ahimsa).* Mohandas Gandhi taught that violence begets more violence, that nonviolence (or *ahimsa*) is a superior moral force, and that humans have a duty to avoid causing harm to other sentient beings whenever possible, and to minimize it when it cannot be avoided. A vegetarian diet minimizes harm to sentient nonhumans.

9. *Religious considerations.* Some religions, notably Jainism, Hinduism, and the Pythagorean cult in ancient Greece, share a belief in reincarnation and in the ensoulment of humans and nonhuman animals. The Pythagoreans held that animals may contain the souls of former humans and thus should not be eaten. Many Hindus, Jains, and Buddhists refrain from eating animals out of respect for kindred beings with souls. Vegetarianism is sometimes advocated for the benefit of abstinence or spiritual purification. Some Jewish and Christian thinkers have taught that God granted humans stewardship rather than dominion over nature. Islam has also been presented as a stewardship religion, with the stronger proviso that causing grievous harm to nature is a direct offense to Allah. Vegetarianism may be seen as required to carry out the task of stewardship. Finally, both Buddhism and the wisdom traditions of Indigenous peoples teach that a spiritual identity and unity bind together all living things. Although this precept most often entails that animals should be killed only out of necessity, reverentially, and without wastefulness, it sometimes issues in a prescription for a vegetarian or semi-vegetarian diet.

Taken in combination, these arguments have considerable persuasive force, and converge on a vegetarian commitment. To many, that commitment focuses attention on our relationship with the rest of nature, as well as on the need to choose a way of life that is morally and ecologically responsible.

See also Veganism

Further Reading

Adams, C. J. 1999. *The sexual politics of meat: A feminist-vegetarian critical theory.* New York: Continuum International.

Fox, M. A. 1999. *Deep vegetarianism.* Philadelphia: Temple University Press.

Gold, M. 2004. The global benefits of eating less meat: A report. Petersfield, Hampshire, UK: Compassion in World Farming Trust: www.ciwf.org.

Hill, J. L. 1996. *The case for vegetarianism: Philosophy for a small planet.* Lanham, MD: Rowman & Littlefield.

Rice, P. 2005. *101 reasons why I'm a vegetarian.* New York: Lantern Books.

Sapontzis, S. F., ed. 2004. *Food for thought: The debate over eating meat.* Amherst, NY: Prometheus Books.

Singer, P., & Mason, J. 2006. *The way we eat: Why our food choices matter.* Emmaus, PA: Rodale Books.

Walters, K., & Portmess, L., eds. 1999. *Ethical vegetarianism: From Pythagoras to Peter Singer.* Albany: State University of New York Press.

Walters, K., & Portmess, L., eds. 2001. *Religious vegetarianism: From Hesiod to the Dalai Lama.* Albany: State University of New York Press.

Young, Richard A. 1998. *Is God a vegetarian? Christianity, vegetarianism, and animal rights.* Chicago: Open Court.

Michael Allen Fox

VETERINARY MEDICINE AND ETHICS

Given the centrality of ethics to the veterinary profession, it is surprising how little attention veterinary medicine has devoted to ethical issues. A study of veterinary practice conducted in the early 1980s showed that veterinarians spend more time managing ethical issues than in any other single activity. It is also arguable that the major challenges facing veterinary medicine in North America are societal ethical questions: What should be done about the welfare of food animals raised in intensive confinement systems? Ought the legal status of animals as property be modified, and if so how? Given the strength of the human-companion animal bond, graphically illustrated during Hurricane Katrina, ought the value of companion animals be raised from mere market worth? How should veterinarians respond to the thinking underlying increasing public demand for non-evidence-based alternative medicine? How does one determine and weigh considerations of animal quality of life in medical decision-making?

Organized veterinary medicine and veterinary educational institutions have exhibited little understanding of or formal training in dealing with ethics. Indeed, historically veterinary ethics amounted to little more than veterinary etiquette, with ethical codes addressing issues like advertising, the size of one's sign, sending Christmas cards, and totally ignoring issues like teaching surgery using multiple animals in sequential procedures, or regulation of the use of animals in research, or the historical absence of analgesia in veterinary teaching, research, and practice.

This disregard for genuine ethical issues came from a variety of sources, including the historical subordination of veterinary medicine to agriculture and the general failure of science and medicine to embrace ethics, captured in the mantra that science is value-free. But as society has become more concerned about animal welfare issues and animal treatment, and has also grown more litigious, ethics is ignored by professions at their own peril. It is thus imperative for nascent veterinarians to enjoy at least a rudimentary understanding of the logical geography of ethics.

At the outset, it is essential to distinguish between Ethics 1 and Ethics 2. Ethics 1 is the set of beliefs about right and wrong, good and bad, just and unjust, fair and unfair, that all persons acquire in society as they grow up. One learns Ethics 1 from a multiplicity of sources—parents, friends, church, media, teachers, and so on. For most people, these diverse teachings are haphazardly stuffed into one's mental hall closet, and are not critically examined or much discussed. Yet the chances of their forming a coherent whole are negligible. Consider, for example, what parents teach about sexual ethics, versus what one learns from friends, college roommates, and films.

Ethics 2, on the other hand, is the systematic study and examination of Ethics 1, addressing such questions as whether the beliefs in question are consistent, why and if one must have ethics, whether there is a coherent way to affirm that some ethics views are better than others, how one justifies Ethics 1, statements, etc. One learns Ethics 2 from philosophers, since philosophy is the branch of knowledge whose purpose is to critically examine what we take for granted.

Some further distinctions must be made. Under Ethics 1, we can distinguish three subclasses, social ethics, personal ethics, and professional ethics. A moment's reflection makes one realize that, if we wish to avoid a life of chaos and anarchy where, as Hobbes put it, life is "nasty, brutish, and short," ethical notions must be binding on everyone in society. That is what one may call the social consensus ethic, and it is most clearly found reflected in the legal system. The social consensus ethic does not dictate all ethical decisions. Much is left to an individual's personal ethic, his or her own beliefs about right and wrong, good and bad. Such ethically-charged issues as what one eats, what one reads, what charities one chooses to support are, in Western democracies, left to one's personal ethic, with the proviso that the societal ethic trumps the personal on matters of general interest.

What is professional ethics? A profession is a subgroup of society entrusted with work society considers essential, and which require specialized skills and knowledge, for example, law, medicine, veterinary medicine, accounting. Loath to prescribe the methods by which a profession fulfills its function, society in essence says to professionals: "You regulate yourselves the way we would regulate you if we understood in detail what you do. If you fail to do so, we will hammer you with draconian regulation." Not to respect this charge is to risk loss of autonomy, as

has occurred in the United States with unethical accounting practices.

Some years ago, Congress became concerned about excessive use of antibiotics in animal agriculture, both as growth promotion and as a way of masking poor husbandry, since such overuse led to the evolution of dangerous antibiotic-resistant pathogens. When it became clear that veterinary medicine was partly responsible, Congress considered withdrawing the privilege of extra-label drug use from veterinarians. Had this indeed transpired, veterinary medicine as we know it would have been dealt a mortal blow, since veterinary medicine relies on human drugs used in an extra-label fashion.

It is important to stress that every area of ethics is subject to being rationally criticized, else one could make no moral progress. For example, U.S. societal ethics was criticized during the Civil Rights era because segregation was logically inconsistent with the fundamental principles of American democracy.

Similarly, though most people don't realize it, personal ethics is also subject to rational criticism. For example, it can be argued that a person cannot logically be both a Christian and an ethical relativist, that is, one who believes that good and bad vary from society to society or person to person.

Finally, professional ethics can be rationally criticized, as when Congress was about to punish veterinary medicine for indiscriminate dispensing of antibiotics despite its commitment to ensuring public health.

But before one can deal with an ethical issue, one must realize that it is an issue, and identify all relevant ethical components, even as in medicine one must diagnose before one can treat. However, identifying all ethically relevant components of a situation is not always easy, as we perceive not only with our sense organs, but also with our prejudices, beliefs, theories, and expectations.

There exists a heuristic device to help veterinarians hone in on all ethical aspects of a case. This involves reflecting on the ethical vectors relevant to veterinary practice, and applying the ensuing template to new situations. Veterinarians have moral obligations to animals, to clients, to peers and their profession, to society in general, to themselves, and to their employees. Ethically charged situations present themselves, where any or all or various combinations of these obligations occur and must be weighed. In every new situation, the veterinarian should consider each of these ethical vectors and see if they apply to the case at hand. In this way, he or she can minimize the chances of missing some morally relevant factor.

The question of a veterinarian's moral obligation to animals is so important to veterinary medicine that I have called it the Fundamental Question of Veterinary Ethics. The issue, of course, is to whom does the veterinarian owe the primary obligation, owner or animal? On the Garage Mechanic Model, the animal is like a car, where the mechanic owes nothing to the car, and fixes it or not depending on the owner's wishes. On the Pediatrician Model, the clinician owes primary obligation to the animal, just as a pediatrician does to a child, despite the fact that the client pays the bills. When I pose this dichotomy to veterinarians, the vast majority profess adherence to the Pediatrician model as a moral ideal. Happily, though animals are property, society's ever-increasing concern with animal welfare is putting increasing limitations on what humans can do with animals.

Leaving obligations to animals aside for the moment, how does one deal with ethical questions regarding people, assuming one has diagnosed all the relevant ethical components? In the simplest cases, of course, the answer is dictated by the social consensus ethic which, for example, prohibits stealing, assault, murder, etc. So, for example, throttling an obnoxious client, however tempting, is not a real option. In other cases, of course, one appeals to one's personal ethic.

None of this, however, helps us to resolve the Fundamental Question of Veterinary Ethics, since the societal ethic has historically been silent with regard to the moral status of animals and our obligations to them, and few people have bothered to develop a consistent personal ethic theory for animal treatment.

However, as society has developed increasing concern for animal treatment, a characterizable ethic has begun to emerge. In essence, society has demanded that we protect animals' basic natures and interests even as we use them, just as we protect humans. This means applying the notion of rights to animals. Though animals are legally property and cannot strictly have rights, the same result is being achieved by a proliferation of laws limiting how people can use animals. Thus U.S. laboratory animal laws require pain and distress control, forbid repeated invasive uses, require exercise for dogs, etc. And some European and U.S. laws have forbidden sow stalls. This mechanism is the root of what I have called animal rights as a mainstream phenomenon. This also explains the proliferation of laws pertaining to animals as an effort to ensure their welfare in the face of historically unprecedented uses.

This new ethic is good news for veterinarians, as they can now expect more and increasing social backing for their priority commitment to animals, which I have called the Pediatrician Model. Veterinary medicine must engage and lead in providing rational answers and laws protecting animal wellbeing in all areas of animal use. Not only will job satisfaction increase, but as the status of animals rises in society, so too does the status of these who care for them.

Further Reading

Rollin, Bernard E. 2007. *Veterinary medical ethics: Theory and cases*, 2nd ed. Ames, Iowa: Blackwell.

Tannenbaum, Jerrold. 1995. *Veterinary ethics: Animal welfare, client relations, competition and collegiality*. St. Louis: Mosby.

Bernard E. Rollin

VIRTUE ETHICS

A virtue ethics is any system, theory, or approach in ethics or morals that regards virtues as a central component. Today, virtue ethics is experiencing a revival. The term virtue refers to traits of character and personality that predispose individuals, including nonhuman animals, to act in good or right ways. In contrast, a vice is a trait inclining them to act in bad or wrong ways. For example, in companion animals as well as people, loyalty and affection are virtues, and meanness and laziness are vices. Due to the influence of Greek, Roman, and Christian thought, virtue ethics dominated Western morals until the 1700s, when it was replaced by approaches based on duties, rights, consequences, utility, and welfare. The latter are centered on externally observable actions and their consequences, rather than on the internally non-observable psychol-

ogy or mindset required by virtues, such as, dispositions, motivations, purposes, intensions, attitudes, and the like.

Today, ethicists agree that virtues are a central component of ethics and morality, but there the agreement ends. The disagreements today concern how virtues are connected to the other central components of ethics. To be complete, a theory of ethics needs three parts: (1) a theory of virtues that explains what kinds of traits morally good agents ought to have, (2) a theory of duties and rights that explains what makes some actions morally required and others morally prohibited, and (3) a theory of the good that explains why some consequences, things, states of being, and conditions are good and others bad. During the ancient and medieval eras, Plato, Aristotle, Thomas Aquinas and most others believed that virtues were directly tied to real human and animal natures, essences, or souls created and implanted by God or nature. Part of this belief was the idea that everything and everyone have real purposes (*telos* in Greek) given by nature or God. Consequently, virtues were the traits that enabled individual persons and animals to achieve their natural or God-given purposes.

Modern science and evolutionary biology refute the old belief in real natural purposes. According to evolution, individuals and species populations result from three interrelated processes: reproductive success, genetic variation, and environmental adaptation. These processes are largely random and unpredictable. Consequently, the ancient and medieval belief connecting virtues to natural or divine purposes is no longer plausible. In response to this objection, religious thinkers have proposed ways of fitting their doctrines into the worldview of contemporary science, and virtue ethicists have attempted to find some alternative foundation for virtues.

In their responses, ethicists have constructed theories of virtue in ways that are either indirect or direct. One indirect way is to derive virtues from the consequences of actions. According to Alasdair MacIntyre, individuals derive their purposes and goals from their social communities, and virtues are the traits needed to achieve those goals. When people possess certain virtues, Julia Driver's theory proposes, they are more likely to produce good than bad. In Thomas Hurka's account, virtues are intrinsically good states of character that result when individuals love that which is intrinsically good and hate what is intrinsically bad. A second possible kind of indirect account would see virtues as the intrinsically good states of character that result when individuals love right actions for their own sake and hate wrong ones (see Copp and Sobel, 2004, pp. 515–516).

The direct way to construct a virtue ethics is to explain and defend virtues without appealing to any other foundation. Rosalind Hursthouse proposes that a virtue is a disposition to act in a characteristic way for characteristic reasons. Calling his theory agent-based, Michael Slote thinks that virtues are admirable traits of character that are morally primitive or fundamental, since virtues are used to derive and explain moral judgments. Finally, Robert Adams argues that virtues are intrinsically excellent on their own independent of other considerations.

Except perhaps for MacIntyre's and Hursthouse's theories, nonhuman animals are conceivably virtuous in all these direct and indirect ways. For instance, even though being loved by a human may be of greater value, the affection of a companion animal is arguably intrinsically good. However, at issue for any theory of virtues, whether or not it attempts to include nonhuman animals, is the extent to which rational reflection and self-consciousness, in contrast to instinctive and conditioned dispositions and behaviors, are necessary for traits of character to be genuinely virtuous. To what extent, for example, does courage require knowledge and assessment of true danger rather than merely an instinctive or conditioned reaction to a stimulus? Is it truly virtuous when a pet loyally serves an abusive master?

See also Telos and Teleology; Utilitarianism

Further Reading

Adams, Robert. 2006. *A theory of virtue*. New York: Oxford University Press.

Copp, David, and Sobel, David 2004. Morality and virtue: An assessment of some recent work in virtue ethics. *Ethics* 114 (April 2004): 514–554.

Driver, Julia. 2001. *Uneasy virtue*. Cambridge: Cambridge University Press.

Foot, Philippa 2001. *Natural goodness*. New York: Oxford University Press.

Hurka, Thomas. 2000. *Virtue, vice, and value*. New York: Oxford University Press.

Hursthouse, Rosalind. 1999. *On virtue ethics*. New York: Oxford University Press.

MacIntyre, Alasdair. 2007. *After virtue,* 3rd ed. Notre Dame, IN: University of Notre Dame Press.

Welchman, Jennifer, ed. 2006. *The practice of virtue: Classic and contemporary readings in virtue ethics*. Indianapolis: Hackett, 2006.

Jack Weir

W

WAR AND ANIMALS

From elephants to pigeons, all manner of animals have been drawn into humanity's wars, to be used as offensive and defensive forms of weaponry or to serve as couriers and, more recently, as disposable subjects for chemical and biological weapons experimentation.

Hannibal of Carthage used Indian elephants in his ambitious plan to defeat the Roman army on their home soil via a journey to Italy over the Alps in 215 BC. With 50,000 foot soldiers as reinforcement, the elephants plowed into the Roman ranks like modern-day tanks, trampling the enemy and creating general chaos.

Horses were perhaps the most commonly employed of wartime animals because of their endurance, agility, and speed. Among the first to launch a war using horses were the tribal Hyksos from modern day Turkey, who conquered Egypt around 2000 BC with horse-drawn chariots, from which their archers could deliver their arrows with deadly accuracy. In 450 BC, Attila the Hun used horses with the addition of saddles and a new invention, the foot stirrup, which gave his warriors superior balance and leverage to accurately fire an arrow, swing a sword, or throw a spear. Horses would continue in much this same capacity in the centuries to follow, serving as a mobile foundation from which strategic assaults could be launched.

With their innate devotion to humans and superior physical senses, dogs have been one of the more easily exploited animals in military history. The ancient Egyptians, Romans, and Greeks depended on barking dogs to give early warning of approaching enemies. Also common were heavy-set dogs trained to maim and kill. Cloaked in armor and wearing collars studded with metal spikes, these Molosers would be unleashed on enemy infantry to tear out the throats and bellies of soldiers and horses. The advantages of using dogs as weapons were not lost on later strategists, either. Upon arriving in Jamaica in 1494, one of Christopher Columbus' first acts was to unleash a large hound on a reception party of ceremonially painted natives, killing six of them within minutes. Subsequent conquerors of the New World brought their own detachments of killer dogs, and easily routed every native community in Latin America.

Like dogs, pigeons have played a recurring role throughout centuries of warfare. News of the conquest of Gaul (modern France and Belgium) in 56 BC by Caius Julius Caesar was dispatched to Rome via a homing pigeon with a papyrus message tied to one of its legs. Similarly trained birds were also present at the battle of Waterloo in 1815, when Wellington used them to convey word of his overwhelming victory against Napoleon's

585

forces. And during the Prussian siege of Paris in 1870, the French depended on pigeons to keep in touch with those inside the city. They devised a means of copying messages onto a primitive version of microfilm, thereby allowing more information to be compacted into a portable size. Over the course of four months, 150,000 official memorandums and one million personal letters were transported by birds.

Commencement of the war to end all wars in 1914 saw the largest mobilization of animals in history. Three million horses, mules and oxen, 50,000 dogs and other creatures were ensnared in this protracted and devastating conflict. World War I would prove fatal for most of them, because for the first time they were pitted against mechanized weaponry and lethal chemical agents.

A dashing cavalry charge typical of earlier wars was impossible given the nature of this new battlefield landscape, which was fraught with artillery craters big enough to contain a house, bottomless pits of sucking mud, and miles of impossibly tangled razor wire. Trapped in this quagmire, whole regiments of horses were easily mowed down with a single machine gun. Eyewitness accounts describe pitiful scenes of horses which, upon hearing the retreat bugle, struggled to return to the defensive line despite being horribly wounded. The bodies of soldiers and horses killed during the day often had to be used as stepping stones to prevent teams of pack animals and their human handlers from being pulled under and smothered by the mud.

Some horses seemed to know in advance when an attack was imminent. One former polo pony on the British side would stamp her feet and neigh loudly a full five minutes before enemy planes appeared overhead. Others heard the faint whistle of incoming mortars and, like the soldiers, dropped to their bellies and pressed their heads to the ground.

Dogs, too, played a key role in this war, although their use as attack animals was no longer needed, given advancements in other forms of weaponry. Swift canines were invaluable for relaying messages in the heat of battle, as were carrier pigeons, and the two often worked in tandem. Of particular note was one greyhound named Satan who turned the tide of the battle for Verdun. The town was being smashed by a German battery when the besieged French spotted the black dog racing toward them. A German bullet caught Satan and sent him crashing to the ground, but moments later he staggered back to his feet. Despite one shattered hind leg, he pressed forward and limped the remaining yards to his friends. The note tied to his collar stated that reinforcements were on the way, and in his saddle pack were two homing pigeons, which the soldiers used to send back the location of the enemy so that artillery could knock out the German position. Thanks to the courageous actions of these animals, Verdun was saved.

Every country had its own Red Cross organization during World War I, and they all trained mercy dogs to locate wounded soldiers lost on the battlefields. Whenever they found a wounded soldier, these dogs collected the soldier's helmet or a piece of uniform and returned to the trench to lead stretcher-bearers back to his location. One such dog named Prusco located more than a hundred wounded soldiers and was strong enough to drag many unconscious men into sheltering craters before fetching the ambulance team.

The years leading up to World War II saw vast improvements in mechanized forms of transportation, weaponry, and wireless communication, thereby reducing the need to conscript so many animals, particularly horses. Even so, dogs continued to be needed to support the troops in various capacities. A civilian organization called Dogs for Defense formed in 1942 to issue a public call for dogs, and Americans donated 40,000 canines, many of them household pets. Those that made it through a basic doggie boot camp went on to work as sentries, patrolling the defensive perimeters of military facilities with an armed human escort. Others worked as scouts in the field with detachments of soldiers, where they alerted to potential ambushes.

The German shepherd Chips was among the most celebrated dogs of this war. He first worked as a tank guard and marched in Patton's Seventh Army through eight campaigns in Africa, the Mediterranean, and Europe. The dog's mettle under fire was further tested on the coast of Sicily where, against the commands of his handler, he bolted down the beach and leapt into what was thought to be an abandoned pillbox. In fact it held six German soldiers poised to open fire with a machine gun. In spite of being wounded in the scuffle, Chips subdued the gunner and frightened the other soldiers into surrendering. For his actions, he received the Purple Heart and the Silver Star, medals usually reserved for humans.

At the end of the war, the public was outraged to learn that the Army planned to euthanize the surviving war dogs rather than return them to their families. Yielding to protests, the Department of Defense agreed to release the dogs following a brief retraining period to reacclimate the animals to civilian life. Several hundred dogs went home, including Chips. His family reported that he didn't seem much changed from his wartime adventures, except that he seemed less interested in chasing the garbage men when they rattled the cans.

Several thousand canines were again deployed in the Korean War (1950–1953) and, as in World War II, they primarily worked as sentries and scouts. DOD strategists determined that whenever the dogs were used in times of imminent contact with the enemy, they reduced casualties by more than 65 percent. A decade later, during the Vietnam conflict, scout dogs in particular were vital in helping soldiers avoid jungle ambushes and hidden explosives. A harmless-looking footpath could harbor spring-loaded poisoned spikes and shrapnel-packed mines, and it was up to the scout dogs to identify these hazards in time to avoid disaster. Walking off-leash in front of the unit, these canines worked in silence and signaled when something was amiss by sitting down or returning to the handler's side. By war's end in 1972, the dog teams had been credited with discovering more than a million pounds of enemy supplies, seven tons of ammunition, and 4,000 enemy booby traps. By some estimates, they saved as many as 10,000 soldiers' lives.

Remembering public protest over the treatment of decommissioned military dogs after World War II, the government reclassified all canines as equipment rather than personnel, meaning that they could be disposed of in any manner. Just as the United States formally announced its withdrawal from Vietnam, orders were issued to leave the dogs behind. Most of them were given to the Army of the Republic of Vietnam, which had little interest in or experience with working with dogs in such a manner. American GIs who credited the dogs with saving their lives thought it the height of betrayal not to bring them home as well. To this day, some combat veterans wonder if their canine comrades perished from neglect or were killed and eaten, as was customary throughout much of Asia at this time.

In the decades since, dogs have continued to be used to patrol airbases and military installations domestically and overseas. Approximately 1,000 canines are currently deployed in Afghanistan and Iraq, working as patrol dogs or in the detection of hidden explosives. The military dogs of today are inducted soon after being weaned and assigned to Air Force-operated kennels to become acclimated to the sound of gunfire and helicopters. After they've mastered the basics, they begin to work alongside human handlers in more specialized training. Belgian Malinois make up the majority of modern military dogs, but even diminutive canines such as beagles are of use, because their size is advantageous for working in close quarters such as submarines. In the Middle East, dogs specializing in explosives detection by scent are in particular demand, because even in this high-tech era, nothing has proven as reliable as the canine nose, which can pick up specific scents at up to a third of a mile.

Today's soldiers feel just as strongly about their canine comrades as their Vietnam-era counterparts. The passage of House Resolution 5314 in November 2000, the first federal law to stipulate an adoption alternative to euthanasia for retiring war dogs, has resulted in many of the animals being able return to America

Spec.4 Rayford Brown of Florence, South Carolina, and his tracker dog relax for a moment at Fire-Base Alpha Four, a U.S. outpost near the DMZ in south Vietnam on January 2, 1971. Such dogs are used to track enemy troops and find booby traps and mines. Many people are concerned with the ethics of using animals in war. (AP Photo/stf)

to live out their remaining days with loving families.

Other animals continue to play a role in the military, although they are not as well publicized as dogs. Each year an estimated 300,000 primates, pigs, goats, sheep, rabbits, mice, cats, and other creatures are experimented on by the U.S. Department of Defense or other contracted private entities. They are subjected to experimental chemical and biological weapons, or purposely shot and burned so their wounds can be studied for weapons efficiency. Animal advocacy groups have demanded greater accountability from these research programs and, with increased media coverage of how the military exploits animals, the public is becoming vocal in its disapproval.

There is not one member of the animal community that has not been affected at one time or another by man's wars. The only way to repay them is to ensure that they are treated with greater respect and kindness in times of peace. Perhaps in the future they will be involved in our conflicts as little as possible, for ultimately the involvement of animals in wars of our own making dehumanizes us all.

Further Reading

Ambrus, Victor. 1975. *Horses in battle.* London: Oxford University Press.

Brereton, J. M. 1976. *The horse in war.* Newton Abbott: David and Charles.

Cooper, Jilly. 1983. *Animals in war.* London: William Heineman Ltd.

Donn, Jeff. 2007. "Dogs of war valuable in detecting bombs, protecting lives." *Austin American-Statesman,* August 15, 2007, B7.

Greene, Gordon. 1994. *A star for Buster.* Huntington, WV: University Editions Ltd.

Grier, John and Varner, Jeanette. 1983. *Dogs of the conquest.* Oklahoma City: University of Oklahoma Press.

Hamer, Blyth. 2001. *Dogs at war: True stories of canine courage under fire.* London: Carlton Publishing Group.

Lemish, Michael. 1996. *War dogs: Canines in combat.* McLean, VA: Brasseys.

Putnam, William W. 2001. *Always faithful: A memoir of the dogs of World War II.* New York: The Free Press.

Redmond, Shirley Raye. 2003. *Pigeon hero!* New York: Aladdin Paperbacks.

Silverstein, Alvin. 2003. *Beautiful birds.* Brookfield, CT: Twenty-first Century Books.

Susman, Tina. 2008. A special bond between soldiers in Iraq. *Los Angeles Times,* February 25, 2008.

Thurston, Mary Elizabeth. 1996. *The lost history of the canine race: Our fifteen thousand-year love affair with dogs.* Kansas City: Andrews and McMeel.

Web sites

www.militaryworkingdogadoptions.com: A nonprofit organization dedicated to promoting the adoption of retiring war dogs.

www.petcem.com: Hartsdale Pet Cemetery is home to the first War Dog Memorial, which was dedicated after World War I.

www.scoutdogpages.com: A Web site devoted particularly to the role of scout dogs in the Vietnam War.

www.uswardogs.org: United States War Dog Association is an online archive of military dog history and current news.

www.vdhaonline.org: This veterans' organization puts veteran dog handlers in touch with one another and educates the public about the role of canines in the Vietnam War.

Mary E. Thurston

WAR: USING ANIMALS IN TRANSPORT

Nonhuman animals have often been exploited by humans and the victims of human conflict. Recent examples include the loss of life suffered by birds in the U.S.-Iraq War, the killing of cats and dogs in London prior to World War II, and the current ongoing impact of the Rwanda and Congo conflicts on Mountain gorillas. Outcomes have been horrendous and are indeed incalculable.

The direct exploitation of animals by humans as tools of war has been particularly extensive. These animals include dogs, pigeons, horses, donkeys, camels, elephants, cats, and dolphins. During the Great War, at least 20,000 pigeons were used and died, as well as over a million horses just in the French campaign alone; few survived. Many individual animals, most given personal names, have been praised and awarded for their bravery and courage under fire, and they are now glorified in statues and other tributes that memorialize their efforts. The most famous of these is the newly erected Brook Gate memorial in Hyde Park in London.

Pigeons have been used, and some would say exploited, by humans for millennia. During the Great War, the U.S. Pigeon Service had some 54,000 pigeons in service, and individuals were given ranks such as captain. In England, the Dicken Medal of Gallantry was awarded to 32 pigeons for their courageous flights under fire. Three horses, 24 dogs, and a cat have also been awarded the Dicken Medal, the most recent a British springer spaniel, for service in Iraq.

Another classic example of the use of animals by humans in war is the case of mules and donkeys during the Great War. Mules, which are a hybrid created by mating a male donkey and a female horse, had been the main means of transport in most theaters of war since the Roman and Greek armies used them for pack work and riding. Animal transport was still vital in many areas, such as the hot, dry, and mountainous conditions at Gallipoli. The British Army turned to those units that had served in India, where mules had proved invaluable on the rugged Northwest Frontier. Well-trained mules had proved their ability to march for over fourteen hours along the most difficult and dangerous paths, especially in mountainous areas.

Mules endured terrible conditions in the trenches of France; the muddy ground was unsuitable for them. Most of the ammunition at Passchendale, for example, was delivered by mules over ground that was hardly passable, transformed into lakes of deep mud. Many hundreds drowned in mud and shell holes. However, their good health and their length of life at the front won accolades from all quarters. Their powers of endurance and resistance to bad conditions were legendary. Unlike horses, few fell sick, and they were incredibly brave under fire. Mules are highly intelligent and have amazing stamina.

Many thousands of donkeys, which are slightly smaller than mules, also served in the Great War. In the East Africa Campaign of 1916–17, over 30,000 died a terrible death from tsetse fly, others from the supposed antidote, arsenic. Donkeys served with all of the Allied armies in France. Small enough to weave their way along the trenches, they carried food and ammunition to the soldiers on the front lines. One account relates how they saved the soldiers at El Salt. Food and ammunition were running out and the troops were stranded. Two hundred donkeys were loaded up and, marching all night over appalling country, they covered the forty miles to save the stranded soldiers.

During the Gallipoli campaign, donkeys were mainly used to carry water to the soldiers, but mules made the most valuable contribution in the transportation of vital materials up the treacherous ravines to the front lines. Each mule carried two boxes of ammunition as they sure-footedly trotted up the steep

The Animals in War memorial, in Hyde Park (London, England) was dedicated in 2004 by Princess Anne. It was designed to honor the animals who have served in wars throughout history. (Mark Bridge)

hillsides under fire. When a mule was hit, it was unhitched, the ammunition removed, and the caravan went on. The challenges of transporting goods from the beaches to the soldiers in the mountains were an ongoing cause for concern during the campaign, and those who served at Gallipoli were aware of how much they owed to the mules and their Indian drivers for the supplies of guns and ammunition, food and water that they carried up razor-sharp cliffs to the front lines.

Mules were beneficial during the Second World War, to Italian forces in the European Alps, the British and other Allied forces in Burma and in China. Hundreds of mules were abandoned by Allied forces at Dunkirk in 1940.

Even before the Carthaginian Hannibal led his war elephants over the Alps to defeat the Romans, elephants were exploited by humans for many purposes. Elephants were first tamed more than 4,000 years ago, and were used for transport and recreation, and also killed for their ivory. Elephants were employed extensively in wars in India and Southeast Asia, as when the Magadha Empire defeated Alexander the Great in 327 BCE, and in the 300-year war between Burma and Thailand up to 1593. Elephants have been used in numerous conflicts, even to modern times, in World War II, and by Saddam Hussein against the Kurds.

Further Reading

Alexander, H. M. 1917. *On two fronts: Being the adventures of an Indian mule corps in France and Gallipoli.* London: Heineman.

Baynes, E. H. 1925. *Animal heroes of the Great War.* London: Macmillan.

Cochrane, P. 1992. *Simpson and the donkey: The making of a legend.* Melbourne: Melbourne University Press.

Cooper, J. 2002. *Animals in war: Valiant horses, courageous dogs, and other unsung animal heroes.* Guilford: Lyon's Press.

Travis, L. 1990. *The mule.* London: J. A. Allen.

Rod Bennison and Jill Bough

WHALES AND DOLPHINS: CULTURE AND HUMAN INTERACTIONS

Culture is seen by many as a uniquely human attribute. But if we define culture in any way that includes the generally-accepted forms of human culture, such as religion, language, art, technology, symbolism, social conventions, political structures, and pop culture, then nonhumans have culture too. The key to culture is social learning, or learning behavior from others. Once behavior is imitated, emulated, taught or transferred between individuals through any form of social learning, culture can happen. With culture, the processes of genetically driven evolution are changed. Behavior can sweep through a population, or be entrenched in it by cultural conservatism. Group-specific badges, such as ethnolinguistic markers, can evolve and drive cooperation within, and competition between, culturally marked groups. These processes have dominated the recent history of humans, but they occur in other species, including oceanic species, as well, and they can affect how these species interact with humans.

In the centuries since humans have traveled the oceans, interactions between humans and whales have mostly involved humans intentionally killing whales. The scale of the slaughter was extraordinary; whaling was the principal cause of death among most large whales in the 20th century. But as whaling ran its course in the 1970s, human-caused deaths did not cease. Whales are killed, often slowly and painfully, by entanglement in fishing gear, by ship strikes and, as has been recently discovered, by noise.

Humans can affect whales in ways other than through a fast or slow death. We can injure them, disturb them, and affect their behavior. Humans' profound alterations of the marine habitat have closed some niches and opened others. In the North Pacific, gray whale calves seem to be an important food for some killer whales. In the North Atlantic there have been no gray whales since their extirpation several hundred years ago. During the course of whaling, killer whales in all oceans scavenged the carcasses of other species of whale killed by whalers. But when whaling virtually stopped in the 1970s, the killer whales moved on. In many parts of the world they have started removing fish from long lines, to the consternation of fishermen. The destruction of sea otter populations along the Alaskan Aleutian archipelago in the 1980s, and consequent restructuring of almost the entire near-shore ecosystem, seems to have been the result of a prey shift by just a few killer whales, perhaps some of those that had subsisted largely on whale carcasses in the heyday of whaling.

That diet shifts by just one nonhuman predator should have such significant conservation and management consequences is partially a tribute to the killer whale's power, size, and intelligence. But as with another voracious predator, the human, there is another important factor: culture.

Culture is defined in many ways, but the essence is that individuals learn their behavior from each other in such a way that groups of individuals acquire distinctive behavior. When behavior becomes determined by culture rather than by genes or individual learning, then it can take some unusual forms and have immense consequences. Humans are the prime example. Human culture includes some wonderfully useful features that enrich our lives. These include language, technology, art, and music. But some forms of culture, such as Kamikaze cults, guns, and fast-food restaurants are harmful to individual humans, and others, such as nuclear weapons, rabid religious beliefs promoting violence, and fossil-fuel burning, threaten us, and in many cases others, as a species.

Because of the capabilities of our brains and the opposability of our thumbs, human culture has reached extraordinary heights and depths, literally and figuratively. But other animals have culture. It has been found in fish, rats, and many other species, but is best known in songbirds, primates, and cetaceans. The cultures of different species vary characteristically. For instance, songbirds seem to be cultural primarily in their songs, whereas culture has a particular role in the foraging and social behavior of chimpanzees. In one important respect, whale and dolphin culture seems closest to that of humans. In several species of whale and dolphin, social groups that use the same habitat behave differently, in an analogous fashion to multicultural human societies.

And as human culture profoundly affects our interactions with others species, their cultures may also influence interspecies relationships. Here are some examples that have arisen over recent years during our dealings with whales and dolphins.

The bottlenose dolphin is the best-studied cetacean. It is found in many parts of the world, and has been studied in several of them. The site of one of the longest and most detailed studies is Shark Bay, Western Australia. The dolphins in Shark Bay have a wide diversity of feeding strategies, ranging from using sponges as tools to probe beneath the surface, to stranding intentionally on beaches, to attacking very large fish. It seems as though these strategies are largely passed on through social learning, perhaps principally from mother to offspring, and so are a form of culture. One of the strategies, begging for fish from beachgoers, has important negative consequences: the calves of the dolphins who exhibit this behavior have higher mortality, and the behavior only involves a few animals. On the other side of Australia, in Moreton Bay, there are two communities of bottlenose dolphins. They use the same waters, but one regularly feeds on discards from prawn-trawlers, probably a cultural behavior. The other does not. The communities rarely interact. They will be differentially affected by human activities, such as changes in trawling activity due to overexploitation of the prawns.

On a more positive note, 25–30 bottlenose dolphins in Laguna, Brazil essentially run a fishing cooperative with local human fishermen, in which the dolphins and fishermen follow a strict protocol, with the dolphins herding the fish into the nets and feeding on the entrapped fish, to the benefit of both. This has been going on for generations, the cooperative fishing culture apparently passed from mother to daughter in the dolphins, and father to son in the humans. There are other dolphins in the Laguna area who do not participate in the cooperative fishing, and sometimes try to disrupt it. There are

In this photo from the U.S. Navy, Sergeant Andrew Garrett watches K-Dog, a bottle nose dolphin attached to Commander Task Unit 55.4.3, leap out of the water while training near the USS *Gunston Hall* in the Persian Gulf on March 18, 2003. Commander Task Unit 55.4.3 is a multinational team from the United States, Great Britain, and Australia conducting deep/shallow water mine clearing operations to clear shipping lanes for humanitarian relief and conducted missions in support of Operation Iraqi Freedom. Many people are concerned with the ethics of using dolphins to detect mines. (AP Photo/ U.S. Navy, Brien Aho, HO)

reports of similar human-dolphin fishing cooperatives in other places, including one involving a different species, the Irrawaddy dolphins in Burma.

In a similar vein, there was for many decades a whaling cooperative in Twofold Bay, Australia. Generations of killer whales would herd baleen whales into the hunting areas of shore-based whalers, and then scavenge the dead animals once the whalers had done their work

From a cultural perspective, the most interesting whales and dolphins may be those which form permanent matrilineal groups. In such species, most female whales swim in the same social unit as their mothers while both are alive. In killer whales and pilot whales this often extends to the males, so that there is no dispersal from the natal social unit by either sex. In such cases the female-dominant pods can develop distinctive cultures. The most easily studied parts of these cultures are vocal repertoires. Pods of killer whales have distinctive dialects and are grouped into clans, which are recognized by vocal similarity, but seem to be based upon common ancestry. Sperm whale social units associate preferentially with other units from their own clan, even though units from two or more clans may share particular waters. In humans, dialects are markers of rich cultural differences between ethno-linguistic groups, and so it seems to be in the whales. The non-vocal cultural differences are those that are most likely to interact with anthropogenic effects on the ocean habitat.

The two principal sperm whale clans off the Galapagos Islands can be distinguished by their codas–Morse code-like patterns of clicks. But they use the waters differently. Groups of the Regular clan (click-click-click-click) primarily use the waters close to the islands, and have convoluted paths as they search these waters for deep water squid. In contrast, the groups of the Plus-one clan (click-click-click-click-pause-click) are generally further from shore and move in straight lines. Under most conditions, the groups of the Regular clan appear to have greater feeding success, but in the years when El Niño strikes and the waters warm, losing much of their productivity, all whales in the area have less success, but the Plus-one clan is less affected and does relatively better. Global warming seems likely to increase the frequency and strength of El Niños as well as the prevalence of El-Niño-like conditions. Preserving the cultural inheritance of the Plus-one clan may be crucial to the survival of the sperm whales in these waters.

More is known about killer whale cultures. Cultural differences have been recognized across several tiers of social structure—matrilineal units, pods, clans, communities, and types—and span a wide variety of behavior. Apart from the vocal dialects evident within all tiers, there are differences in foraging behavior, social behavior, and play behavior. The southern resident community has a ritualized greeting ceremony when pods meet, and is known for breaching and leaping from the water. For a short while, its members had a strange distinctive fad: pushing dead salmon around. In contrast, the northern residents do not show the greeting ceremony and rarely breach, but have a rubbing beach that they use regularly.

Some of these differences interact with human behavior. Most dramatically, when killer whales were captured for the display industry, they were fed fish. This was fine for the residents, who eat fish. But transient killer whales primarily eat mammals, and a transient who was also caught with the residents died of starvation rather than eat fish.

This is an example of cultural conservatism taken to the extreme, but culture can play it either way. Sometimes it promotes conservative behavior, preventing adaptive responses to changed circumstances, but in other situations culture can allow a species to quickly adopt to new environments as animals learn new ways of life from one another. The spread of scavenging from human whalers and feeding from long lines by killer whales noted previously are two examples in which social learning likely helped spread an activity which some humans found extremely annoying.

When culture becomes a major determinant of behavior, as it appears to have done with killer whales and sperm whales, it can take dramatic forms, as a look at human behavior so clearly shows. Cultural conservatism and cultural opportunism are joined by group-specific cultural badges and maladaptive behaviors. We do not know why groups of apparently healthy whales and dolphins mass strand on beaches, but it seems likely that a usually sensible cultural imperative such as stay with the group whatever happens plays a part. Thus we need to view the behavior of cultural animals with a different perspective, and this must carry through when we are implementing conservation measures.

This came to a head with the transborder southern resident killer whales, whose small population in the inland waterways between British Columbia and Washington was declining. They differ from the northern residents north of British Columbia, a healthier population, by only one known base-pair of genes in their genetic code, but also by a host of cultural traits. Should the southern residents be specifically protected under endangered species legislation? The Canadian listing committee (COSEWIC) thought so, and listed them as endangered. The U.S. equivalent, the National Marine Fisheries Service (NMFS), thought not, and listed all killer whales in the area as depleted. However, after protests and legal chal-

lenges, the NMFS changed its perspective and upgraded the southern residents to endangered in 2006.

Cultural species may, through a new and rapidly spreading form of behavior, quickly become embroiled in a conflict with humans or, through their cultural conservatism, they may not react appropriately when we change their environment. But cultures, like genes, have evolved through natural selection, and they mostly have an important role in allowing animals to live their lives. Just as we seek to preserve genetic biodiversity, we must preserve the cultural diversity of such species, so that the cultured species of the ocean, like sperm and killer whales, have the knowledge to survive when we change their habitat.

Further Reading

Chilvers, B. L., and Corkeron, P. J. 2001. Trawling and bottlenose dolphins' social structure. *Proceedings of the Royal Society of London* B 268:1901–1905.

Rendell, L., and Whitehead, H. 2001. Culture in whales and dolphins. *Behavioral and Brain Sciences* 24:309–324.

Richerson, P. J., and Boyd, R. 2004. *Not by genes alone: How culture transformed human evolution.* Chicago, IL: University of Chicago Press.

Whitehead, H., Rendell, L., Osborne, R. W., and Würsig, B. 2004. Culture and conservation of non-humans with reference to whales and dolphins: review and new directions. *Biological Conservation* 120:431–441.

Hal Whitehead

WHALES AND DOLPHINS: SENTIENCE AND SUFFERING

Whales, dolphins and porpoises, collectively known as cetaceans, are remarkable nonhuman animals that exhibit complex social lives. Recent research on this large group of animals not only informs us about their conservation status and provides fascinating insights into their unique ways of life, but also tells us a great deal about their capacity to suffer, both as individuals and as groups.

The size range of this group of marine mammals is quite extraordinary, ranging from the colossal blue whale, the largest animal ever to have lived on the Earth, to the small, rare, and critically endangered Vaquita and Maui's dolphin. Cetaceans have adapted to life in a range of habitats, from the murky river and estuarine waters of the Ganges and the Amazon to busy coastal areas, right out past the edge of the continental shelf to the almost extraterrestrial remoteness and depths of the Earth's great oceans.

Their capacity to communicate, navigate, migrate, find a mate, feed, and give birth in some of the oceans' more challenging environments has given rise to a number of unique and amazing adaptations. Most notable is cetaceans' use of sound. They use sound not only to communicate with those in close proximity but, for some species, over much greater distances, even across ocean basins. In the toothed cetacean species (*Odontocetes*), sound is also used to echolocate, helping them to find food and also providing three-dimensional information about the world around them.

Since the nature of the aquatic world which they inhabit is largely alien and inaccessible to humans, there are some significant challenges associated with studying cetaceans; researchers often have to piece together information about the complex lives of these marine mammals from fleeting glances of surface behavior, underwater encounters, or stranding events.

However, dedicated long-term studies of particular groups and individuals in the wild are starting to reveal some intriguing insights into the fascinating lives of whales and dolphins. For example, scientists recently discovered that the male boto, a South American river dolphin, uses seaweed, a stick, or a lump of clay to attract a mate (Martin and da Silva, 2008), a unique activity among mammals. Humpbacks, fin whales, orcas, and sperm whales all have a large number of spindle neurons, special cells previously believed to be unique to humans and other great apes, which are found in the areas of the brain associated with social organization, empathy, and speech (Hof and Van der Gucht, 2007). Certain cetaceans may have very specific roles within societies (Lusseau and Newman, 2004) and, perhaps most remarkable of all, there is now compelling evidence that some cetaceans societies exhibit cultural transmission between groups, transmitting knowledge on hunting skills or foraging methods, and some show evidence of having developed distinctive dialects (Whitehead et al., 2004).

The evolving field of cetacean science is beginning to unravel some of the mysteries of the lives of these great river and ocean dwellers and, in turn, inform us a great deal about their sentience and even their sapience. An important consequence of such insights is that the scientific community now has a duty to ask itself some searching question about the ability of these animals to suffer as a result of human activities. We are the human guardians of cetaceans and their habitats, and our improved understanding of their lives heralds the realization that we have even greater responsibility towards protecting them.

A southern right whale cow swims with a calf. Right whales, who can weigh up to 100 tons, got their name from whalers, who said they were the "right" whales to hunt. Their numbers were greatly reduced by the whaling industry. (Photos.com)

So, how do cetaceans suffer as a result of human activities? The answer is: in very many ways. Besides the more obvious acute threats such as hunting, trade in cetacean products, which drives hunting practices, and the estimated 300,000 cetaceans that die annually worldwide as a result of becoming entangled in fishing gear, cetaceans also provide a major attraction in one of the world's captive entertainment businesses: dolphinaria. Despite the apparent smile of the bottlenose dolphin, the most common dolphin species in captivity, the very medium in which captive dolphins exist is oftentimes barely recognizable as the sea environment it is intended to imitate; captive dolphins are usually trapped inside sanitized and chemically controlled tanks, perhaps the human equivalent of solitary confinement, rather than freeswimming in ocean currents, with tides and waves, surrounded by other species. Evidence suggests that captive dolphins are far from content and often have significantly shorter lives in captivity than in the wild (Rose et al., 2006). Ironically, it is the public's desire to get close to cetaceans in captivity that is one of the driving factors behind the brutal dolphin drive hunts in Japan (Vail and Risch, 2006) and it is this same desire that stimulates dolphin-assisted therapy, an expensive and expanding industry which has no scientifically proven benefits over any other animal-assisted therapies (Marino and Lilienfeld, 2007; Brakes and Williamson, 2007).

There are also many more insidious, but no less threatening, pressures on cetaceans and their wellbeing. There is considerable evidence that these threats are increasing and they include the degradation of dolphin habitat from ocean noise, chemical pollution, marine debris, fisheries extraction, harassment, disturbance, and ship strikes, to the many and varied threats which may result from climate change, the resultant changes in ocean circulation, temperature, salinity, prey availability, and ocean acidification (Simmonds, 2006).

Scientists have developed methods for evaluating the various threats to cetaceans and continue to attempt to quantify how these threats will impact cetaceans at the population and species level. However, further effort needs to be expended to determine how these myriad threats will act synergistically and how they will influence the lives of cetaceans at the individual level, which will of course also better inform us about wider population and species implications. Some methods have been developed for evaluating physical injuries incurred, for example, from ship strikes or entanglements, but there has been little attempt to truly quantify the psychological suffering resulting from various threats. For example, the short- and long-term effect on conspecifics of the removal of a particular individual from a population has to date received little attention. This is in part due to the difficulty associated with collecting and interpreting these types of data. However, our growing understanding of the complex lives of these animals suggests that removal of key individuals may play an important roll in both the welfare of the remaining individuals and/or the long-term conservation status of a population (Lusseau and Newman, 2004).

Limited data are available on the suffering of whales and dolphins during hunting. The data show that during Japanese whaling operations whales take, on average, several minutes to die. These data also demonstrate that in some instances death takes a great deal longer than a few minutes; some whales last an hour or even longer (Brakes et al., 2004). This does not compare well with other forms of commercial meat production. For example, in many countries it is a legal requirement that an animal slaughtered in a slaughterhouse should be irreversibly stunned before being slaughtered so that it does not regain consciousness. Hunted whales and dolphins are not afforded anything close to the same level of protection at the time of slaughter, despite the considerable profit that the sale of their meat can yield. Furthermore, very little is known about the deaths of most of the cetaceans that die as a result of entanglement, as it often takes place below the ocean surface and usually beyond regulatory oversight. Postmortem data from recovered cadavers indicate that the diving reflex, which enables a dolphin to hold its breath, is so powerful that dolphins tend to asphyxiate rather than drown in nets.

Our growing understanding that cetaceans are sentient, sapient animals with complex social lives also engenders a moral responsibility for researchers to ensure that their research efforts do not harm their subject animals. Moreover, there is also arguably an experimental imperative to ensure accuracy, and that researchers do not inadvertently influence the very systems or variables they are attempting to measure. However, conflicting pressures on scientists to collect and publish data, which in the case of cetacean research is often in difficult and/or remote areas which can be expensive and logistically challenging to monitor, has driven a burgeoning interest in the use of telemetry data, where satellite or radio tags are attached to cetaceans, often with varying degrees of invasiveness, to collect data remotely. These methods tend to collect data on just a few specific variables in isolation, without recording the wider context of the animal's behavior, environment or interactions with conspecifics and other species. There is some concern that the instrumentation effect may not always be taken into consideration, thus confounding the interpretation of some of the results and potentially limiting their value.

Furthermore, a growing understanding of cetacean cognitive ethology and the complexity of cetacean societies is accompanied by an ever more pressing need to recognize the interests and intrinsic rights of these intelligent animals. In his book *In Defense of Dolphins*, Thomas I. White asks whether dolphins qualify as persons and, therefore, whether they should be afforded the level of protection associated with personhood. As with the great ape debate, such a paradigm shift in the philosophy of our relationship with other animals would alter not only the way in which the global community views cetaceans, but also require fundamental legislative reform to meet the responsibilities associated with such a revelation.

In the meantime, this leaves us with some important practical questions for the protection of whales, dolphins, and their habitats. How human cultures protect them as individuals, populations, and species, and how we work to recognize their intrinsic rights as sentient individuals deserving of the status of personhood is also likely to influence their conservation status as well as their individual and group wellbeing. One thing is certain; there is a great deal more for us to learn about these amazing animals. As we consider how we treat our marine cousins and what we need to do to protect them in the coming centuries, we must also overlay the likely growing pressures from offshore power production, climate change, fishing activities, and the other myriad threats which are likely to further impinge upon their livelihoods and habitats.

It is perhaps easier for us to empathize with the great apes with whom we share a more recent ancestral lineage; it may be more challenging to make the same required leap in thinking for us to welcome cetaceans and other species under this mantle. Although cetaceans are, like us, warm-blooded mammals who suckle their young, their environment is almost entirely alien to us, as is their method of seeing the world through sound. Through our growing understanding of their complex lives, we can perhaps slowly begin to gain better insights into the true implications of our actions on them as individuals, groups, populations and, perhaps, even as cultures.

See also Affective Ethology; Consciousness, Animal

Further Reading

Brakes, P., Butterworth, A., Simmonds, P., & Lymbery P. 2004. Troubled Waters: A review of the welfare implications of modern whaling activities. World Society for the Protection of Animals, London. Available at: http://www.wdcs.org/submissions_bin/troubledwaters.pdf.

Brakes, P., & Williamson, C. 2007. Dolphin Assisted Therapy: can you put your faith in DAT? Whale and Dolphin Conservation Society, London. Available at: http://www.wdcs.org/submissions_bin/datreport.pdf.

Hof, P. R., & Van der Gucht, E. 2007. The structure of the cerebral cortex of the humpback whale, *Megaptera novaeangliae* (Cetacea, Mysticeti, Balaenopteridae). *Anatomical Record,* 290, 1–31.

Lusseau, D., & Newman, M.E.J. 2004. Identifying the role that animals play in their social networks. *Proceedings of the Royal Society of London B (Suppl.),* 271, S477-S481.

Marino, L., & Lilienfeld, S. O. 2007. Dolphin Assisted Therapy: more flawed data and more flawed conclusions. *Anthrozoos,* 20(3), 239–249.

Martin, A. R., da Silva, V.M.F., & Rothery, P. 2008. Object carrying as socio-sexual display in an aquatic mammal. *Biology Letters,* 4(3), 243–245.

Rose, N. A., Farinato, R., & Sherwin, S. 2006. *The case against marine mammals in captivity*, 3rd ed. The Humane Society of the United States and World Society for the Protection of Animals. Available at: http://www.hsus.org/web-files/PDF/MarMamCptvtyBklt.pdf.

Simmonds, M. P. 2006. Into the brains of whales. *Applied Animal Behaviour Science,* 100, 103–116.

Vail, C. S., & Risch, D. 2006. Driven By Demand: dolphin drive hunts in Japan and the involvement of the aquarium industry. Whale and Dolphin Conservation Society, London. Available at: http://www.wdcs.org/submissions_bin/drivenbydemand.pdf.

White, T. I. 2007. *In defense of dolphins: the new moral frontier*. Madden, MA: Blackwell Publishing.

Whitehead, H., Rendell, L., Osborne, R. W., & Würsig B. 2004. Culture and conservation of non-humans with reference to whales and dolphins: review and new directions. *Biological Conservation,* 120(3), 427–437.

Philippa Brakes

WHALES AND DOLPHINS: SOLITARY DOLPHIN WELFARE

Dolphins are extraordinary, intelligent, and undoubtedly self-aware (Simmonds, 2006). Biologist Philippa Brakes puts forward the proposition that we should change our relationship with them, and there is a compelling case to recognize them as nonhuman persons and award them rights under our laws (White, 2007). Dolphins are immensely popular, which should ensure their welfare and conservation, but this popularity actually creates problems. The difficulty of balancing our enthusiasms for interacting with these wonderful animals and protecting them from harm is particularly marked in the case of solitary-sociable dolphins.

Many solitary-sociable dolphins are young animals. They may have lost their mother and become detached from their school. In the UK, it is easy to see how this might happen. A dolphin swimming south, away from the Moray Firth population in Scotland, could travel hundreds of miles with little chance of meeting others of its own kind. Alternatively, it is possible that some dolphins naturally disperse away from their natal population to spend some time living alone.

What is arguably unnatural, however, is the relationship that wild solitary dolphins can develop with humans. A number of stages in the development of such relationships have been proposed (abridged from Wilke et al., 2005):

Stage 1. The dolphin first appears and remains in a new home range, sometimes restricting itself to a small area, often less than 1 km^2. The dolphin does not yet approach humans.

Stage 2. Local people become aware of its presence and attempt to swim with it. Dolphin appears curious but remains at a distance from swimmers.

Stage 3. The dolphin becomes familiar with the presence of one or more people and interacts with only a limited number. Its behavior may include swimming in close proximity or diving side by side, and it allows itself to be touched, including having its dorsal fin held to allow swimmers to be pulled along.

Stage 4. The presence of the animal becomes widely known. Inappropriate human behavior may provoke unwanted and possibly dangerous behavior by the dolphin, including dominant, aggressive, and sexual behaviors directed at humans.

However, not everyone agrees that solitary sociable dolphins are created by such a process or that they are at risk. There are web sites and organizations dedicated to promoting swimming with solitary dolphins. They argue that the dolphins' behavior is entirely natural. The term ambassador dolphin has been coined for them, encompassing the notion that they are emissaries for their species. This line of argument can also draw on the long history of positive interactions between human and dolphin, with examples ranging from our earliest cultures to very recent stories of dolphins rescuing people from drowning or from sharks, accounts that are too detailed and frequent to be dismissed as flights of fancy.

Many of the historical stories appear to involve what we would now recognize as solitary-sociable dolphins. A recent example of one such dolphin was Dave, a bottlenose dolphin, later determined to be a juvenile female, who arrived on the coast of Kent in the southeast of the UK in April, 2006 (Simmonds and Stansfield, 2007; Eisfeld et al., 2008). She adopted a small range, a few kilometers long, close inshore on a coastline that is one of the most densely populated in the UK. Many people regularly bathe here, and leisure boating and recreational angling are also popular. At first, Dave was wary of people. Then she started to associate with one or two regular swimmers. As the presence of a small friendly dolphin became more widely known, more people came to seek her out, and she was even promoted by the local Chamber of Commerce as a tourist attraction. Many people, perhaps inspired by what they had seen on TV or in dolphinaria, tried to interact with her and, over time, she increasingly allowed this.

By September, 2007, she was avidly seeking out swimmers and kayaks to play with, and there were incidents where the robustness of her play caused alarm, including a couple of cases where she prevented people from leaving the sea. Dave also had some shallow wounds on her body by this time. In October, 2007, Dave received a life-threatening wound, with about a third of one tail fluke torn away, probably because of entanglement in fishing line. She was treated with antibiotics and monitored carefully. Her swimming seemed to become stronger over the next few days, but then she disappeared. It seems likely that she died.

The life history of another of these animals, a small female known as Marra, who was first noticed trapped in a dock in Maryport in Cumbria, northeast England, in January, 2006, was strikingly similar to that of Dave. Marra also adopted a small range, and again, the same basic stages in habituation to people followed. Over time she too was seen to be increasingly wounded, ultimately suffering an untimely death from an infection likely caused by her wounds and exposure to contaminated near-shore waters. Three out of the four UK solitary dolphins monitored since 2006 have now almost certainly died as a result of their friendly behavior; the third animal was killed by a boat propeller, and the fate of the fourth animal is unknown (see Simmonds and Stansfield, 2007, and www.wdcs.org/solitaries).

A coalition of welfare groups and local volunteers attempted to keep Marra and Dave safe. The coalition consulted widely around the world and concluded that it would be better if the young dolphins did not become habituated. Outreach programs were developed to try to persuade the local community and visitors to leave the dolphins alone, but they failed. The allure of increasingly friendly dolphins was too great, and people could not resist joining them in the water, thereby largely unwittingly changing the behavior of these animals, making them more vulnerable to harm.

The stories of Dave and Marra are being repeated across the world. There are many other solitary-sociable dolphins. Many are bottlenose dolphins, but there are also solitary-sociable Risso's dolphins and belugas. Orcas can also show this behavior, and the famous case of Luna, a juvenile living in Nootka Sound in Canada, killed in March, 2006, by a tugboat propeller, is now the subject of a remarkable award-winning film. Friendly whales and dolphins are vulnerable not only to accidental harm, but also to deliberate attack. There are at least four examples of such animals being deliberately killed by humans, while others have mysteriously disappeared (Samuels et al., 2000; Frohoff et al., 2006). On the other hand, it is suggested that some solitary-sociables reunite with their own kind, and there is at least one famous example of a solitary-sociable who has mainly survived apart from his species for several decades.

A case can be made that a fully habituated solitary dolphin benefits from the presence of its human friends, because they may be meeting its social needs. But they may actually be inhibiting the animal from seeking its own kind, and are almost certainly putting it at risk from other dangerous human interactions.

Dolphins deserve their reputation for being gentle and friendly. Very few dolphin-human interactions have ended in serious harm to people, all the more remarkable bearing in mind the animals' size and superior aquatic abilities. The weight of scientific opinion is that we are generating increasing numbers of animals that lose their natural fear and seek to interact with us. The fact that we are increasingly invading their environment may be facilitating this. Thus, we need to work out how to offer them better protection or how to prevent this habituation to humans.

Further Reading

Eisfeld, S. M., Simmonds, M. P., Stansfield, L. R. (in press). Behavior of a solitary sociable female bottlenose dolphin (*Tursiops truncatus*) off the coast of Kent, SE England. *Journal of Applied Animal Welfare Science.*

Frohoff, T., Vail, C. S, & Bossley, M. 2006. Preliminary Proceedings of the Workshop on the Research and Management of Solitary Sociable Odontocetes convened at the 16th Bienniel Conference on the Biology of Marine Mammals, San Diego, California, December 10, 2005.

Samuels, A., Bejder, L., and Heinrich, S. 2000. A review of the literature pertaining to swimming with wild dolphins. Prepared for the US Marine Mammal Commission.

Simmonds, M. P. 2006. Into the brains of whales. *Applied Animal Behaviour* 100: 103–116.

Simmonds, M. P., and Stansfield, L. 2007. Solitary-sociable dolphins in the UK. *British Wildlife* 19: 96–101.

White, T. I. 2007. *In defense of dolphins: The new moral frontier.* Madden, MA: Blackwell Public Philosophy Series.

Mark P. Simmonds

WILD ANIMALS AND ETHICAL PERSPECTIVES

Few ethicists today doubt that humans have duties toward domestic animals, but the question of duties to wild animals is more vexing. Some of the leading issues are hunting and trapping, animal suffering, appropriate levels of management intervention, poisoning, habitat degradation, feral animals, restoration, and endangered species.

Duties to wild animals, if they involve care, also involve noninterference, sometimes called hands-off management. In 1988, with the world watching on news media over a two-week period, two gray whales were rescued from winter ice off Point Barrow Alaska. A Russian icebreaker opened a path to the sea; considerable time and expense was required. But perhaps there is no duty to save stranded whales; human compassion may have become exaggerated.

In February 1983, a bison fell through the ice into the Yellowstone River and struggled to get out. Snowmobilers looped a rope around the animal's horns and attempted a rescue. They failed, and the park authorities ordered them to let the animal die, and refused even to mercy-kill it. "Let nature take its course," is the park ethic.

In 1981–82, bighorn sheep in Yellowstone developed conjunctivitis or pinkeye. Partial blindness often proves fatal on craggy slopes. More than 300 bighorns perished, over 60 percent of the herd. Wildlife veterinarians might have treated the disease, as they would have with any domestic herd, but the Yellowstone ethicists claimed that the disease should be left to run its natural course. Humane caring was not a criterion for decision. Rather, the sheep were left to be naturally selected for a better adapted fit.

If suffering is a bad thing for humans, who seek to eliminate it, then suffering is also a bad thing for wild animals. Some respond that here human nature urges compassion, and why not let human nature take its course? Do unto others as you would have them do unto you. But compassion is not the only consideration, and in environmental ethics it plays a different role than in a humanist ethics. Animals live in the wild, subject to natural selection, and the integrity of the species is a result of these selective pressures. To intervene artificially is not to produce any benefit for the good of the kind, although it may benefit an individual bison or whale. Human beings, by contrast, live in a culture where the forces of natural selection are relaxed, and a different ethic is appropriate.

Wild animals are often impacted by human-introduced changes, which can change the ethic. Colorado wildlife veterinarians have made extensive efforts to rid the Colorado bighorns of a lungworm disease, in contrast to the Yellowstone authorities who refused to treat their bighorns with pinkeye. Arguments were that the lungworm parasite was contracted, some think, from imported domestic sheep, or that, even if it is a native parasite, the bighorns' natural resistance is weakened because human settlements in the foothills deprive sheep of their winter forage and force them to winter at higher elevations. There, undernourished, they contract the lungworm first and later die of pneumonia.

The difference is that with the introduced parasite, or the disrupted winter range, or both, natural selection is not taking place. Letting the lungworm disease run its course would not be an instance of letting nature take its course and, both in concern for the species and in concern for suffering individuals, treatment was required.

The ethic changes again where an endangered species is involved. In the spring of 1984, a sow grizzly and her three cubs walked across the ice of Yellowstone Lake

A gray wolf, also known as a timber wolf, remains alert. The habitat of wolves throughout Eurasia and North America continues to dwindle for these predators. (Photos.com)

to Frank Island, two miles from shore. They stayed several days to feed on two elk carcasses, when the ice bridge melted. Soon afterward, they were starving on an island too small to support them. This time park authorities rescued the mother and her cubs and released them on the mainland.

The relevant difference was a consideration for an endangered species, much interrupted by humans who have long persecuted grizzlies. The bears were saved lest the species be imperiled. Duties to wildlife are not simply at the level of individuals; the ethic is that one ought to rescue individual animals in trouble where they are the last tokens of a type.

Wolves have recently been reintroduced to Yellowstone National Park, having been exterminated there early in this century. The restoration earned protests from some in the ranching community.

Such restoration arises, according to most advocates, from a duty to the wolf as a species, coupled with the fact that the wolf was historically, and ought to be again, the top predator in the Yellowstone ecosystem. Conservationists also realize that problem wolves will have to be relocated, sometimes killed, and believe this is an acceptable killing of individuals in order to have the wolf species present. It removes wolves who turn to killing sheep or cattle, not their natural prey; it also protects ranchers against losses. In the recommended mix of nature and culture, if we are to have wolves, we must kill wolves.

Duties to animals can conflict with concern for endangered animal or plant species. In a 1996 case, the U.S. Fish and Wildlife Service moved to poison 6,000 gulls at Monomoy National Wildlife Refuge off Cape Cod, in order to save

35 piping plovers, an endangered species. A U.S. District Court rejected an appeal by the Humane Society of the United States to stop the killing.

San Clemente Island, off the coast of California, has both endemic plant species and a population of feral goats, introduced by Spanish sailors two centuries ago. To protect plants numbering in the few hundreds, the Fish and Wildlife Service and the U.S. Navy have shot tens of thousands of feral goats. The Fund for Animals protested that it is inhumane to count a few plant species more than many mammal lives. But again the ethic of species triumphed.

Further Reading

Armstrong, Susan, and Botzler, Richard G., eds. 2008. *The animal ethics reader,* 2nd ed. New York: Routledge.

Hargrove, Eugene C., ed. 1992. *The animal rights/environmental ethics debate.* Albany: State University of New York Press.

Mighetto, Lisa. 1991. *Wild animals and American environmental ethics.* Tucson: University of Arizona Press.

Rolston, Holmes, III 1988. Higher Animals: Duties to Sentient Life, Chapter 2, in *Environmental ethics.* Philadelphia: Temple University Press.

Rolston, Holmes, III. 1992. Ethical responsibilities toward wildlife. *Journal of the American Veterinary Medical Association* 200(1992):618–622.

WILDLIFE ABUSE

Although not all hunters engage in excessive or especially cruel methods of hunting, some do, leading to slaughters and endangerment of animals such as passenger pigeons and buffalo. Methods of hunting captive animals, moreover, are seen by most to be not only inhumane but also unsporting. Hunting seems to be lessening in popularity in the United States.

According to figures from the U.S. Fish and Wildlife Service, the number of hunters 16 and older declined by 10 percent between 1996 and 2006—from 14 million to about 12.5 million (U.S. Department of Commerce, 2006, p. 22) Hunting organizations point to changing demographics, urbanization, and decreased access to hunting land as the cause for the shift. However, with the growing increase in outdoor activity overall, a possible cause is a change in values.

Membership in humane-based organizations is at an all-time high, and in the last several years, states have passed a record number of animal-friendly laws. At the same time, the number of Americans who participate in other humane wildlife activities like bird watching and photography continues to rise. Wildlife watching appears to be increasing at a rate of about 16 percent from 1996 to 2007 (U.S. Department of the Interior and U.S. Department of Commerce, 2006, p. 36).

The Changing Face of Wildlife Management

Organized wildlife management in this country dates back to the early 20th century, and was partially a reaction to the wholesale slaughter of many of the country's most prolific species. At the end of the 19th century, the commercialization of wildlife was reaching epidemic proportions. Hunters were killing large numbers of animals for their fur, skins and parts; the most notorious case was the American buffalo, which hunters brought to the brink of extinction.

A second, even more troubling massacre followed. The passenger pigeon was once the most common bird on the continent, perhaps in the world—migrating in flocks that took days to pass overhead.

Seals being clubbed during a cull on the Namibian coast. The practice generates a war of words between the government and animal rights activists who object to the practice, which will see 80,000 pups clubbed to death in a year. (AP Photo/STR/Seal Alert)

Thanks to shooters' unwavering zeal, the last known passenger pigeon, "Martha," died in Ohio on September 1, 1914.

Conservation leaders at the time developed a new set of ideals, and slowed the commercialization of wildlife. Most notably, Aldo Leopold suggested a wildlife management model, later labeled the North American Wildlife Conservation Model, that held the fundamental principle that wildlife belong to all people and are managed in trust by the government. According to his model, wildlife managers have a responsibility to sustain animal populations forever.

The North American wildlife conservation model sets limits by restricting activities as a means of protecting wildlife. The conservation model commands thorough consideration in the killing of wildlife, although it views wildlife as resources and the killing of wildlife as harvesting, ideas that fewer and fewer Americans agree with.

Although the North American wildlife conservation model has been the dominant philosophy for the past century, the system shows signs of inevitable unraveling. The model sits squarely on the shoulders of consumptive users of wildlife, that is, hunters and fishermen, because license sales directly fund state agencies responsible for wildlife management.

Wildlife Abuse

Those responsible for the buffalo and passenger pigeon massacres are rightly notorious as people who simply didn't understand the power that humans wield over nature. The tragic stories of these animals are chalked up as examples of ecological ignorance and unknowing people unable to control their impulses. These stories from the past are more than history, though. They are windows into something else, perhaps something darker in the human character, the inexplicable lust to kill.

Captive Hunting

Often occurring at places with names like game ranches and shooting preserves, captive hunts are actually commercial killing fields where customers pay large sums to kill animals inside enclosed areas.

The victims, ranging from zebras to Himalayan mountain sheep to endangered species like the scimitar-horned oryx, are bred on ranches or purchased from dealers. Sometimes dealers visit petting zoos and roadside zoos looking for living targets like warthogs, rhinos, and exotic deer.

Whether bred or bought, they are typically semi-tame, perhaps hand-reared animals who have lost most of their fear of humans. They might look like wild animals, but they are domesticated enough to trust people, and that trust makes them particularly easy targets. Some captive hunts utilize tiny pens, while others convey the illusion of more space by covering hundreds of acres. But the size does not matter. Ranch hands, who call themselves guides, know all of the haunts and hiding places. They can always lead the customer straight to the target. In some cases, animals are killed at their scheduled feeding time, which is why the operators of even the largest canned shoots can advertise with perfect confidence, "No kill, no bill!"

Killing Contests

In wildlife killing contests, participants attempt to kill as many animals as possible for money, hundreds or even thousands of dollars, and prizes. The events conclude at a checkout station where participants pile up the dead animals for photographs before dumping the bodies elsewhere.

Pigeon Shoots   In pigeon shoots, tame birds are released from boxes called traps to be shot from 30 yards away. Nine traps are lined up in front of the competitor. Sometimes electrified to shock the birds into flight, the traps pop open one at time in a random sequence, with each pigeon on the receiving end of two rounds of shot. The shooter gets points for each shot bird that lands within a large ring. Often, wounded birds escape the ring to the surrounding area and suffer for days before succumbing to their injuries. At the end of the day, prizes are awarded based upon who shot down the most birds into the ring.

Coyote Contests   Coyote calling contests, in which contestants compete for prizes to see who can kill the most coyotes in a specified period of time, are found across the West and Midwest. Coyote hunters sometimes gather at bars the night before a hunt to bet on winning teams and who will kill the biggest animal. Contestants use two basic techniques, both involving mechanical, commercially-manufactured calls: imitating coyote distress cries, and those of downed prey animals, usually deer or rabbits. The coyote then comes to investigate. Instead of finding a fellow coyote in trouble or a meal, the coyote instead encounters a two-person team of hunkered-down, camouflaged killers. One is a shooter with a high-powered, long-range, tripod-balanced, scope-mounted rifle, often equipped with an electronic range finder. The other is a spotter, using powerful binoculars to search the countryside for any signs of a coyote on a mission of mercy or in search of dinner. It is not unusual for several hundred coyotes to be killed in the course of a three- or four-day contest. How many are wounded by difficult, long-range shots and left to wander off and die slow, painful deaths is something that contest aficionados never talk about.

Prairie Dog Contests   At prairie dog-killing contests, participants set up shooting benches at varying distances from a prairie dog colony, and they each fire as many as a thousand rounds of ammunition at the unsuspecting animals to see who can kill the most in a specified period of time. The kills glorify the cruelty they inflict, with contestants typically cheering the explosion of varmint vapor with each shot. Contestants receive extra points when a prairie dog flies into the air upon impact. Some shooters aim for specific body areas, hoping to throw the animal in a certain direction, and some kill multiple animals with one shot. Shooters have a number of phrases to characterize the slaughter, including Montana mist, Dakota droplets, red mist, and dog popping.

Stocking

State wildlife agencies should be stewards of the environment. But some agencies raise and release non-native ring-necked pheasants for target practice. Native to China, these pheasants don't thrive everywhere in the United States. Yet to meet hunter demand, wildlife agencies release hundreds of thousands of birds each year that have little chance of survival. Because they are pen-raised, stocked pheasants often lack the skills necessary to fend for themselves. In some states, hunters wait in parking lots for trucks bringing crates of these birds, or line up before release for the first shot. The pheasants who survive this initial gauntlet usually succumb to harsh weather, starvation, or predators.

Wildlife Penning

Although dog fighting and cockfighting are illegal in every state, the cruel practice of penning wildlife for fighting may be found across the Southeast. Coyotes and foxes are caught by the heavy steel jaws of leghold traps, often suffering excruciating pain and terror. Traps can tear flesh, cut tendons and ligaments, and break bones. When the animals struggle to free themselves, they aggravate their injuries. Trapped animals have even chewed or twisted limbs off in an effort to escape. Later, the animal is removed from the trap and packed into a cramped cage with other injured animals to be sold and transported, often across state lines. Untreated for their injuries, the coyotes and foxes are released into an enclosure. In the pens, packs of hound dogs are released to chase the animals, which are exposed to repeated, prolonged, and unavoidable pursuit. Even though some pens have escape shelters, the trapped animals often meet an agonizing and terrifying end when torn apart by packs of dogs.

Targeting Bears

In many states, bears are targeted with some of the most inhumane practices, particularly baiting, hound hunting, and spring hunting. Bear baiting involves using piles of donuts, rotten meat, or other garbage to lure bears into the shooters' sights. As the bears eat, hunters in nearby tree stands pick them off at close range. Hound hunters sometimes use packs of GPS-equipped hound dogs to chase bears until they're so exhausted they have no choice except to climb a tree in a futile attempt to escape their pursuers. Once the bear is treed, the so-called hunter simply shoots him down. Spring bear hunting involves shooting bears when they have just come out of hibernation and sows are still nursing dependant cubs. When their mothers are killed, the cubs are left to die from starvation, exposure, or predation.

Poaching

Many poachers kill animals solely for trophies to hang on their walls. A poacher may kill an elk or deer, chop off the head and antlers, and then simply leave the rest of the body behind. Some stockpile the antlers or send trophy hunting magazines macabre photographs of the bodies. Officials estimate that for each one of the tens of millions of wild animals killed legally every year, another is killed illegally. And with scarce wildlife enforcement resources and countless acres of open land, only a miniscule percentage of poachers are ever caught and punished for their crimes.

Doves

Mourning doves are the traditional bird of peace, and to many a welcome backyard visitor. They delight millions of birdwatchers and people who simply attract these gentle birds to their backyard feeders. But a minority of Americans view mourning doves as nothing more than live targets, sometimes referring to them as cheap skeet. More than 20 million doves are killed each year, earning them the tragic distinction of the most-hunted animal in the United States. Studies consistently reveal that, after being shot, nearly one in three birds is wounded and simply left to die. Because they are so small, and at their lightest weight during the shooting season, many hunters don't even bother eating them.

Since doves are not overpopulated and do not damage crops or property, hunters can't even claim that there is any excuse to kill them. Shooting doves also damages the environment, since dove shooters favor cheap lead shot. Most of the shot falls to the ground, where it quickly accumulates and poisons the soil and the groundwater. Doves and other birds frequently ingest lead pellets, which are toxic to them and the birds who prey upon them.

Further Reading

U.S. Department of Interior and U.S. Department of Commerce. Bureau of Census. 1996. *National survey of fishing, hunting, and wildlife-associated recreation, 1996.*

U.S. Department of Interior and U.S. Department of Commerce. Bureau of Census. 2006. *National survey of fishing, hunting, and wildlife-associated recreation.*

Andrew Page

WILDLIFE CONTRACEPTION

Interest in wildlife contraception has grown sharply among animal advocates, since it was demonstrated in the late 1980s that contraceptive vaccines could be used safely to prevent pregnancy in free-roaming wild horses. These so-called immunocontraceptives were first injected into the famous wild horses of Assateague Island National Seashore, Maryland, by researchers using darts and dart guns (Kirkpatrick et al., 1991). The ability to administer contraceptives to free-ranging wildlife by dart, without capturing or handling them, raised the possibility of using such agents to manage populations of deer, elephants, and other species that are often controlled by sport hunting or other forms of systematic killing often referred to as culling.

Many animal advocates feel that such a nonlethal means of wildlife population control would be a very desirable alternative to hunting and culling. Indeed, various forms of wildlife contraception have now been applied, usually as part of a research study, to free-ranging white-tailed deer, elk, African elephants, African lions, prairie dogs, coyotes, pigs, kangaroos, koalas, Canada geese, pigeons, and many other species. Yet many questions remain about the ethics of contraceptive use on free-ranging wildlife, ranging from narrower questions about the effects of specific contraceptive treatments on the health, behavior, and genetics of treated wildlife, to questions about when, where, and whether contraceptives should be used on wildlife in the first place.

In the policy arena, the debate over the merits of wildlife immunocontraception is most commonly waged in strictly human-centered utilitarian terms; arguments weigh the costs and benefits of contraception to the human community. Can contraception reduce the number of deer-vehicle collisions on the roads, or the damage to backyard shrubs and gardens? How fast? At what cost? Can this be accomplished faster by culling? Is contraception safer than hunting or culling? The interests of the animals themselves may receive lip service, or not be recognized at all.

One important ethical argument rooted in human-centered values is whether contraception will reduce the availability for human use of an important wildlife resource. A common criticism of proposals to manage African elephant populations with contraception, which is now feasible in many circumstances (Delsink et al., 2007) is that native peoples would thereby lose the economic opportunities provided by hunting or culling, such as selling ivory, imposing fees for trophy hunting, and consuming, distributing, or selling meat. In that context, opponents may characterize contraception as a foreign concept being forced on native peoples by Westerners whose attitudes towards animals differ markedly from those of the natives. This can be a very effective argument in policy debates, regardless of whether it is supported by facts.

The questions shift when the interests of the animals themselves are incorporated into the ethical calculations. In an expanded utilitarian discussion, the interests of the human community are balanced against the interests of the animals. How grave are the consequences of wildlife overpopulation for the human community? If the impacts can be characterized as trivial or frivolous, perhaps minor damage to lawns or ornamental shrubs, or the presence of fecal matter on lawns or sidewalks, does this really justify major intervention such as contraception or killing in the lives of wild animals? Or perhaps conflicts between people and wildlife can be resolved without wildlife population control, by excluding wildlife from sensitive areas, modifying the behavior of wildlife, or encouraging people to make simple changes in their own behavior. This living with wildlife view is promoted by many animal protection organizations such as the HSUS and MSPCA (Hadidian, 2007). For example, securing food, removing bird feeders, using bear-proof trash containers, and applying aversive conditioning make human settlements less attractive to black bears, and may reduce or eliminate the need to control bear population size. In this expanded utilitarian approach, people benefit by the reduction of impacts from wild animals, and wild animals are spared suffering or death.

In the case of more serious conflicts between wildlife and people, a utilitarian approach may justify the use of contraception or even killing of wildlife to protect human interests or a broader ecological community of plants and animals. To take one well-known example, the release of rabbits, red foxes, housecats, and other animals of European origin into Australia has had dramatic and harmful impacts on the native wildlife of Australia, causing extinctions of some species and threatening the existence of many more. Wildlife managers have responded by killing these introduced animals on a massive scale, using shooting, trapping, explosives, poison, infectious disease, and other techniques, many of which are widely perceived to be extremely cruel (Oogjes, 1997). While the control of such species might be justified by utilitarian calculations, the associated animal suffering and death weigh heavily against current control practices.

Threats to public health and safety associated with wildlife populations at high densities also push utilitarian calculations toward active control of wildlife numbers. Personal injuries and property damage associated with deer-vehicle collisions, as well as the injuries and deaths sustained by the animals themselves, may justify the application of population control, especially if such methods are perceived as more humane than death by vehicle. Wildlife population control in response to wildlife impacts that are strictly economic, such as limited crop damage or livestock depredation, is ethically problematic, since the utilitarian calculations differ sharply depending on whether or not it is you who are experiencing the damage. Those who work with livestock generally like the idea of government-sponsored predator control; taxpayers are less enthusiastic since they bear the cost but share little of the benefit, and some may even favor the interests of predators over those of stockmen. Of course the predators themselves favor it least of all.

The principal utilitarian argument for the use of contraception is that, in case of serious human-wildlife conflicts or harmful ecological impacts, contraception may provide a more humane and less invasive method of wildlife population control than other management alternatives. Here we must enter the tricky ethical ground of deciding on the wild animal's behalf what course of human action is in his or her best interest. The common assumption, on which the utilitarian case for contraception is made, is that from the animal's perspective, foregoing reproduction is preferable to death and the suffering that may be associated with death. Many animal welfarists are comfortable with this argument, since it is also a fundamental ethical assumption of spay-neuter campaigns advanced for cats and dogs; the invasiveness of sterilization is justified by the prevention of suffering and death that would have been experienced by dogs and cats for whom good homes could not be found.

Not everyone accepts this argument. Just as many Europeans consider surgical sterilization of companion animals to be an unethical mutilation (Salmeri et al., 1991), it has been argued that contraception deprives female wild animals of activities fundamental to their natures; the ability to try to carry out those activities should not be sacrificed even at the risk of early death. When reversible, of course, contraception does not pose such a stark choice; contraception may cause only a delay of reproduction, not a lifetime deprivation, which weakens the case in opposition.

A related ethical argument in opposition to wildlife contraception is that it is unnatural or playing God. Often this argument is voiced by sport hunters, who feel that they themselves are the natural population control agents for wildlife, and that they fill the ecological niche left empty by the natural predators that have been displaced by modern civilization (Porton, 2005). They perceive contraception as yet another agent of the civilization that has made such a mess of nature in the first place.

Sometimes it may be possible to obtain an answer of sorts to the question of the animals' interest in contraception or death from the behavior of the animals themselves. For example, female white-tailed deer will generally abandon their young to their fates when threatened by predators. This suggests a preference for delaying reproduction rather than risking death, which one might argue supports the view that the animal would prefer reversible contraception to death. Whether the behavioral expression of preference is intentional, incidental, or a product of Darwinian natural selection or something else is another question.

The best interest of the animal may also depend on the nature of the contraceptive.

The attractiveness of the contraceptive first used on wild horses at Assateague Island (PZP, or porcine zona pellucida) rests not only on a relatively high level of effectiveness, but on the capacity to administer it without handling the animal, which is typically stressful to wildlife, and on the absence of serious side effects with respect to the health and behavior of the treated animal (Kirkpatrick & Rutberg, 2001). Contraceptives that pose health risks to the animal, as some steroid contraceptives do for cats, or that change natural behavior in important ways, will compare less favorably to alternatives (Munson, 2006). As implied above, permanent sterilization might be considered more invasive than reversible contraceptives, and therefore less preferred from the viewpoint of the animal's interest. If conserving genetic variability in the population is an important value, permanent sterilization is also less desirable than reversible contraceptives because it effectively removes the treated individual from the gene pool.

Thus, the ethical logic of all of the arguments presented above rests on a utilitarian foundation, that is, weighing the interests of people and animals against each other to reach an outcome that produces the most good for the most community members. Even when the interests of animals are weighted equally with those of people, however, this does not constitute an animal rights view in the strict sense. Rather, the animal rights view sets firm limits on what is right to do to a wild animal, and what is wrong, just as the Bill of Rights in the U.S. Constitution was written to limit what the government can do to restrict the behavior of individual citizens, regardless of the will of the majority. Depending on exactly what rights one believes animals should have, what boundaries cannot be transgressed, one may or may not oppose wildlife contraception. One strongly held rights view is that free-ranging wild animals should be able to live free of systematic manipulation by people, to fulfill their basic natures and experience their lives on their own terms (Porton, 2005; Hammer, 2006). Contraception of free-ranging wild animals is such a systematic manipulation, at minimum denying animals the experience of reproduction and its consequences, or at least denying the animal the control over its own reproductive schedule. Consequently, wildlife contraception is not justified under the ethical premise of no systematic manipulation. In an odd parallel with sport hunters, those who advocate this right perceive contraception as an unnatural intervention; of course, rather than advancing hunting as the alternative, they argue against any intrusive intervention.

The animal rights position against systematic manipulation of free-ranging wild animals raises some difficult ethical questions. It is easier to make a no intervention argument for free-ranging wild animals living in relatively natural habitat, large tracts of land where the human footprint is shallow. It is more difficult to advance this argument for free-ranging wildlife living in cities, towns, and suburbs, where human actions and activities dominate the environment. Even from a rights perspective, intervention in the lives of animals that thoughtfully considers their interests might appear more ethical than the indifference to the interests and rights of wild animals which commonly prevails in the day to day activities of human communities.

Other framings outside the formalized structures of utilitarianism and rights language may provide a more robust ethical foundation for guiding the use of contraception on free-ranging wild animals. The language of guardianship is advancing as a way to think about the respectful treatment of companion animals, and may also help guide human relationships with wild animals. This ethic shares features with relationship- and context-based feminist ethics of care, and may offer a platform that extends beyond the single-species focus of traditional wildlife management to broader aspects of the biological community.

In particular, a guardianship ethic may help resolve ethical paradoxes within the rights concept. Compassion, care, connectedness, and responsibility are built into the common notion of guardianship, but so also are respect for and recognition of the interests and autonomy of the object of guardianship. Because guardians ideally encourage the autonomy and independence of their charges, a guardianship ethic could minimize or preclude human intervention, including contraception, in the lives of wild animals occupying habitats where human impacts are minimal. In a guardianship framing, animals that live in human-dominated environments might require, and deserve, thoughtful humane intervention, potentially including contraception, to reduce animal suffering and facilitate amiable coexistence with people.

See also Animal Reproduction, Human Control of

Further Reading
Porton, I. J. 2005. The ethics of wildlife contraception. In C.S. Asa and I. J. Porton, eds. *Wildlife contraception: Issues, methods, and applications*, 3–16. Baltimore: Johns Hopkins University Press.

Delsink, A. K., van Altena, J. J., Grobler, D., Bertschinger, H., Kirkpatrick, J., & Slotow, R. 2007. Implementing immunocontraception in free-ranging African elephants at Makalali Conservancy. *Journal of the South African Veterinary Association* 78, 25–30.

Donovan, J., & Adams, C., eds. 2007. *The feminist care tradition in animal ethics.* New York: Columbia University Press.

Grandy, J., & Rutberg, A. T. 2002. An animal welfare view of wildlife contraception. *Reproduction Supplement* 60, 1–7.

Hadidian, J. 2007. *Wild neighbors: The humane approach to living with wildlife.* Washington, DC: Humane Society Press.

Hammer, D. 2006. Putting other animals on the pill: Should we or shouldn't we? ActionLine Spring 2006. http://www.friendsofanimals. org/actionline/spring-2006/animals-on-the-pill.php (accessed 19 September 2008).

Kirkpatrick, J. F., Liu, I.K.M., & Turner, J. W. Jr. 1991. Remotely-delivered immunocontraception in feral horses. *Wildlife Society Bulletin* 18, 326–330.

Kirkpatrick, J. F., & Rutberg, A. T. 2001. Fertility control in animals. In D. J. Salem and A. N. Rowan, eds. *State of the Animals 2001*, 183–198. Washington, DC: Humane Society Press.

Munson, L. 2006. Contraception in felids. *Theriogenology* 66, 126–134.

Oogjes, G. 1997. Ethical aspects and dilemmas of fertility control of unwanted wildlife: An animal welfarist's perspective. *Reproduction, Fertility & Development* 9, 163–167.

Rutberg, A. T., ed. 2005. *Humane wildlife solutions: The role of immunocontraception.* Washington DC: Humane Society Press.

Salmeri, K. R., Olson, P. N., & Bloomberg, M. S. 1991. Elective gonadectomy in dogs: A review." *Journal of the American Veterinary Medical Association* 198, 1183–1192.

Allen T. Rutberg

WILDLIFE SERVICES

Wildlife Services, a program of the U.S. Department of Agriculture and part of the Animal and Plant Health Inspection Service, spends more than $100 million annually to kill more than one million animals, primarily birds, and hundreds of thousands of mammals such as black and grizzly bears, beavers, mountain lions, coyotes, and wolves (USDA-APHIS-WS 2008a). Wildlife Services was a major force in eliminating wolf and grizzly bear populations in the continental United States by 1940 (Robinson, 2005; Mighetto, 1991; Dunlap, 1988).

Wildlife Services aerial guns, traps, and snares animals, and broadcasts a panoply of dangerous toxicants that harm a variety of taxa, for the purported benefit of the agricultural industry. Between 2004 and 2007, Wildlife Services killed 8,378,412 animals (USDA-APHIS-WS, 2008a). The numbers of mammals in the overall kill has increased in recent years. In 2004, for instance, the agency killed 179,251 mammals, compared with 207,341 in 2006 (USDA-APHIS-WS, 2008a). Wildlife Services has escalated the numbers of endangered species it killed in recent years for a total 2,481 individuals, primarily gray wolves, since 1996 (USDA-APHIS-WS, 2008a). The average number of endangered species killed between 1996 and 2004 was 177.5. In comparison the average number of endangered species killed between 2005 and 2007 was 294.3, representing a 66 percent increase in the numbers of endangered species killed in the past three years (2005–2007), as compared to the previous nine (1996–2004) (USDA-APHIS-WS, 2008a).

Yet Wildlife Services cannot accurately count each poisoned individual. Many toxic bait sites go undocumented. Grizzly bears may trigger an M-44, a device that expels deadly sodium cyanide, only to die unnoticed in the wilderness. Numerous family dogs have been exposed to M-44s, as have people (Keefover-Ring, 2007). Tens of thousands of birds, poisoned by DRC-1339, an avian toxicant, rain down from the sky, forcing some homeowners to scoop them up with pitchforks (Antone, 2008; Slabaugh, 2008). Because the toxicant can take three days to act, many birds are not found and included in the agency's statistics (see Johnston et al., 2005). Wildlife Services sprays pesticides from helicopters onto cattails in wetlands to reduce breeding sites for migratory blackbirds to benefit the sunflower industry (USDA-APHIS-WS, 2008b). These treatments likely cause harm to wetland functionality, water quality, and wildlife habitats.

Why the killings? Wildlife Services is designed to help agribusiness reduce losses caused by wildlife. Because its focus is on utilitarian values (USDA-APHIS-ADC, 1994), little energy is afforded to conservation concerns, people's diverse values for wildlife (Kellert, 1996), or even an emphasis on non-lethal wildlife controls (US GAO, 1995, 2001). Biologists, economists, and federal oversight agencies have, however, criticized the efficacy of Wildlife Services. Biologists have dubbed the agency's predator-control program the sledgehammer approach to wildlife management because of the breadth of extermination (Treves and Karanth, 2003; Stolzenburg, 2006; Mitchell et al., 2004). Large-scale predator-killing programs are unsustainable and environmentally harmful. Few livestock producers actually experience predator problems, because most unintended cattle and sheep deaths come from birthing problems, disease, or weather, but not predation (Keefover-Ring, 2008). An economic study shows that lamb prices, wages and hay costs, but rarely predators, harm sheep producers (Berger, 2006). More ominous to many, several federal agencies have determined

that Wildlife Services' practices prove hazardous.

Wildlife Services presents a national security threat, according to federal oversight agencies. In a series of audits since 2001, the USDA's Office of Inspector General has sanctioned Wildlife Services for its unsafe handling of lethal biological agents, toxins that could be used in biological warfare (Fleischman, 2002; USDA-OIG, 2004a, b, 2005, 2006), particularly sodium cyanide and Compound 1080, both of which can be used in chemical warfare and are extremely toxic to humans. In March, 2008, the Environmental Protection Agency issued a notice of warning letter to Wildlife Services for its illegal and unsafe placement of M-44s that resulted in the injury of a U.S. Fish and Wildlife Service biologist and the death of his hunting dog. In November, 2007, Wildlife Services itself admitted that it had experienced a series of accidents that involved its aerial gunning program, its hazardous chemicals inventory, and more (USDA-APHIS-WS, 2007). The aerial gunning program, for instance, caused ten fatalities and 28 injuries to federal employees and contractors in the years between 1979 and 2008 (Keefover-Ring, 2008).

Despite this track record, Wildlife Services skirts around disclosure laws. For instance, in July 2000, WildEarth Guardians, a nonprofit organization whose mission is devoted to protecting and restoring wildlife in the American West, requested documents pursuant to the Freedom of Information Act concerning aircraft accidents. The response arrived in October, 2007, seven years late, and incomplete. A major report was missing, and 82 of 400 pages were redacted. Wildlife Services finds federal disclosure laws inconvenient. Despite its public status and funding sources, Wildlife Services, according to critics, remains publicly unaccountable.

Most of Wildlife Services' budget comes from federal tax dollars, but states and counties also contribute. The agency also receives funding from private contributors such as the Woolgrowers Association and the Cattlemen's Association (USDA-APHIS-WS, 2008a). This biologically and fiscally expensive program burdens taxpayers.

To many, Wildlife Services appears to kill America's wildlife in order to benefit agribusiness. In fact, it is the mission of Wildlife Services' parent agency, the Animal and Plant Health Inspection Service (APHIS), to "protect the health and value of American agriculture and natural resources" (USDA-APHIS, 2008). It argues that the government's role "in preventing and controlling damage caused by wildlife is sensible" because "wildlife belong in common to the country's citizens" (USDA-APHIS-ADC, 1994, Chapter 3, p. 51). Yet taxpayers are unwittingly funding the deaths of hundreds of thousands of animals each year. Those deaths are conducted in ways that are harmful to the environment, the public, protected species, and family pets.

Viable nonlethal alternatives to using dangerous toxicants, traps, and aerial gunning are available but go unused. While practical and time-tested nonlethal aids are available to the livestock industry and farmers, the federal government neither actively uses them, nor does it appear to spend significant resources developing new ones. To some, Wildlife Services appears to shoot first and deflect questions later. In 2008, WildEarth Guardians released a report to Congress calling upon it to defund Wildlife Services' lethal animal control measures.

Further Reading

Animal Damage Control. 1994/1997. Final Environmental Impact Statement.

Antone, R. 2008. Birds by the bagful a surprise. *Yakima Herald-Republic* http://www.yakima-herald.com/stories/2008/03/15/birds-by-the-bagful-a-surprise.

Berger, K. M. 2006. Carnivore-livestock conflicts: Effects of subsidized predator control and economic correlates on the sheep industry. *Conservation Biology, 20,* 751–761.

Dunlap, T. R. 1988. *Saving America's wildlife.* Princeton, NJ: Princeton University Press.

Fleischman, J. N. 2002. Statement of Joyce N. Fleischman, Acting Inspector General, U.S. Department of Agriculture. *Subcommittee on Agriculture, Rural Development, Food and Drug Administration, and Related Agencies.* http://www.aphis.usda.gov/about_aphis/

Johnston, J. J., Holmes, M. J., Hart, A., Kohler, D. J., & Stahl, R. S. 2005. Probabilistic model for estimating field mortality of target and non-target bird populations when simultaneously exposed to avicide bait. *Pest Management Science, 61,* 649–659.

Keefover-Ring, W. 2007. Sinapu et al.'s Petition to the Environmental Protection Agency to Ban Sodium Cyanide (M-44) and Sodium Flouroacetate (Livestock Protection Collars), Environmental Protection Agency Docket Number: EPA-HQ-OPP-2007-0944.

Keefover-Ring, W. 2008. AGRO: A Coalition to End Aerial Gunning of Wildlife: www.goAGRO.org.

Kellert, S. R. 1996. *The value of life: Biological diversity and human society.* Washington, DC: Island Press.

Mighetto, L. 1991. *Wild animals and American environmental ethics.* Tucson: University of Arizona Press.

Mitchell, B. R., Jaeger, M. M., & Barrett, R. H. 2004. Coyote depredation management: current methods and research needs. *Wildlife Society Bulletin, 32,* 1209–1218.

Robinson, M. J. 2005. *Predatory bureaucracy: The extermination of wolves and transformation of the West.* Boulder: University Press of Colorado.

Slabaugh, S. 2008. Bird die-off causes a flap in Winchester: Man who discovered starlings lives next door to CAFO. http://www.wildearthguardians.org/Portals/0/support_docs/report_WOWR_2_09.pdf.

Stolzenburg, W. 2006. Us or Them. *Conservation in Practice,* 7, 14–21.

Treves, A., & Karanth, K. U. 2003. Human-carnivore conflict and perspectives on carnivore management worldwide. *Conservation Biology,* 17, 1491–1499.

U.S. Department of Agriculture—Animal and Plant Health Inspection Service—Wildlife Services. 2008a. Wildlife Damage: Program Data Reports, 1996–2007. http://www.aphis.usda.gov/wildlife_damage/prog_data_report.shtml.

U.S. Department of Agriculture—Animal and Plant Health Inspection Service—Wildlife Services. 2008b. FY 2007 Monitoring Report and Amendment to the EA for Management of Blackbird Species to Reduce Damage to Sunflower, Corn, and Other Small Grain Crops in the Prairie Pothole Region of North Dakota and South Dakota.

U.S. Department of Agriculture—Animal and Plant Health Inspection Service—Wildlife Services. 2007. Wildlife Services Stakeholder's Newsletter: 2007 Fall Edition.

U.S. Department of Agriculture—Office of Inspector General. 2004a. Audit Report: Animal and Plant Health Inspection Service, Wildlife Services' Controls Over Hazardous Materials Inventory.

U.S. Department of Agriculture—Office of Inspector General. 2004b. Audit Report: Security Over Animal and Plant Health Inspection Service's Owned and Leased Aircraft.

U.S. Department of Agriculture—Office of Inspector General. 2005. Animal and Plant Health Inspection Service, Evaluation of the Implementation of the Select Agent or Toxin Regulations, Phase I. *Report No. 33601-2-AT.*

U.S. Department of Agriculture—Office of Inspector General. 2006. Audit Report: Animal and Plant Health Inspection Service, Evaluation of the Implementation of the Select Agent or Toxin Regulations, Phase II. *Report No. 33601-3-AT.*

U.S. Department of Agriculture, Animal and Plant Inspection Services. "About APHIS."

U.S. General Accounting Office. 1995. Animal Damage Control Program: Efforts to Protect Livestock from Predators.

U.S. General Accounting Office. Nov. 2001. Wildlife Services Program: Information on Activities to Manage Wildlife Damage. Washington, D.C.: GAO.

Wendy Keefover-Ring

WOLVES AND ETHICAL PERSPECTIVES

Religious and ethical perceptions of wolves are unsurprisingly intertwined with the ways that wolves come into conflict or cohesion with human interests. From ravenous beasts, to protective gods, to wildlife superstars, wolves have played various symbolic roles throughout history. Because the human imagination is entangled with the physical landscape, wolves have alternatively been decimated, persecuted, respected, or allowed to flourish based on the degree to which humans have considered them a part of their moral and sacred communities.

Wolves as Kin

As a species particularly well equipped for symbolic thought, humans have long looked to other animals for their behavioral cues, adapting and adopting various nonhuman animals as social models. For many small-scale societies that depended on coordinated hunting as a means of subsistence, wolves were often treated with admiration and seen as teachers and masters of the hunt. Recognition of the similarities between wolves and humans was often reflected positively through a kinship-based ethic and oral narratives describing the manner in which wolves aided hunters, religious specialists, and warriors in times of need.

Kinship relations, based on physical proximity and mythic importance (the two often being related), between humans and nonhumans were and remain important for many Amerindian peoples. Spiritual power could be given or withheld by animals, and was believed to be dependent on individual and corporate rituals that ensured proper respect toward particular animals. The Skidi Pawnee are perhaps best remembered for their social correspondences with wolves, but other plains-based tribes, such as the Tonkawa and the Cheyenne, ritually reenacted oral narratives through elaborate dances that explained their origins as hunting peoples, expressed their cultural dependence on wolves, and were intended to ensure productive hunts. Origin myths of the Paiute, Cree, Blackfoot, and Arikara recall how a wolf helped to form the Earth itself. For indigenous peoples in northwestern North America, such as the Nootka, Kwakiutl, and Quillayuk, wolf people played a special role in initiation ceremonies that served to ritually incorporate young people as members of their respective societies.

For cultures such as these, an emphasis on the permeable boundaries between humans and nonhuman animals was and is common. In many ways, wolves' high degree of sociability makes them likely candidates for special attention. To name just a few characteristics, wolves have elaborate systems of communication; they are socialized and learn valuable skills through play; they coordinate their movements and hunts to accomplish goals that could not be accomplished in isolation; they interact in ways that increase intragroup bonding while regulating distances between other wolf populations; breeding adults form strong pair bonds, and they spend extended periods of time caring for their young. Shared social relationships between wolves and humans

may also lead to a historical identification with their fate as a species. In recent years, certain native peoples, including the Nez Perce, have identified their own historical persecution with that of wolves, and therefore have welcomed the reintroduction of wolves as a symbol of renewed tribal strength.

Wolves as Outlaws

Prior to concentrated eradication efforts by humans, wolves occupied territory stretching throughout nearly all of the northern hemisphere, from Mexico City and southern India, northward all the way to the arctic extremities. Yet the same evolutionary adaptations that made wolves one of the most successful and wide-ranging carnivores also brought them into conflict with human communities. In various parts of the world, herding communities that depend upon domestic stock for their livelihood often fear the damage that wolves can incur upon their flocks and/or herds. In such a context, wolves are frequently labeled in negative terms as thieves, varmints, villains, or attributed preternatural powers. The Abrahamic religions, Judaism, Christianity, and Islam, for example, arose in a predominantly pastoral context, and in these traditions wolves are commonly metaphors of destruction or deception (for biblical examples, see Gen 49:27, Jer 5:6, Matt 7:15, John 10:12, Acts 20:29). Even when pastoral economies and lifeways were left behind, wolves' metaphorical roles as sources of pollution or agents of evil persisted as a way of categorizing spiritual and physical threats.

Especially in central and northern Europe, perhaps because of their association with scavenging human corpses on medieval battlefields, wolves were depicted as unwelcome transgressors of the boundaries between civilization and wilderness. This is imaginatively embodied in werewolf folklore, fairy tales that cast wolves as cunning predators, like "Little Red Riding Hood," popular bestiaries of the Middle Ages (books that assigned specific human characteristics, such as greed or valor, to various animals) in which wolves were depicted as symbols of humankind's baser instincts; and epic literature, such as *Beowulf*. These mythic and popular images served as moralistic warnings to humans, while also incarnating dark fears of the uncultivated forest.

It is difficult to gauge the precise impact of such tales upon actual wolf populations, but there is evidence that the fears expressed in these stories served to justify acts of retributive justice in both Europe and, later, in North America. Convicted criminals in 10th-century England, for example, could avoid jail by delivering a prescribed number of wolf tongues to authorities. In France, beginning in the ninth century and continuing well into the 19th century, special groups of wolf hunters were organized to exterminate wolves for payment. In short, wolves were understood as the epitome of the outlaw creature, unable to remain in their proper place away from domesticated property, and therefore were the frequent target of vigilante justice.

Wolves as Deities

If wolves have been the ultimate criminals to some, they have been an object of reverence and even worship for others. Ancient gods like the Greek huntress Artemis or the Teutonic war-god Odin had powerful wolf companions. According to legend, Rome was founded by twin boys nursed by a she-wolf. Likewise, in Inner Eurasia, the Turks and the Mongols believed themselves to be descended from a wolf. Permeable lines were also sometimes believed to exist between deities and wolves themselves, as in the case of the shape-shifting sun god Apollo, the patron of shepherds, who took the form of a wolf in some Greek legends, signaling perhaps the dual capacities of the gods in Hellenistic culture to protect and destroy.

Though agriculturally-based societies have typically had ambivalent relationships with wolves, the worship of wolves in Japan was widespread among mountain farmers up until the 19th century. According to historian Brett Walker, the wolf was known as the Large-Mouthed Pure God and, when properly treated, was believed to protect the people's crops from the ravages of wild boars and deer. The power of wolves could also be harnessed in talismans and charms that served to protect their wearers from disease and infertility, among other misfortunes. The Ainu, an indigenous Japanese tribe, worshipped wolves as their divine ancestors. The modernization of Japan in the late 19th century, however, led to the waning of wolves' sacred status. In a span of a few decades, the two subspecies of wolves in Japan were eradicated, vividly demonstrating how changing ideologies can be expressed on the physical landscape.

Wolves as Symbols of Wilderness

Japan was certainly not the only country to experience a dramatic reduction of wolves. Much has been written on the Puritan encounter with the howling wilderness of New England and, for most of the early settlers in America, wolves figured predominantly as treacherous actors on a divine stage, harassing livestock that were allowed to roam free outside of colonial settlements. Economic interests often mixed with biblical injunctions to protect the flock, and wolf bounties were legislated early to fulfill a dual purpose: secure economic prosperity and perform a spiritual catharsis on the land by clearing it of unwanted threats.

In the colonial context, the means of wolf eradication, though lethal, were geographically limited. In North America at large, the scale of this eradication became magnified over the 19th century as advances in technology and a growing government bureaucracy linked progressive ideals with national economic interests. Despite early calls for animal protection and conservation in the late 19th century, wild predator animals remained ensconced in the category of the unworthy. Wolves in particular represented the epitome of the bad animal, a quintessential varmint with neither sporting manners nor moral qualms about their violent acts.

Cultural and religious constructs, however, are not static. A sense of loss, better ecological understandings of the importance of wolves to their habitats, and support for biodiversity, have led to calls for the reintroduction of wolves in selected areas. As wild places and creatures diminished in an increasingly urbanized United States, old myths began to lose their weight, and new values began to emerge. Ecological studies played no small part in such changing views. One notable conversion experience comes from Aldo Leopold, an early 20th-century government forester whose writings had a tremendous influence on ecological discourse and the field of environmental ethics. Leopold recalled in his posthumously published *A Sand County Almanac* (1949/1987) that when he was young and "full of trigger itch," he once shot a mother wolf during a timber survey assignment in the American Southwest, arriving in time to see the "green fire" dying in the wolf's eyes. This moment etched itself upon his memory and altered his view of the human place in the biotic community. In order to "think like a mountain," Leopold later argued, one had to consider the wolf's integral place in the larger ecosystem. In the absence of natural predators, deer would denude the mountain, encouraging erosion and degrading the entire ecosystem. Humans, Leopold concluded, have a great responsibility, not to be superior, but merely a "plain member and citizen" of the natural world.

Since Leopold's time, the status of wolves has undergone a substantial shift in North America, and even worldwide. In the affluent and increasingly urban and suburban context of a post-World War II America, people were becoming more receptive to ideas like Leopold's, more willing to question the role of the government in controlling wildlife according to progressive-era management philosophies, and more interested in visiting the wildlands that constituted America's natural heritage. The immediate threat of wolves, both real and perceived, had largely passed into legend. A trickle of disapproval from select scientists over indiscriminate predator control would turn into a flood of public sentiment in the latter part of the 20th century. During this period, wolves became the icon of choice to represent endangered species, ecologically threatened lands, and a vision of humanity that placed less emphasis on dominance over the nonhuman world.

Wolves now grace the publications and websites of numerous environmental advocacy groups, and the proliferation of wolf images in the media oftentimes indicates an empathetic stance toward what was once an object of derision. Even the howls that were once considered portents of death and evil are assuming new associations, and listening for wolf howls with park rangers has become a popular nighttime tourist activity at several Canadian national parks and wolf education facilities in the United States. Preeminent wolf biologist L. David Mech remarked that since the wolf has come to symbolize disappearing wilderness, "the creature now symbolizes [all] endangered species and has become the cause célèbre of numerous animal-interest groups," which has resulted in "wolf deification" (1995, p. 271). This deification does not have the same connotations as it formerly did in the context of Japan or ancient Greece, but it does perhaps signal a growing appreciation for, and an extension of, religious and ethical concern to wolves.

Wolf Reintroduction in the United States

In human relationships with nonhuman animals, religious and cultural narratives may serve to reinforce kinship relations and concomitant ethical obligations with specific species or individual animals. Religious and cultural narratives may also, inasmuch as they indicate what is outside the realm of sacred consideration, reinforce the unworthiness or the object status of certain or all animals. In the context of gray wolf reintroductions in North America, which began in Yellowstone and Idaho in 1995 and was followed by the reintroduction of Mexican gray wolves to the Southwest in

1998, religious rhetoric was sometimes used to capture the sense that wolf recovery may signal a rapprochement between humans and nature. As Hal Clifford, executive editor of *Orion* magazine, expressed it,

This is the renaissance of the land. The wolf sings it into being. The wolf is all the connections of the land, and that includes our connection, too. As we make room for the wolf we take another step toward embracing the complexity of the world—the glorious, magical complexity that is the expression of God in all things—and we begin to stitch ourselves into the fabric of place. (Clifford, 2005, p. 194)

As Clifford indicated, for some people wolf restoration signaled the fruition of Leopold's "green fire" experience: a willingness to accept a humbler human role as part of a larger biotic community. Other people, particularly livestock owners and those living near rural reintroduction sites, are much less enthusiastic about the prospect of wolves close to home, and regard wolves as an unwelcome animal unnecessarily foisted upon struggling rural communities. For people who are against wolf reintroductions, the idea of going backward, reviving the presence of animals that were intentionally trapped, poisoned, and shot out of existence, constitutes a regressive plunge that decivilizes the land and threatens to disrupt humans' position as nature's rightful manager. Moreover, inasmuch as wolves may require changes in human uses of land, their presence threatens the religious and cultural narratives that encourage, or are interpreted as encouraging, the idea of human dominion.

Rami, a gray wolf from Mission Wolf, a sanctuary located in Silver Cliff, Colorado (http://www.missionwolf.com) patiently stands on a table during a news conference as handler Kent Weber talks about the animal and why we should respect wolves and other animals. (AP Photo/David Zalubowski)

The reintroduction of wolves thus highlights a collision of narratives, in which different groups assert alternative visions of humans and their relationships to and within the natural world. It could be argued that wolf reintroduction has exposed conflicting priorities over inherited traditions and stories, ways of life, and notions of what makes such a life worth living, and the authority to enact this way of life as people work toward an ideal vision of the future. Particularly in the context of wolf reintroduction areas, in which people must negotiate not only the symbolic meaning of wolves, but their tangible impacts on local human communities and the larger biotic community, the iconic status of wolves brings variant views of the natural world forward for necessary discussion. Sophisticated treatments of the ethical factors involved in wildlife management are also becoming more prevalent, and one can expect to see more work in this area as ethicists call attention to and offer prescriptions for bridging the gaps between ideas, policy, and practice (see, for example, Jickling and Paquet, 2005; Lynn, 2002, 2006).

Once and Future Wolves

Wolves have clearly been symbolically powerful in various ways throughout human history, and they continue to be so for many people. Historically, across cultures, humans have treated wolves in radically diverse ways depending on their social and geographical contexts. Wolves have been magnets for expressions of loathing and devotion and, in various regions where they are now recovering, they have been iconic animals that illuminate social divisions and conflicting suppositions about shared relationships between humans and the natural world. For the foreseeable future, the prospect of wolf recovery is likely to challenge various groups and individuals to grapple with their relationships to one another, their local landscapes, and why it might be of value to adjust human lifestyles and livelihoods so that wolves may repopulate portions of their former historical ranges.

Further Reading

Coleman, J. T. 2004. *Vicious: Wolves and men in America*. New Haven, CT: Yale University Press.

Clark, T. W., Rutherford, M.B., & Casey, D. 2001. *Coexisting with large carnivores: Lessons from Greater Yellowstone*. Washington D.C./Covelo, CA: Island Press.

Clifford, Hal. 2005. Saved by wolves. In G. Wockner, G. McNamee, & S. Campbell, eds. *Comeback wolves: Western writers welcome the wolf home*, 190–194. Boulder, CO: Johnson Books.

Dunlap, T. 1988. *Saving America's wildlife: Ecology and the American mind, 1850–1990*. Princeton: Princeton University Press.

Hampton, B. 1997. *The great American wolf*. New York: Henry Holt and Company.

Jickling, B., & Paquet, P.C. 2005. Wolf stories: Reflections on science, ethics, and epistemology. *Environmental Ethics*, 27/2, 115–134.

Leopold, A. 1949/1987. *A Sand County almanac* and *Sketches here and there*. New York: Oxford University Press.

Lopez, B. 1976/2004. *Of wolves and men*. New York: Scribner.

Lynn, W. S. 2002. "*Canis lupus cosmopolis*: Wolves in a cosmopolitan worldview." *Worldviews*, 6/3, 300–327.

Lynn, W. S. 2006. Between science and ethics: What science and the scientific method can and cannot contribute to conservation and sustainability. In D. Lavigne, ed. *Wildlife Conservation: In Pursuit of Ecological Sustainability*, 191–205. Limerick, IRL: University of Limerick Press.

McIntyre, R., ed. 1995. *War against the wolf: America's campaign to exterminate the wolf*. Stillwater, MN: Voyageur Press.

Mech, L. D. 1995. The challenge and opportunity of recovering wolf populations. *Conservation Biology*, 9/2, 270–278.

Mech, L. D., & Boitani, Luigi. 2003. *Wolves: Behavior, ecology, and conservation*. Chicago and London: University of Chicago Press.

Schlesier, K. H. 1987. *The wolves of Heaven: Cheyenne shamanism, ceremonies, and prehistoric origins*. Norman, OK: University of Oklahoma Press.

Sharpe, V. A., Norton, B.G., & Donnelley, S. 2001. *Wolves and human communities: Biology, politics, and ethics*. Washington D.C./Covelo, CA: Island Press.

Walker, B. L. 2005. *The lost wolves of Japan*. Seattle: University of Washington Press.

Gavin Van Horn

X

XENOGRAFT

The demand for transplantable tissues and organs is much greater than the supply. Many people on transplant waiting lists die every year; in the United States alone one person dies approximately every 90 minutes waiting for an organ transplant (Satel, 2006). Physicians and medical researchers have long been fascinated by the idea that nonhuman animals might become an appropriate source for organs, and that xenografts, organs or tissues transplanted between animals of different species, could even solve the organ scarcity problem. Supporters of this idea have imagined setting up farms on which animals would be kept at the ready for human beings who need new hearts, livers, kidneys, lungs, or other body parts.

The idea that no one need die waiting for an organ is an attractive one, but there are many obstacles, both technical and ethical, in the way of xenografts becoming the solution to this problem. Technically, organs from nonhumans have not yet been shown to be viable for use in humans. In fact, every effort of this kind, from the implantation of a chimpanzee heart into a 68-year-old man in 1964, through the transplantation of a baboon's heart into the infant Baby Fae twenty years later, to the 1994 attempt to transplant a pig's liver into a 26-year-old woman, has ended dismally. In every case, the patient died shortly after receiving the xenograft.

Yet even if the technical problems are solved someday, the moral problems would remain. The central ethical challenge to xenografting concerns whether taking organs from healthy animals for use in human beings can be justified. A number of serious moral arguments conclude that animals may not be treated in this way, even if doing so would offer a human being a considerable chance of living longer. For example, Tom Regan's claim that many animals, including those which might become attractive organ sources for humans, "have a distinctive kind of value in their own right, if we do; therefore, they too have a right not to be treated in ways that fail to respect this value" would, if correct, imply that xenografting is immoral. An allied view, based on the argument from marginal cases, would also condemn xenografting unless we were willing to regard the mentally handicapped or other marginal members of our species as potential sources of transplant organs as well.

Those who favor developing xenografting as a reliable method of obtaining organs often point out that we take animal lives for many less serious reasons than obtaining organs for people who will die without them. For example, we eat and wear animal products when there is no real life or death need to do so. Further, xenografting is just a particularly visible way in which animals are used in medical research, education, and therapy; a great deal of what happens to any patient in very many medical encounters involves the suffering and death of animals, in drug testing, or as research subjects for physicians and surgeons. Finally, there is great interest among those who are involved in xenograft research in using pigs rather than primates as sources of organs. Whereas primates are scarce, expensive, and disturbingly humanlike, pigs are breakfast food; if it is morally legitimate to raise pigs in confinement settings and then eat sausage, why is it not morally legitimate to genetically engineer pigs in laboratories and then use their organs for people who may die without them? The answer to this question may simply be that it is not morally legitimate to use animals for food and clothing, even though people commonly do, and not defensible to use animals as we have done in medical research, testing, and education.

Recently, medical researchers have shown interest in placing human organs into animals. A goal of such reverse xeno-grafting might be to preserve reproductive organs that would be otherwise damaged by, for example, cancer treatments. This work remains highly experimental, and raises ethical questions similar to those discussed above.

See also Genetic Engineering; Genethics; Marginal Cases

Further Reading

Caplan, Arthur. 1992. "Is xenografting morally wrong?" *Transplantation Proceedings* 24: 722–727.

Kushner, Thomasine, and Belliotti, Raymond. 1985. Baby Fae: A beastly business. *Journal of Medical Ethics* 11.

Nelson, James Lindemann. 1988. "Animals as a source of human transplant organs." In J. Humber and R. Almeder (eds.), *Biomedical ethics reviews 1987*. Totowa, NJ: Humana Press.

Nelson, James Lindemann. 1992. Transplantation through a glass darkly. *Hastings Center Report* 22: 6–8.

Regan, Tom. 1983. *The case for animal rights*. Berkeley: University of California Press.

Satel, Sally. 2006. Death's waiting list. *New York Times*, May 15, 2006.

James Lindemann Nelson

Z

ZOOS: HISTORY

When measured by today's standards, zoos of the 1800s and early 1900s are often said to have been collections of newly-caught wild animals that lived short lives in prison-like barred cages for the pleasure of the paying public. When taken in the context of their times, however, those zoos developed many of the philosophies and husbandry practices of today's professionally managed zoos.

With few exceptions, the earliest collections of captive wild animals were privately-held menageries that were symbols of wealth and power. The ancient Egyptians are thought to be the first people to keep collections of wild animals. Animals of religious significance were kept as representatives of gods. In 1490 BC, Queen Hatshepsut directed an animal collecting trip through Africa to fill her royal menagerie and to trade with neighboring countries. Chinese emperor Wen Wang, of the Chou dynasty, kept a variety of plants and animals in a 1,500-acre Intelligence Park around 1100 BC. Like the menageries in Egypt, it was intended primarily to show the wealth of the empire.

By the third century BC, private collections of animals in Greece were used for study, experimentation, and as pets. Alexander the Great opened the first public menagerie in Alexandria in Egypt.

Wealthy Romans kept small menageries and aviaries in their villas. By the second and first centuries BC, most captive animals were kept on exhibit in imperial menageries until they were sent into the arena to fight people or other animals, or killed for food. The public was charged an admission fee to see them.

In the 1200s, Kublai Khan's collection in Asia held elephants, monkeys, fish, hawks, and other species found in his vast empire. In 1519, conquistador Hernando Cortes visited a large menagerie held by the Aztec King Montezuma in Mexico, which was staffed by 300 keepers. The collection included exhibits featuring American animals as well as human dwarfs and slaves. As in many of today's exhibits, the animals were exhibited in barless, moated enclosures.

By the 1600s, foreign conquests, trade, and the spread of agriculture and industry into undeveloped lands brought tales of great beasts and occasionally living specimens to Western nations. Because collections were still mostly private, the demand for animals in traveling menageries that could be seen by the public grew.

The first modern zoos were European zoological collections like Tierpark Schoenbrunn in Austria, which opened in 1765, Menagerie du Jardin des Plantes in Paris, which opened in 1793, and the London Zoological Garden, which opened in 1828. Like modern zoos, they were open to the public. Animal exhibits were surrounded by exotic plants in a gardenlike setting. These zoos/gardens (hence the term zoological garden) differed from earlier menageries, in that closely related species were exhibited near each other. In keeping with the scientific spirit of the age and the growth of Darwinism, they were established for scientific studies and education.

Exhibit techniques reflected what public attitudes towards animals, technology, space, and resources of the day would allow. Zoo managers and most visitors believed the animals to be in spacious enclosures that resembled, and even improved upon, their natural habitat. Compared with the cramped cages of the more familiar menageries, that was probably true. A landscaped garden surrounding the outside of the enclosure was viewed as representative of a tame jungle. Zoos were living museums.

In the United States, having a zoo in your town became as important as having a museum or art gallery. Many zoos' first animals, directors, curators, and experienced animal keepers were from circuses. The first true European-style zoo in the United States was the Philadelphia Zoo, opened in 1874, which was modeled after the London Zoological Garden. Animals were housed in permanent ornate buildings. The zoo was supported by a zoological society, and it was managed by a director knowledgeable about wildlife. Zoos also began to formally include scientific research as part of their mission. The National Zoo in Washington, D.C. was established in 1891 "for the advancement of science and the instruction and recreation of the people."

As more municipal zoos emerged, there was a competition among zoos to have as many different kinds of animals as possible represented. This is often compared to a stamp collection, because the emphasis was on a great variety of species. Often social animals were not exhibited in groups or pairs, so that the public would not be offended by witnessing breeding behavior. Without breeding in captivity, there was a need to replenish the supply of captive animals. Expeditions were organized to trap and transport wild animals to the zoo. Animal mortality during capture, transport, and at the zoo was high. Since little was known about wild animal care, many exhibits were small and barren, in the belief that this would minimize disease risk. Exhibits were typically barred cages for the safety of the visitors and the animals, and to allow visitors to see the animals as close as possible. Animal buildings were designed for the pleasure of the visitor, and included art such as fine sculpture or tile mosaics.

Around 1907, some zoos began to take advantage of the Hagenbeck Revolution. At his zoo in Hamburg, Germany, Carl Hagenbeck Tierpark, animal supplier Carl Hagenbeck designed concrete moats around exhibits which kept animals in, visitors out, and eliminated the need for bars. His exhibits were recreations of nature as he saw it during his world travels. Exhibit illusions such as a lion sharing space with antelope were created by a moat separating the two animals that was hidden from the visitor's view. The zoological garden had spread from the public walkways into the exhibits. Since Hagenbeck, many zoos have moved from prison-like cages to more naturalistic enclosures.

As the sciences of zoo biology, animal behavior, veterinary medicine, genetics, and animal nutrition grew in the 20th century, animal management improved, more species bred in captivity, and emphasis was no longer on large collections of many species, but on fewer species exhibited in larger, more naturalistic enclosures. There were more mixed-species exhibits and exhibits with social groups of one species. Animals could be exhibited by themes like species relatedness, geographic zone, or habitat. Some zoos chose to focus on local or regional wildlife species. Some zoos, like the Durrell Wildlife Conservation Trust in the United Kingdom, maintain and breed only species that are endangered and can benefit from zoo and field research. New exhibit technologies, coupled with greater knowledge of animal behavior and husbandry, have led to a surge in immersion exhibits that allow visitors to enter the habitat by means of, for example, acrylic tunnels, safari vehicles, and boat rides. Some have even blurred the lines between zoo and aquarium by integrating terrestrial and aquatic exhibits.

With the recognition that many species of animals were threatened with extinction due to human activities, zoos have also become major centers of conservation and public education. Instead of a staff of mostly animal collection managers, modern zoos have added veterinary, nutrition, conservation and research, and education departments. A few animals are bred specifically for reintroduction to places where their numbers are few or they have disappeared completely. The New York Zoological Park is widely credited for rescuing the American bison in the early 1900s through captive breeding and reintroduction. Some zoos maintain their own offsite breeding and research centers. Many zoos have major field research programs. Professional zoo societies have emerged worldwide and have facilitated zoos' working with each other and partnering with conservation groups to address local and international wildlife issues.

Finally, concern about the wellbeing of animals in zoos, particularly since the 1980s, has resulted in increased oversight. This guidance includes self-regulation, requirements for accreditation in professional zoo organizations, and local, state, and national laws.

To the scholar, early zoos were living museums, places of scientific and artistic opportunity. To the visitor, they were urban retreats, gardens of novelty, entertainment, and education. Those roles have not changed, although emphasis is now on education and conservation. Zoos' continued popularity makes them ideal venues for these missions.

Further Reading

Bell, C.E. (ed.) (2001). *Encyclopedia of the world's zoos.* Chicago and London: Fitzroy Dearborn Publishers.
Hanson, E. (2002). *Animal attractions: Nature on display in American zoos:* Princeton, NJ: Princeton University Press.
Hoage, R. J., and Deiss, W. A. (eds.) (1996). *New world, new animals: From menagerie to zoological park in the nineteenth century.* Baltimore: Johns Hopkins University Press.
Maier, F., and Page, J. (1990). *Zoo: The modern ark.* New York: Facts on File.
Mullan, B., and Marvin, G. (1987). *Zoo culture.* London: Weidenfeld and Nicholson.

Michael D. Kreger

ZOOS: ROLES

If animals have a right to freedom, zoos seem to infringe on that right, and therefore to be questionable on welfare grounds. Today's thousands of zoos, attracting millions of visitors worldwide, vary enormously from so-called roadside zoos, which are condemned outright by reputable ones, to zoological parks whose animals, many of them in large, naturalistic, and/or behaviorally enriched enclosures, often give every indication of being in a state of wellbeing.

The question remains whether it is still misguided, as some feel, to maintain wild animals, how ever well cared for, outside their natural habitats, to which millions of years of evolution have adapted them. Zoos and their critics agree now that wild species must be protected, and reputable zoos now take very few animals, especially mammals, from the wild, though they need to do this occasionally for serious conservational reasons. If it is acceptable to keep domesticated animals, perhaps it is not wrong to keep what can only be relatively wild animals in zoos. Indeed, some of them could be argued to be slightly domesticated because of their individual adjustment to zoo conditions, or because of some perhaps unavoidable selective breeding. It is true that many domesticated animals, such as intensively reared hens and pigs, are kept in appalling conditions, but this is because of economic greed, not because they cannot be kept humanely. Zoo animals' captive environments can similarly be vastly improved by studying their behavioral requirements.

The degree to which animals show their natural behavior is a main criterion for judging their wellbeing or otherwise, as well as a guide to how their facilities may be improved. Other criteria include their degree of physical health, their readiness to breed, and the degree to which they show abnormal behavior such as the stereotyped weaving of some captive polar bears.

If animals in zoos are only relatively wild or even slightly domesticated, this makes keeping them more acceptable, but at the same time it casts doubt on zoos'

claim to maintain truly wild animals, and on whether these animals or their descendants could successfully be reintroduced to the wild. This is one of many real problems for zoos, and some critics deny their ability to save animals who are wild in any meaningful sense. On the other hand, zoos now have elaborate conservational arrangements to help to maintain their animals' wildness, at least genetically. These include studbooks for many endangered species and computerized, linked animal records (part of ISIS, the International Species Information System, started thirty years ago) to assist in the management of zoo animals as members of total captive populations with minimal inbreeding and maximal genetic diversity, as in a wild population. Enlightened zoo conditions help to maintain behavioral wildness also. Successful reintroductions have already occurred, such as the reintroduction of the Arabian oryx. However, just how successful some reintroductions have been, for example, that of the golden lion tamarin, is arguable. Thus zoos' ability to save, or at least reintroduce, many wild species remains unproven. However, threats face many wild species, from the hunting of rhinos and tigers to the threats to almost all wild habitats from the exploding human population, and zoos can help considerably. Again, some critics see a concentration on captive breeding as a dangerous distraction from the primary conservational task of protecting actual wild habitats. But zoos see their captive breeding as merely complementing this, and some zoo scientists assist greatly in the protective management of actual wild populations. Many more zoos help to educate the public about threats to wild habitats. Zoos' conservational roles also bring their own moral problems, such as whether saving endangered species can justify killing surplus animals, for example, nearly eighty hybrid orangutans in American zoos who are unsuitable for reintroductions.

Serious zoos are in many ways allies of all those who care about animals as individuals and about their survival as species. Apart from their conservational captive breeding, zoos constitute a kind of powerhouse of ordinary people's fondness and concern for animals. Though zoo critics tend to see zoos as demonstrations of domination over nonhumans, many of the millions who visit zoos probably do it because of animals' appeal to them. These people are potentially a huge body of support for conservation and animal protection. A first step here is the introduction of legislation to regulate zoos, which already exists in Britain and some other European Union countries. In Britain also, detailed welfare and conservational requirements are laid down in the Secretary of State's Standards of Modern Zoo Practice. In the United States, zoos are licensed and inspected by the U.S. Department of Agriculture and other agencies.

See also Enrichment and Wellbeing for Zoo Animals

Further Reading

Bostock, S. St.C. 1993. *Zoos and animal rights: The ethics of keeping animals*. London and New York: Routledge.

Broom, D. M., and Johnson, K. G. 1993. *Stress and animal welfare*. London: Chapman and Hall.

Margodt, K. 2000. *The welfare ark: Suggestions for a renewed policy in zoos*. Brussels: VUB University Press.

Norton, B. G., Hutchins, M., Stevens, E. F., and Maple eds. T. L., eds. 1995. *Ethics on the ark: Zoos, animal welfare, and wildlife conservation*. Washington, DC: Smithsonian Institution Press.

Secretary of State's Standards of modern zoo practice. London: Department of the Environment, Transport and the Regions, 2000; www.defra.gov.uk/wildlife-country side/gwd/zooprac/pdf.

Shepherdson, D. J., Mellen, J. D., and Hutchins, M., eds. 1998. *Second nature: Environmental enrichment for captive animals*. Washington, DC: Smithsonian Institution Press.

Tudge, C. 1992. *Last animals at the zoo*. Oxford: Oxford University Press.

World Association of Zoos and Aquariums (WAZA), *Building a future for wildlife: The world zoo and aquarium conservation strategy*. Bern, Switzerland: WAZA Executive Office; www.waza.org/conser vation/wzacs.php.

Stephen St.C. Bostock

ZOOS: WELFARE CONCERNS

In recent years there has been a great deal of discussion about the welfare of animals raised for food, used in research, and confined in zoos. This has led to discussion of what welfare consists of, attempts at behavioral enrichment, and debate about whether adequate levels of animal welfare can ever be secured in zoos, laboratories, and slaughterhouses.

In addition to these concerns about welfare, another critique has developed that appeals to a wide range of animal interests. Some critics have argued that the fact that animals may suffer in zoos and laboratories is only part of what makes these facilities unjust. To use a common but controversial analogy, what was wrong with American slavery was not only that slaves suffered, but that slavery systematically violated virtually every significant interest of those who were legally defined as slaves. Similarly, what is wrong with zoos, in this view, is not only that they cause animal suffering, but that they violate a range of interests that are central to the lives of many animals. Just as happy slaves do not justify slavery, so behaviorally enriched animals do not justify zoos.

This second critique can only have moral force among people who already believe that animals have significant moral standing. Once this is granted, zoos become morally suspect, since virtually all creatures with significant moral standing have an interest in directing their own lives. If animals are to be confined in zoos, then the moral claim in favor of respecting this interest will have to be overcome.

Some, such as Tom Regan, argue that this moral claim cannot be overcome. Humans and many nonhumans enjoy an equal moral status that manifests in rights. Fundamental rights, in his view, can almost never be infringed. Since zoos violate the rights of many animals, they are thus morally indefensible.

Others, such as Dale Jamieson, believe that this presumption can in principle be overcome, if there are weighty enough reasons for keeping animals in captivity. In recent years, education and conservation have been invoked most frequently in attempts to justify zoos.

While an appeal to education may carry some weight, it is alarming that there is so little empirical evidence about what zoos actually achieve in their educational efforts. Anecdotes are available, but reliable data are hard to come by. But even if we grant that zoos are successful in educating the public in some positive way, given the technological resources (webcams, virtual reality, etc.) that are now coming online, it is far from clear that holding animals in captivity is necessary for delivering positive educational results.

Conservation is the justification most often appealed to by scientists in the zoo

A rhinoceros sleeps against a barred window in an impoverished cage in a zoo. (Dreamstime.com)

community. There are variations on the theme. Some want to use zoos as bases for captive breeding and reintroduction. Others want to use the economic and political power of zoos to protect habitat. Still others would be satisfied if zoos could be constituted as genetic libraries for animals that no longer exist in viable populations. Sometimes it seems that conservation is so highly valued that any activity directed toward this end is thereby considered justified.

Despite the rhetoric, most zoos have no habitat conservation programs, and among those that do, it is rare when more than one to two percent of the budget is spent on them. Reintroduction has been a mixed success. Benjamin Beck, formerly Chair of the American Zoo and Aquarium Association's Reintroduction Advisory Group, rather understates his conclusion when he writes, ". . . we must acknowledge frankly at this point that there is not overwhelming evidence that reintroduction is successful." David Hancocks, who directed zoos in both the United States and Australia, writes that "[t]here is a commonly held misconception that zoos are not only saving wild animals from extinction but also reintroducing them to their wild habitats."

Whatever the role of captive breeding and reintroduction in species preservation, an inconsistency arises when it is enlisted as a justification for zoos. Zoos are places where people come to see animals. They are places to take children on Sunday afternoons. They are urban amenities like football and baseball teams, part of what makes a city big league. Increasingly, zoos are even the sites of rock concerts and fundraisers. However, the best institutions for captive breeding and reintroduction would not play these roles. They would remove animals from

excessive contact with people, give them relatively large ranges, and prepare them for reintroduction in ways that zoo visitors might find shocking, for example by developing their competence as predators. What the importance of captive breeding and reintroduction justifies, if anything, is not the existing system of zoos, but a different kind of institution entirely, one that protects animals from people rather than putting them on display. Thus, whatever the power of the appeal to conservation, the present zoo system seems to be unjustified.

Arguments whose conclusions diverge from what people want to believe are not always greeted with enthusiasm. Yet, in 1994, the citizens of Vancouver, Canada voted to close the Stanley Park Zoo, and San Francisco, Chicago, Detroit, Philadelphia, and the Bronx Zoo in New York City have all agreed to stop exhibiting elephants. However, since most zoos will continue to exist for the foreseeable future, we should ensure that the highest standards of welfare are maintained.

Further Reading

Beck, Benjamin. 1995. Reintroduction, zoos, conservation, and animal welfare. In B. Norton, M. Hutchins, E. Stevens, and T. Maple eds., *Ethics on the ark: Zoos, animal welfare, and wildlife conservation,* 155–163. Washington: Smithsonian Institution Press.

Hancocks, David. 2001. *A different nature.* Berkeley, University of California Press.

Hancocks, David. 1995. An introduction to reintroduction. In B. Norton, M. Hutchins, E. Stevens, and T. Maple eds., 181–183. *Ethics on the ark: Zoos, animal welfare, and wildlife conservation.* Washington: Smithsonian Institution Press.

Jamieson, Dale. 1985. "Against zoos." In Peter Singer, ed., *In defense of animals*, 108–117. New York: Basil Blackwell.

Jamieson, Dale. 1995. Zoos revisited. In B. Norton, M. Hutchins, E. Stevens, and

T. Maple eds., *Ethics on the ark: Zoos, animal welfare, and wildlife conservation,* 52–66. Washington: Smithsonian Institution Press.

Norton, B., Hutchins, M., Stevens, E., and Maple eds., T. 1995. *Ethics on the ark: Zoos, animal welfare, and wildlife conservation.* Washington: Smithsonian Institution Press.

Regan, T. 1983. *The case for animal rights.* Berkeley: University of California Press.

Regan, T. 1995. Are zoos morally defensible? In B. Norton, M. Hutchins, E. Stevens, and T. Maple eds., *Ethics on the ark: Zoos, animal welfare, and wildlife conservation,* 38–51. Washington: Smithsonian Institution Press.

Dale Jamieson

Chronology of Historical Events in Animal Rights and Animal Welfare

Below is a brief chronology of some historical events in the United States, unless otherwise indicated, related to the use of animals and to animal rights and animal welfare. "UK" stands for the United Kingdom.

Globally, there is a lot of ongoing legislation concerning animal protection. Details about the federal Animal Welfare Act can be found at http://www.nal.usda.gov/awic/legislat/usdaleg1.htm. A general search on Google for "history animal protection" will lead to numerous web sites on legislation in specific countries and for specific species. The Animal Welfare Information Center (AWIC) Newsletter updates information in its "Congress in Action" section (see also http://awic.nal.usda.gov/nal_display/index.php?info_center=3&tax_level=1).

For more information, see the first edition of the *Encyclopedia of Animal Rights and Animal Welfare* (Greenwood Publishing Group, 1998) and also Michigan State University College of Law: Animal Legal & Historical Web Center (www.animallaw.info); Animal Law Review (http://www.lclark.edu/org/animalaw/); National Center for Animal Law (http://www.lclark.edu/org/ncal/); Society for Animal Protective Legislation (http://www.saplonline.org/); A Chronology of Key Events in the Scientific Use of Chimpanzees in the United States (http://www.releasechimps.org/pdfs/chronology-of-key-events.pdf); the National Association for Biomedical Research: Animal Law Section (http://www.nabr.org/AnimalLaw/contactUs.htm); http://worldanimal.net/const-leaflet.htm for a global summary of animal protection legislation; the website for the Animal Welfare Institute (http://www.awionline.org/legislation/index.htm; http://capwiz.com/compassionindex/issues/bills; http://www.awionline.org/links.html); and the website for the Animal Welfare Information Center (http://awic.nal.usda.gov/nal_display/index.php?info_center=3&tax_level=1).

A summary of regulations on the use of animals in research in European countries can be found at http://arbs.biblioteca.unesp.br/viewissue.php in the essay by Annamaria Passantino titled "Application of the 3Rs Principles for Animals Used for Experiments at the Beginning of the 21st Century."

A summary of species used in research can be found at http://www.hsus.org/animals_in_research/species_used_in_research/ and A Timeline of Progress for Farm Animals can be found at http://www.americanhumane.org/site/PageServer?pagename=pa_farm_animals_timeline.

1822   Ill-Treatment of Cattle Act

1822   Martin's Anticruelty Act (UK)

635

1824   Society for the Prevention of Cruelty to Animals (SPCA) (UK) founded

1826   Bill to Prevent the Cruel and Improper Treatment of Dogs

1832   Warburton Anatomy Act (UK)

1840   SPCA becomes the Royal Society for the Prevention of Cruelty to Animals (RSPCA) with patronage of Queen Victoria (UK)

1866   American Society for the Prevention of Cruelty of Animals (ASPCA) founded

1868   Massachusetts Society for the Prevention of Cruelty to Animals (MSPCA) founded

1875   Victoria Street Society for the Protection of Animals from Vivisection (UK) founded

1876   Cruelty to Animals Act (UK)

1877   American Humane Association founded

1883   American Anti-Vivisection Society founded

1889   American Humane Education Society (AHES) founded

1891   The Humanitarian League founded

1895   New England Anti-Vivisection Society founded

1898   British Union for the Abolition of Vivisection (UK)

1906   Animal Defence and Anti-Vivisection Society (UK) founded

1911   Protection of Animals Act (England, UK)

1912   Protection of Animals Act (Scotland, UK)

1925   The Performing Animals (Regulations) Act (UK)

1926   University of London Animal Welfare Society founded (name changed to Universities Federation for Animal Welfare [UFAW] in 1938) (UK)

1929   Victoria Street Society for the Protection of Animals from Vivisection changes its name to National Anti-Vivisection Society (UK)

1946   National Society for Medical Research founded

1948   Morris Animal Foundation founded

1949   The Docking and Nicking of Animals Act

1950   Animal Protection Law (covers farm animals and bans battery cages) (Denmark)

1951   Animal Welfare Institute founded

1952   Institute for Animal Laboratory Resources founded

1954   Humane Society for the Unite States (HSUS) founded

1954   The Protection of Animals (Anaesthetics) Act (UK)

1955   Society for Animal Protective Legislation founded

1957   Friends of Animals founded

1958   Humane Slaughter Act

1959   Beauty without Cruelty (UK) founded

1959   Wild Horses Act

1959   Catholic Society for Animal Welfare (now International Society for Animal Rights) founded

1960   The Abandonment of Animals Act (UK)

1961   Lawson-Tait Trust (UK) founded

1962   Bald and Golden Eagle Protection Act

1962   The Animals (Cruel Poisons) Act (UK)

1963   British Hunt Saboteurs Association (UK) Act

1965   Brambell Report on Farm Animal Welfare (UK)

1965   Littlewood Report (UK, a discussion of alternatives to the use of animals)

1965   American Association for Accreditation of Laboratory Animal Care founded

1966   Laboratory Animal Welfare Act

1967   Fund for Animals (UK) founded

1967   Farm Animal Welfare Advisory Committee (UK) founded

1968   Animal Protection Institute founded

1969   Council of Europe Convention on Animals in Transport

1969   International Fund for Animal Welfare (IFAW) founded

1969   Endangered Species Act

1969   Fund for the Replacement of Animals in Medical Experiments (FRAME) (UK) founded

1969   International Association against Painful Experiments on Animals (UK) founded

1970   Laboratory Animal Welfare Act broadened and renamed Animal Welfare Act; legislation extended to include all warm-blooded animals (including pet and exhibition trades)

1970   Dr. Hadwen Trust for Humane Research (UK) founded

1971   Greenpeace (now International) founded

1971	Wild Free-roaming Horse and Burro Act
1971	Law requiring approval of new buildings for animal protection (Sweden)
1972	American Zoo and Aquarium Association accreditation standards and code of professional ethics
1972	Marine Mammal Protection Act
1972	Animal Protection Act (Germany)
1973	International Primate Protection League founded
1973	National Antivivisection Society founded
1973	Endangered Species Act
1973	Convention on International Trade in Endangered Species (CITES) of wild fauna and flora (international) is signed by 167 countries
1976	Animal Rights International founded by Henry Spira
1976	Animal Welfare Act broadened to cover, among other things, transportation and prohibitions against dog fighting and cockfighting
1976	Horse Protection Act
1976	Fur Seal Act
1976	Protest at American Museum of Natural History (Henry Spira)
1976	The Dangerous Wild Animals Act (UK)
1977	First International Conference on the Rights of Animals, Trinity College, Cambridge, England (organized by Andrew Linzey and Richard Ryder)
1978	Humane Slaughter Act broadened
1978	Animal Legal Defense Fund founded
1978	Swiss Animal Welfare Act
1979	Association for Biomedical Research (founded as Research Animal Alliance) founded
1979	Coalition to Abolish the Draize Test founded by Henry Spira
1979	First European Conference on Farm Animal Welfare, the Netherlands
1979	Packwood-Magnuson Amendment to the International Fishery Conservation Act
1980	People for the Ethical Treatment of Animals (PETA) founded
1980	Psychologists for the Ethical Treatment of Animals (PsyETA) founded
1981	Association of Veterinarians for Animal Rights (AVAR) founded
1981	Johns Hopkins Center for Alternatives to Animal Testing founded

1981   Silver Spring monkeys case, which led to the 1985 revision of the federal Laboratory Animal Welfare Act

1981   The Zoo Licensing Act (UK)

1981   Foundation for Biomedical Research founded

1982   Marine Mammal Protection Act reauthorized

1982   World Women for Animal Rights/Empowerment Vegetarian Activist Collective founded

1983   In Defense of Animals founded

1984   Humane Farming Association founded

1984   Performing Animal Welfare Society founded

1984   Break-in, Head Injury Clinical Research Laboratory, University of Pennsylvania

1985   Improved Standards for Laboratory Animals Act (an amendment of the Animal Welfare Act) to include requirements for psychological enrichment for non human primates. It mandates minimal cage size (for chimpanzees: 5 x 5 x 7 feet).

1985   Head Injury Clinical Research Laboratory closed

1985   National Association for Biomedical Research founded (merger of National Society for Medical Research, Association for Biomedical Research, and Foundation for Biomedical Research)

1985   Jews for Animal Rights founded

1986   Farm Animal Reform Movement (FARM) founded

1986   Animal Welfare Information Center founded

1986   European Directive Regarding the Protection of Vertebrate Animals Used for Experimental and Other Scientific Purposes (Council of Europe)

1988   Swedish Animal Welfare Act

1989   Veal Calf Protection Bill hearings (U.S. Congress)

1990   Veal Crate Ban (UK)

1990   Pet Theft Act, amendment to the Animal Welfare Act

1990   Rutgers Animal Rights Law Center founded

1991   The Ark Trust, Inc., founded

1991   Americans for Medical Progress founded

1991   European Union Regulation against Leghold Traps

1992   Czechoslovakian Law against Cruelty on Animals (first welfare legislation in former Communist countries)

1992   Wild Bird Conservation Act

1992   International Dolphin Conservation Act

1992   Driftnet Fishery Conservation Act

1992   Protection of Animal Facilities Act

1992   Animal Enterprise Protection Act

1993   National Health Revitalization Act

1993   First World Congress on Alternatives and Animals in the Life Sciences held in Baltimore, Maryland

1993   European Centre for the Validation of Alternative Methods (ECVAM)

1995   Second World Congress on Alternatives and Animals in the Life Sciences, Utrecht, Netherlands

1996   House of Representatives holds a hearing on the Society Animal Protection Legislation (SAPL)-supported Pet Safety and Protection Act

1996   The Pet Protection Act is considered

1997   Dolphin Protection Consumer Information Act makes it a violation of the Federal Trade Commission Act for any producer, importer, exporter, distributor, or seller of any tuna product that is exported from or offered for sale in the United States to include on the label of that product the term "dolphin safe" or any other term or symbol that falsely claims or suggests that the tuna contained in the product were harvested using a method of fishing that is not harmful to dolphins if the product was obtained by tuna harvesting

1998   Multinational Species Conservation Fund was created to carry out the African Elephant Conservation Act, the Asian Elephant Conservation Act, and the Rhinoceros and Tiger Conservation Act

1999   New legislation bans the creation, sale, and possession with intent to sell, of animal crushing or stomping films

1999   Depiction of Animal Cruelty: This statute makes it a crime knowingly to create, sell, or possess any visual or audio depiction of animal cruelty with the intention of placing that depiction in interstate or foreign commerce for commercial gain. It provides an exception for "any depiction that has serious religious, political, scientific, educational, journalistic, historical, or artistic value."

1999   The New Zealand Animal Welfare Act becomes law. Great apes are banned from use in research, testing, or teaching.

2000   The United States passes the Chimpanzee Health Improvement, Maintenance, and Protection (CHIMP) Act. The CHIMP Act provides for the retirement and lifetime care of chimpanzees not in active protocols, prohibits euthanasia and breeding, but allows for them to be recalled back into research

2000   The Great Ape Conservation Act establishes a $5 million conservation fund to assist in global projects to conserve great ape populations.

2000   New legislation requires the immediate termination of the Department of Defense practice of euthanizing military working dogs at the end of their useful working life, and facilitates the adoption of retired military working dogs.

2000   The Dog and Cat Protection Act of 2000 makes it unlawful to import into or export from the United States any dog or cat fur product, or to engage in the commerce of any dog or cat fur product.

2000   ICCVAM Authorization Act of 2000 provides that the Interagency Coordinating Committee on the Validation of Alternative Methods (ICCVAM) shall among other things review and evaluate new or revised or alternative test methods; the ICCVAM was established by the Director of the National Institute of Environmental Health Sciences.

2002   Senate passes the Bear Protection Act; the bill eliminates the incentive to kill bears for their gallbladders by making it illegal to sell, import, or export the internal organs of a bear, as well as products containing bear parts.

2002   Farm Bill includes a Resolution dictating that the Humane Methods of Slaughter Act of 1958 should be fully enforced, preventing the needless suffering of animals; it also requires the Secretary of Agriculture to track violations of the Act and report results to Congress.

2002   Rhinoceros and Tiger Conservation Act, African Elephant Conservation Act, and Asian Elephant Conservation act are all reauthorized.

2002   Four new laws are enacted, and cruelty to animals is now a felony in 37 states.

2002   Sweden bans the use of great apes in biomedical research including a ban on the lesser apes, gibbons, and siamangs.

2003   The Captive Wildlife Safety Act prohibits the interstate transport of exotic big cats for private ownership as pets.

2004   The Marine Turtle Conservation Act states that its purpose is to assist in the conservation of marine turtles and the nesting habits of marine turtles in foreign countries by supporting and providing financial resources for projects to conserve the nesting habitats, conserve marine turtles in those habitats, and address other threats to the survival of marine turtles.

2005   The House of Representatives passes an amendment to stop the use of taxpayers' dollars to fund horse slaughterhouse inspection, effectively banning horse slaughter for one fiscal year if passed.

2006   The Supreme Court of India bans breeding of animals in zoos.

2006   Germany bans imported seal products from the country.

2006   Arizona and Michigan pass animal protection laws in their respective states. Arizona prohibits confining calves in veal crates and confining breeding pigs in gestation crates. Michigan rejected a proposal that would have permitted a target-shooting season on the mourning dove, the state's official bird of peace, a protected species there since 1905.

2008   Colorado bans the veal crate and the gestation crate.

2008   The Spanish Parliament extends rights to great apes.

2008   The U.S. House of Representatives passes a bill to halt the interstate primate pet trade.

2008   Bullfighting is banned by the mayor in the Portuguese town of Viana do Castel.

2008   Proposition 2 passed in California. This law phases out some of the most restrictive confinement systems used by factory farms—gestation crates for breeding pigs, veal crates for calves, and battery cages for egg-laying hens—affecting 20 million farm animals in the state by simply granting them space to stand up, stretch their limbs, turn around, and lie down comfortably. (http://www.farmsanctuary.org/mediacenter/2008/pr_prop2_victory.html)

2009   On March 14, U.S. Department of Agriculture Secretary Tom Vilsack amended federal meat inspection regulations to completely ban the slaughter of downer cows, those cattle that become non-ambulatory disabled after passing initial inspection.

Resources on Animal Rights
and Animal Welfare

SEARCHING THE LITERATURE ONLINE

Information on both animal welfare and animal rights is abundant. Accessing the Internet and typing a few words in a Google search box retrieves more results than there is time to read. Refining a search in order to retrieve fewer but more relevant results requires consideration of both how and where to search.

The search terms used directly affect retrieved results, no matter where one searches. For example, searching by each of the following six terms individually—*euthanasia, endpoint, slaughter, sacrifice, kill,* or *death*—will produce six lists with very different results. A search using the term vivisection and another using the terms animal research will also retrieve two sets of unique results. This is because Google does not interpret what is meant by a term; rather, it searches literally, letter by letter. Therefore, the term must be used in the search in order to identify any sites using that same term; consider your search terms carefully.

The source of the site, the citation, or the information retrieved in the search results are further essential considerations. Searchers should review who has posted the resource, and evaluate their qualifications and level of expertise. GoogleScholar is one way to limit a search to the scholarly literature, such as peer-reviewed papers, theses, books, abstracts, and articles from academic publishers, societies, and universities. The Advanced Search option is particularly useful in narrowing a search to results that are precise and authoritative.

The free databases published by the U.S. government, PubMed (published by the U.S. National Library of Medicine), http://www.ncbi.nlm.nih.gov/pubmed/, and Agricola (published by the National Agricultural Library), http://agricola.nal.usda.gov/, allow users to access research publications in the medical and agricultural fields, respectively. Academic and research institutions and libraries subscribe to proprietary databases, such as PsycInfo and Web of Science, which offer additional avenues to search for and locate useful and reliable information.

Finally, the general resources listed in "Books, Essays, and Organizations" below, as well as the "Further Resources" sections at the ends of the entries in the encyclopedia, provide valuable tools to identify and obtain information of interest.

Resources for Searching Animal Rights Literature

Allen, T., & Jensen, D. 2006. Searching bibliographic databases for alternatives. *Animal Welfare Information Center Bulletin,* 12, 1–16.

Hart, L., Wood, M., & Weng, H. 2005. Effective searching of the scientific literature for alternatives: search grids for appropriate databases. *Animal Welfare,* 14, 287–289.

Smith, A., & Allen, T. 2005. The use of databases, information centres and guidelines when planning research that may involve animals. *Animal Welfare,* 14, 347–359.

Wood, M. 2006. Techniques for searching the AAT literature. In *Handbook on Animal-Assisted Therapy: Theoretical Foundations and Guidelines for Practice*, 2nd ed. (ed. by A. Fine), 413–423. Boston: Elsevier/Academic Press.

Wood, M. 2007. Education: Information resources on humans and animals. In *Encyclopedia of Human-Animal Relationships: A Global Exploration of Our Connections with Animals*, ed. by M. Bekoff, 678–680. Westport, CT: Greenwood Press.

Mary W. Wood

GENERAL ONLINE AND PRINT PUBLICATIONS

This list of general sources will provide excellent references for a wide variety of issues centering on animal rights and animal welfare. For extensive references, please see the list of sources accompanying each essay in this encyclopedia, the web sites that are included in the list of contributors, and the following web sites and publications.

Web Sites

- Animal Ethics. A Philosophical Discussion of the Moral Status of Nonhuman Animals. http://animalethics.blogspot.com/. A blog spot that lists organizations and books, and provides blogs and discussion on animal rights and welfare.

- Animal People Online. http://www.animalpeoplenews.org/. An online news source for information about animal protection worldwide.

- Animal Welfare Institute. http://www.awionline.org/pubs/online.html; Lists Animal Welfare Institute online publications.

- Center for Environmental Philosophy. University of North Texas. http://www.cep.unt.edu/iseebooks.html. Selected Environmental Ethics Books.

- GEARI. Group for the Education of Animal Related Issues. http://www.geari.org/animal-rights-organizations.html. Provides a list of animal rights organizations.

- Google.com. Books on Animal Ethics. http://www.google.com/products?client=safari&rls=en&q=books+on+animal+ethics&ie=UTF-8&oe=UTF-8&um=1&sa=X&oi=product_result_group&resnum=4&ct=title; Provides a list of books on animal ethics, which link to commercial book ordering Web sites.

- Green People. http://www.greenpeople.org/animalrights.htm. Includes updated information about animal rights organizations in the United States, Canada, and other countries.

- Speak Out for Species at the University of Georgia: http://www.uga.edu/sos/organizations.html. Features a list of animal rights organizations.

- U.S. Department of Agriculture. National Agriculture Library. Animal Welfare Information Center. http://www.nal.usda.gov/awic/pubs/bulletin.shtml; Provides access to online information sources available through the U.S. National Agriculture Library, including all government-funded reports and databases. Searchable access points include research animals; farm animals; zoo, circus, and marine animals; companion animals; government and professional resources; alternatives; literature searching and databases; pain and distress; and "humane endpoints and euthanasia."
- WordPress.com. http://wordpress.com/tag/animal-protection-organizations/. "Blogs about Animal Protection Organizations."
- World Animal Net. http://www.worldanimal.net/. World Animal Net is the world's largest network of animal protection societies with consultative status at the UN.
- Yahoo. Directory. Animal Rights Organizations. http://dir.yahoo.com/Science/Biology/Zoology/Animals__Insects__and_Pets/Animal_Rights/Organizations/. Lists more than 100 organizations.

Publications (Print and Online)

Adams, Carol J. 1994, *Neither man nor beast: Feminism and the defense of animals.* New York: Continuum Publishing Company.

Adams, Carol J. 1999. *The sexual politics of meat: A feminist vegetarian critical theory* (10th Anniversary Edition). New York: Continuum.

Anderson, Allen, and Anderson, Linda. 2006. *Rescued: Saving animals from disaster.* Novato, CA: New World Library.

Anderson, Virginia DeJohn. 2004. *Creatures of empire: How domestic animals transformed early america.* New York: Oxford University Press.

Anthony, L. 2007. *Babylon's ark: The incredible wartime rescue of the Baghdad zoo.* New York: Thomas Dunne Books.

Appleby, M. C., Mench, J. A., and Hughes, B. O. 2004. *Poultry behaviour and welfare.* Cambridge, MA: CABI Publishing.

Arluke, A. 2004. *Brute force: Animal police and the challenge of cruelty.* West Lafayette, IN: Purdue University Press.

Armstrong, S., ed. 2003. *The animal ethics reader.* New York: Routledge.

Balcombe, J. P. 2006. *Pleasurable kingdom: Animals and the nature of feeling good.* London: Macmillan.

Bateson, P.P.G. 1991. Assessment of pain in animals. *Animal behaviour* 42, 827–839.

Baur, G. 2008. *Farm sanctuary: Changing hearts and minds about animals and food.* New York: Touchstone.

Beck, Alan M., and Bekoff, M. 2002. *Minding animals: Awareness, emotions, and heart.* New York: Oxford University Press.

Bekoff, M. 2006. "Animal emotions and animal sentience and why they matter: Blending 'science sense' with common sense, compassion and heart." In *Animals, Ethics, and Trade,* J. Turner and J. D'Silva, eds., 27–40. London: Earthscan Publishing.

Bekoff, M. 2006. *Animal passions and beastly virtues: Reflections on redecorating nature.* Philadelphia: Temple University Press.

Bekoff, M. 2006. "Animal passions and beastly virtues: Cognitive ethology as the unifying science for understanding the subjective, emotional, empathic, and moral lives of animals." *Zygon (Journal of Religion and Science)* 41, 71–104.

Bekoff, M. 2006. "The public lives of animals: A troubled scientist, pissy baboons, angry elephants, and happy hounds." *Journal of Consciousness Studies* 13, 115–131.

Bekoff, M. 2007. *Animals matter: A biologist explains why we should treat animals with compassion and respect.* Boston: Shambhala.

Bekoff, M. 2007. *The emotional lives of animals.* Novato, CA: New World Library.

Bekoff, M. 2007. Why "good welfare" isn't "good enough": Minding animals and increasing our compassionate footprint. Available at http://arbs.biblioteca.unesp.br/viewissue.php.

Bekoff, M., and Jamieson, D. 1996. "Ethics and the study of carnivores: Doing science while respecting animals." In J. L. Gittleman, ed., *Carnivore behavior, ecology, and evolution,* Vol. 2, 15–45. Ithaca, NY: Cornell University Press.

Bekoff, M., and Pierce, J. 2009. *Wild justice: The moral lives of animals.* Chicago: University of Chicago Press.

Bradshaw, G., Schore, A. N., Brown, J. L., Poole, J. H., and Moss, C. 2005. Elephant breakdown. *Nature* 433: 807.

Brakes, P., Butterworth, A., Simmonds, M., and Lymbery, P. 2004. *Troubled waters: A review of the welfare implications of modern whaling activities.* World Society for the Protection of Animals, London. http://www.wdcs.org/submissions_bin/troubledwaters.pdf.

Broom, D. M. 2008. Welfare assessment and relevant ethical decisions: Key concepts. Available at http://arbs.biblioteca.unesp.br/viewissue.php.

Burgess, C. and Dubbs, C. 2007. *Animals in space.* New York: Springer.

Caras, Roger. 2002. *A perfect harmony: The intertwining lives of animals and humans throughout history.* New York: Simon & Schuster.

Carbone, L. 2004. *What animals want: Expertise and advocacy in laboratory animal welfare policy.* New York: Oxford University Press.

Cavalieri, P., and Singer, P., eds. 1993. *The Great Ape Project: Equality beyond humanity.* London: Fourth Estate.

Clubb, R., Rowcliffe, M., Lee, P., Mar, K. U., Moss, C., and Mason, G. 2008. "Compromised survivorship in zoo elephants." *Science* 322, 1649.

Cooper, Jilly. 1983. *Animals in war.* London: Heinemann.

Crist, E. 1999. *Images of animals: Anthropomorphism and animal mind.* Philadelphia: Temple University Press.

Davis, K. 2001. *More than a meal: The turkey in history, myth, ritual, and reality.* New York: New Lantern Books.

Davis S. G. 1997. *Spectacular nature: Corporate culture and the Sea World experience.* Berkeley, CA: University of California Press.

Dawn, K. 2008. *Thanking the monkey: Rethinking the way we treat animals.* New York: HarperCollins.

Eisner, G. A. 1997. *Slaughterhouse.* New York: Prometheus.

Foltz, Richard C. 2005. *Animals in Islamic tradition and Muslim cultures.* Oxford: Oneworld.

Forthman, D., Kane, L. F., Hancocks, D., and Waldau, P. F., eds. 2009. *An elephant in the room: The science and well-being of elephants in captivity.* Tufts Center for Animals and Public Policy, Tufts University, North Grafton, Massachusetts.

Fox, C. H., and Papouchis, C. M., eds. 2004. *Cull of the wild: A contemporary analysis of wildlife trapping in the United States.* Sacramento, CA: Animal Protection Institute.

Francione, G. L. 2000. *Introduction to animal rights: Your child or the dog?* Philadelphia: Temple University Press.

Francione, G. L. 2008. *Animals as person: Essays on the abolition of animal exploitation.* New York: Columbia University Press.

Fraser, D. 2008. *Understanding animal welfare: The science in its cultural context.* Sussex, UK: Wiley-Blackwell.

Garner, R. 1998. "The economics and politics of animal exploitation." In *Political animals: Animal protection politics in Britain and the United States.* New York: St. Martin's Press.

Goodall, J., and Bekoff, M. 2002. *The ten trusts: What we must do to care for the animals we love.* HarperCollins, San Francisco.

Greek, C. R., and Greek, J. S. 2000. *Sacred cows and golden geese: The human cost of experiments on animals.* New York: Continuum.

Greek, C. R., and Greek, J. S. 2002. *Specious science: How evolution and genetics explain why medical research on animals kills humans.* New York: Continuum.

Hall, Lee. 2006. *Capers in the churchyard: Animal rights advocacy in the age of terror.* Darien CT: Nectar Bat Press.

Hatkoff, A. 2009. *The inner world of farm animals.* New York: Stewart, Tabori, and Chang.

International Society for Anthrozoology. "ISAZ: International Society for Anthrozoology". ISAZ. http://www.vetmed.ucdavis.edu/CCAB/isaz.htm.

Irvine, L. 2004. *If you tame me: Understanding our connection with animals.* Philadelphia, PA: Temple University Press.

Irvine, L. 2009. *Filling the ark.* Philadelphia: Temple University Press,.

Jamieson, D. 2008. *Ethics and the Environment.* New York: Cambridge University Press.

Lawrence, E. A.1982. *Rodeo: An anthropologist looks at the wild and the tame.* Knoxville: University of Tennessee Press.

Lawrence, E. A. 1985. *Hoofbeats and society: Studies of human-horse interactions.* Bloomington, IN: Indiana University Press.

Mack, A., ed. 1999. *Humans and other animals.* Columbus: Ohio State University Press.

Manning, A., & Serpell, J., eds. 1994. *Animals and human society changing perspectives.* London: Routledge.

Midgley, M. 1983. *Animals and why they matter.* Athens, GA: University of Georgia Press.

Midgley, Mary. 1995. *Beast and man: The roots of human nature.* New York: Routledge.

Midkiff, K. 2004. *The meat you eat.* New York: St. Martin's Griffin.

Newkirk, I. 2005. *Making kind choices: Everyday ways to enhance your life through earth- and animal-friendly living.* New York: St Martin's Griffin.

Niman, N. H. 2009. *Righteous porkchop: Finding a life and good food beyond factory farms.* New York: Collins Living.

Ogorzaly, M. A. 2006. *When bulls cry: The case against bullfighting.* Bloomington, IN: Author-House.

Peterson, D. 2003. *Eating apes.* Berkeley, CA: University of California Press.

Phelps, N. 2004. *The great compassion: Buddhism and animal rights.* New York: Lantern Books.

Phelps, N. 2007. *The Longest struggle: Animal advocacy from Pythagoras to PETA.* New York: Lantern Books.

Pickover, M. 2005. *Animal rights in South Africa.* Cape Town, SA: Double Storey.

Poulsen, E. 2009. *Smiling bears: A zookeeper explores he behavior and emotional life of bears.* Vancouver, BC: Greystone Books.

Regan, Tom. 1984. *The case for animal rights.* New York: Routledge.

Regan, T. 2004. *Empty cages: Facing the challenge of animal rights.* New York: Rowman & Littlefield.

Renhardt, V., and Renhardt, A. 2008. *Environmental enrichment ad refinement for non-human primates kept in research laboratories.* Washington, DC: Animal Welfare Institute.

Ryder, R. D. 1989. *Animal revolution: Changing attitudes towards speciesism.* Cambridge, MA: Basil Blackwell.

Salem, D. J., and Rowan, A. N., eds. 2007. *The state of the animals IV, 2007.* Washington, DC: Humane Society Press.

Scholtmeijer, Marion. 1993. *Animal victims in modern fiction: From sanctity to sacrifice.* Toronto: University of Toronto Press.

Scully, M. 2002. *Dominion: The power of man, the suffering of animals.* New York: St. Martin's Press.

Serpell, J. 1986. *In the company of animals.* New York: Basil Blackwell.

Serpell, J. 1996. *In the company of animals: A study of human–animal relationships.* Cambridge: Cambridge University Press.

Shrouded by the sea, http://www.wdcs.org/submissions_bin/wdcs_bycatchreport_2008.pdf.

Singer, Peter. 1991. *Animal liberation,* 2nd ed. London: Thorsons.

Singer, P., and Mason, J. 2006. *The way we eat: Why our food choices matter.* Emmaus, PA: Rodale.

Smith, E., and Dauncey, G. 2007. *Building an ark: 101 solutions to animal suffering.* Vancouver, BC: New Society Publishers.

Sneddon, L. U. 2003. "The evidence for pain in fish: the use of morphine as an analgesic." *Applied Animal Behaviour Science* 83, 153–162.

SPEAK—promoting humane education; http://www.speakonline.org/about.html.

Tobias, M., and Morrison, J. 2008. *Sanctuary: Global oases of innocence.* San Francisco: Council Oak Books. (http://epublishersweekly.blogspot.com/2008/05/sanctuary-by-michael-tobias-jane.html)

Turner, J., and D'Silva, J., eds. 2006. *Animals, ethics, and trade.* London: EarthScan Publishing.

Weil, Zoe. 2003. *Above all, be kind: Raising a humane child in challenging times.* Gabriolo Island, British Columbia: New Society Publishers.

White, T. I. 2007. *In defense of dolphins: The new moral frontier.* Malden, MA: Blackwell Publishing.

Williams, E., and DeMello, M. 2007. *Why animals matter: The case for animal protection.* Amherst, NY: Prometheus Books.

Wilson, E. O. 1984 *Biophilia.* Cambridge, MA: Harvard University Press.

Wise, Steven M. 2000. *Rattling the cage: Toward legal rights for animals.* Cambridge, MA: Perseus Books.

Wise, S. 2003. "The evolution of animal law since 1950." In *The state of the animals II,* ed. Deborah Salem and Andrew Rowan. Washington, DC: Humane Society Press. At http://www.hsus.org/press_and_publications/humane_bookshelf/the_state_of_the_animals_ii_2003.html.

Woodroffe, R., Thirgood, S. and Rabinowitz, A., eds. 2005. *People and wildlife, conflict or coexistence?* Cambridge: Cambridge University Press.

About the Editor and Contributors

EDITOR

Marc Bekoff is Professor Emeritus of Ecology and Evolutionary Biology at the University of Colorado, a Fellow of the Animal Behavior Society, and a former Guggenheim Fellow. In 2000 he was awarded the Exemplar Award from the Animal Behavior Society for major long-term contributions to the field of animal behavior. Marc is also an ambassador for Jane Goodall's Roots & Shoots program, in which he works with students of all ages, senior citizens, and prisoners. Marc has published more than 200 scientific and popular essays and twenty-two books, including *The Emotional Lives of Animals, Animals Matter, Animals at Play: Rules of the Game* (winner of the outstanding children's book award from the Animal Behavior Society), *Wild Justice: The Moral Lives of Animals* (with Jessica Pierce), and the *Encyclopedia of Human-Animal Relationships* (Greenwood, 2007), the *Encyclopedia of Animal Behavior* (Greenwood, 2004), and the first edition of the *Encyclopedia of Animal Rights and Animal Welfare* (Greenwood, 1998). In 2005, Marc was presented with The Bank One Faculty Community Service Award for the work he has done with children, senior citizens, and prisoners. His websites are http://literati.net/Bekoff and, with Jane Goodall: www.ethologicalethics.org.

CONTRIBUTORS

The following information about the contributors to this encyclopedia includes relevant Web sites and links that are outstanding interdisciplinary and international resources, containing details about the authors and various educational programs, projects, and organizations concerned with animal rights, animal welfare, and human-animal interactions.

Candace S. Alcorta, Research Scientist, Department of Anthropology, University of Connecticut, Storrs, Connecticut. **Signals and Rituals of Humans and Animals** (by Candace S. Alcorta and Richard Sosis)

Sky Alibhai, Director, WildTrack, Non-invasive Wildlife Monitoring, Monchique, Portugal: www.wildtrack.org. **Field Studies: Animal Immobilization** (by Zoe Jewell and Sky Alibhai)

Colin Allen, Professor, History & Philosophy of Science and Cognitive Science, Indiana University, Bloomington: http://mypage.iu.edu/~colallen/. **Consciousness, Animal**

Arnold Arluke, Professor of Sociology and Anthropology, Northeastern University, Boston, Massachusetts: http://www.socant.neu.edu/faculty/arluke. **Cruelty to Animals: Enforcement of Anti-Cruelty Laws; Laboratory Animal Use—Sacrifice**

Melissa Bain, Assistant Professor, Clinical Animal Behavior Service, University of California-Davis, School of Veterinary Medicine: www.vetmed.ucdavis.edu/vmth/small_animal/behavior/default.cfm www.dacvb.org; www.avsabonline.org. **Companion Animals, Welfare, and the Human-Animal Bond**

Jonathan Balcombe, Senior Research Scientist, Physicians Committee for Responsible Medicine, Washington, DC: www.pleasurablekingdom.com; www.jonathanbalcombe.com. **Pleasure and Animal Welfare; Rats; Stress and Laboratory Routines**

Ann Baldwin, Research Professor, Physiology, University of Arizona, Tucson: http://www.physiology.arizona.edu/index.php/baldwin_lab; www.mind-body-science.com. **Stress, Assessment, Reduction, and Science**

Marsha L. Baum, Professor of Law, University of New Mexico, Albuquerque: http://lawschool.unm.edu/faculty/baum/index.php. **Disasters and Animals: Legal Treatment in the United States**

Tom L. Beauchamp, Kennedy Institute of Ethics and Department of Philosophy, Georgetown University, Washington, DC. **Moral Standing of Animals**

Alan M. Beck, Professor and Director, Center for the Human-Animal Bond, Purdue University, West Lafayette, Indiana: http://www.vet.purdue.edu/chab/. **Humane Education Movement: The Humane University**

Piers Beirne, Professor of Sociology and Legal Studies, Dept. of Criminology, University of Southern Maine, Portland, Maine. **Bestiality**

Anne Bekoff, Professor, Integrative Physiology, University of Colorado, Boulder. **Embryo Research**

Marc Bekoff, **Introduction: Why Animal Rights and Animal Welfare Matter; Deep Ethology; Dogs; Field Studies and Ethics; Human Effects on Animal Behavior**

Marjorie Bekoff, Independent Scholar, Weybridge, Vermont. **Institutional Animal Care and Use Committees: Non-Affiliated Members**

Beth Bennett, Associate Research Professor, Institute for Behavioral Genetics, University of Colorado, Boulder. **Genethics**

Rod Bennison, Conjoint Academic, Geography and Environmental Science, University of Newcastle, Newcastle, Australia. **Ecological Inclusion: Unity Among Animals; War: Using Animals in Transport**

Sarah M. Bexell, Director of Conservation Education and Communications, Chengdu Research Base of Giant Panda Breeding, Chengdu, China: www.panda.org.cn. **Humane Education, Animal Welfare, and Conservation**

Lynda Birke, Professor, Anthrozoology Unit, Department of Biology, University of Chester, UK. **Ecofeminism and Animal Rights**

Stephen St. C. Bostock, Honorary Research Fellow in Philosophy, University of Glasgow, Glasgow, UK. **Zoos—Roles**

Jill Bough, Humanities and Social Sciences, University of Newcastle, Australia. **Donkeys; War: Using Animals in Transport**

Carol Bradley, Independent Scholar and Author, Harvard University Neiman Fellow, Great Falls, Montana. **Puppy Mills**

G.A. Bradshaw, Director, The Kerulos Center, Jacksonville, Oregon: www.kerulos.org. **Conservation Ethics, Elephants**

Philippa Brakes, Senior Biologist, WDCS, the Whale and Dolphin Conservation Society: www.wdcs.org. **Whales and Dolphins: Sentience and Suffering**

Donald M. Broom, Centre for Animal Welfare and Anthrozoology. Department of Veterinary Medicine, University of Cambridge, Cambridge, UK. **Animal Welfare: Coping; Animal Welfare: Freedom; Stereotypies in Animals**

Joseph Bruchac, Director, The Greenfield Review Literary Center, Greenfield Center, New York: josephbruchac.com: ndcenter.org. **Native Americans' Relationships with Animals: All Our Relations**

Gordon M. Burghardt, Alumni Distinguished Service Professor, Departments of Psychology and Ecology & Evolutionary Biology, University of Tennessee, Knoxville. **Anthropomorphism: Critical Anthropomorphism; Reptiles**

Nedim C. Buyukmihci, Emeritus Professor of Veterinary Medicine, University of California, Davis. **Association of Veterinarians for Animal Rights (AVAR)**

Larry Carbone, Senior Veterinarian, Laboratory Animal Resource Center, University of California San Francisco. **Animal Models and Animal Welfare; Euthanasia; Experimentation and Research with Animals; Laboratory Animal Welfare**

Matt Cartmill, Professor, Department of Anthropology, Boston University, Boston, Massachusetts. **Hunting, History of Ideas**

Paola Cavalieri, Milan, Italy, *Etica & Animali* (International Philosophy Journal). **Animal Liberation Ethics; Speciesism—Biological Classification**

Anna E. Charlton Adjunct Professor of Law, Rutgers University School of Law—Newark: http://www.AbolitionistApproach.com; http://law.newark.rutgers.edu. **Abolitionist Approach to Animal Rights; Student Rights and the First Amendment**

Una Chaudhuri, Collegiate Professor, English and Drama, New York University, New York. **Entertainment and Amusement: Animals in the Performing Arts**

Stephen R. L. Clark, Professor of Philosophy, University of Liverpool: http://pcwww. liv.ac.uk/~srlclark/srlc.htm. **Species Essentialism**

Juliet Clutton-Brock, South Barn, Cambridge, UK. **Domestication**

Nicole Cottam, Tufts Cummings School of Veterinary Medicine: www.tufts.edu/vet/behavior. **Toxicity Testing and Animals**

Debbie Coultis, President and CEO, People, Animals, Nature, Naperville, Illinois: www.pan-inc.org. **Blood Sports**

Eileen Crist, Science and Technology, Virginia Polytechnic University, Blacksburg, Virginia. **Evolutionary Continuity**

David DeGrazia, Professor, Department of Philosophy, George Washington University: http://www.gwu.edu/~philosop/faculty/degrazia.htm. **Autonomy of Animals; Equal Consideration**

Margo DeMello, Director, House Rabbit Society, Placitas, New Mexico. **Animal Body, Alteration of; Rabbits; Rescue Groups**

Mark Derr, Independent Scholar and Author, Miami Beach, Florida. **Dog fighting**

Rebecca Dresser, Daniel Noyes Kirby Professor of Law, Professor of Ethics in Medicine Washington University Law School, St. Louis, Missouri. **Institutional Animal Care and Use Committees** (IACUCS): **Regulatory Requirements**

Feng Rui Xi, Conservation Education Program Manager, Chengdu Research Base of Giant Panda Breeding, Chengdu, China: http://www.panda.org.cn/english/index.htm. **Humane Education, Animal Welfare, and Conservation**

Richard Foltz, Associate Professor, Religion, Concordia University, Montreal, Canada: http://artsandscience1.concordia.ca/religion/FoltzR.html. **Religion and Animals: Islam**

Camilla H. Fox, Founding director of Project Coyote (www.ProjectCoyote.org), a project of Earth Island Institute, wildlife consultant for the Animal Welfare Institute (www. awionline.org). **Predator Control and Ethics; Trapping, Behavior and Welfare; Urban Wildlife**

Michael Allen Fox, Professor Emeritus of Philosophy, Queen's University (Canada) and Adjunct Professor, School of Humanities, University of New England (Australia): http://www.une.edu.au/staff/mfox3.php: http://www.queensu.ca/philosophy/faculty. html (Scroll to emeritus professors). **Anthropocentrism; Antivivisectionism; Vegetarianism**

Michael W. Fox, veterinarian, bioethicist, and syndicated columnist: www.doctormw fox.org. **Genetic Engineering and Farmed Animal Cloning**

Gary L. Francione, Distinguished Professor of Law, and Nicholas de B. Katzenbach Scholar of Law and Philosophy, Rutgers University School of Law—Newark: http://www.AbolitionistApproach.com: http://law.newark.rutgers.edu. **Abolitionist Approach to Animal Rights** (by Gary L. Francione and Anna Charlton); **Animal Rights Movement, New Welfarism; Law and Animals**

David Fraser, Professor, Animal Welfare Program, University of British Columbia, Canada: http://www.landfood.ubc.ca/animalwelfare/. **Animal Welfare**

Carol Freeman, Research Associate, Geography and Environmental Studies, University of Tasmania, Hobart: http://fcms.its.utas.edu.au/scieng/geog/pagedetails.asp?

lpersonId=1350; http://www.utas.edu.au/library/exhibitions/thylacine/index.html. **Extinction and Ethical Perspectives**

Chance French, Executive Director, APES: http://www.a-p-e-s.org/. **Sanctuaries, Ethics of Keeping Chimpanzees in** (with Lee Theisen-Watt)

R. G. Frey, Professor, Philosophy Department, Bowling Green State University, Bowling Green, Ohio. **Quality of Life for Animals**

Antoine F. Goetschel, Animal attorney, Zurich, Switzerland: http://www.afgoetschel.com/en/; http://www.swissinfo.ch/eng/swissinfo.html?siteSect=882&sid=10206656. **Law and Animals, European Union**

Alan Goldberg, Professor of Toxicology; Director, Center for Alternatives to Animal Testing(CAAT); Johns Hopkins Bloomberg School of Public Health, Baltimore: http://caat.jhsph.edu: http://altweb.jhsph.edu. **Alternatives to Animal Experiments: Reduction, Refinement, and Replacement**

Jane Goodall, D.B.E., Founder, The Jane Goodall Institute and U.N. Messenger of Peace: www.janegoodall.org. **Sanctuaries**

John Goodwin, Manager of Animal Fighting Issues, The Humane Society of the United States: www.humanesociety.org. **Cockfighting**

Ray Greek MD, Independent scholar, President, Americans for Medical Advancement (www.curedisease.com), Science Advisor, National Anti-Vivisection Society, Chicago, IL: www.navs.org. **Medical Research with Animals**

Michael Greger, MD, Director of Public Health and Animal Agriculture, Humane Society of the United States: DrGreger.org; AtkinsFacts.org; BirdFluBook.org Farm AnimalWelfare.org. **Factory Farms and Emerging Infectious Diseases**

Lori Gruen, Associate Professor, Philosophy and Feminist, Gender, and Sexuality Studies; Director, Ethics in Society Project, Wesleyan University, Middletown, Connecticut: http://first100chimps.wesleyan.edu; http://lgruen.web.wesleyan.edu. **Chimpanzees in Captivity; Ecofeminism and Animal Rights** (by Lori Gruel and Lynda Birke)

John Hadidian, Director Urban Wildlife Program, The Humane Society of the United States: www.wildneighbors.org. **Urban Wildlife** (by John Hadidian, Camilla Fox, and William S. Lynn)

Marlene Halverson, Senior Farm Animal Policy Specialist, Animal Welfare Institute: www.awionline.org. **Food Animals: A Comparison of Methods of Raising Animals**

Lynette Hart, Professor, Department of Population Health and Reproduction, School of Veterinary Medicine, University of California, Davis, California: http://www.vetmed.ucdavis.edu/Animal_Alternatives/main.htm, http://www.lib.ucdavis.edu/dept/animal alternatives/, http://www.vetmed.ucdavis.edu/CCAB/humananimalinteractions.

html, http://www.vetmed.ucdavis.edu/Animal_Alternatives/appendices.html. **Animal-Assisted Therapy; Dissection in Science and Health Education; Mice**

Harold Herzog, Professor, Department of Psychology, Western Carolina University: http://paws.wcu.edu/herzog/. **Sociology of the Animal Rights Movement**

Ned Hettinger, Professor, Philosophy Department, College of Charleston, SC: http://www.cofc.edu/hettinger/. **Environmental Ethics**

Chris Heyde, Deputy Director, Government and Legal Affairs, Animal Welfare Institute: www.awionline.org; www.compassionindex.org. **Horse Slaughter** (with Liz Clancy Ross)

Laura Hobgood-Oster, Professor, Religion and Environmental Studies, Southwestern University, Georgetown, Texas: http://southwestern.edu/departments/religionphiloso phy/; www.hsus.org/religion/ **Blessing of the Animals Rituals**

Alexandra Horowitz, Term Assistant Professor, Barnard College, New York: http://www.columbia.edu/~ah2240/. **Anthropomorphism**

Leslie Irvine, Associate Professor of Sociology, University of Colorado, Boulder: http://socsci.colorado.edu/SOC/People/Faculty/irvine.html. **Disasters and Animals**

Jennifer Jackman, Assistant Professor of Political Science, Salem State College, Salem, Massachusetts; Humane Society University: http://www.hsus.org. **The Gender Gap and Policies toward Animals**

Robert G. Jaeger, Carencor, Louisiana. **Amphibians**

Dale Jamieson, Director of Environmental Studies, Professor of Environmental Studies and Philosophy, New York University: http://philosophy.fas.nyu.edu/object/daleja mieson.html. **Zoos—Welfare Concerns**

Olga S. Jarrett, Associate Professor, Department of Early Childhood Education, Georgia State University, Atlanta, Georgia. **Humane Education, Animal Welfare, and Conservation**

Zoe Jewell, Director, WildTrack, Non-invasive Wildlife Monitoring, Monchique, Portugal: www.wildtrack.org. **Field Studies: Animal Immobilization** (by Zoe Jewell and Sky Alibhai)

Nick Jukes, InterNICHE Coordinator, Leicester, UK: http://www.interniche.org/. **Alternatives to Animal Experiments in the Life Sciences**

Wendy Keefover-Ring, Carnivore Protection Director, WildEarth Guardians: http://www.wildearthguardians.org/; http://www.goagro.org/. **Wildlife Services**

Mark G. E. Kelly, Lecturer in Philosophy, Middlesex University: http://www.mdx.ac.uk/www/crmep/STAFF/MarkKelly.htm. **The Political Subjectivity of Animals**

Lisa Kemmerer, Assistant Professor, Montana State University, Billings: http://www.msubillings.edu/CASFaculty/Kemmerer. **Religion and Animals: Daoism; Religion and Animals: Pantheism and Panentheism; Religion and Animals: Veganism and the Bible**

Josphat Ngonyo Kisui, Director, Africa Network for Animal Welfare, Nairobi, Kenya: www.anaw.org. **Kenya: Conservation and Ethics**

Michael Kreger, Biologist, Laurel, Maryland. **Zoos: History**

S. Chinny Krishna, Chairman, Blue Cross of India, Chennai, India: http://www.blue cross.org.in/; http://www.bluecross.org.in/founder.html. **India: Animal Experimentation**

Hugh LaFollette, Cole Chair in Ethics, University of South Florida St. Petersburg: http://www.stpt.usf.edu/hhl/. **Krogh Principle** (by Hugh LaFollette and Niall Shanks); **Utilitarianism and Assessment of Animal Experimentation** (by Hugh LaFollette and Niall Shanks)

James LaVilla-Havelin, Poet, Museum Educator, Arts Educator, Lytle, Texas. **Museums and Representation of Animals; Poetry and Representation of Animals**

Berel Dov Lerner, Lecturer, Philosophy, Western Galilee College, Israel: http://wgalil. academia.edu/BerelDovLerner. **Religion and Animals: Judaism**

Peter J. Li, Associate Professor of Chinese Politics: http://www.uhd.edu/academic/ colleges/humanities/sos/political_science/lip.htm. **China: Animal Rights and Animal Welfare**

Andrew Linzey, Director of the Oxford Centre for Animal Ethics and a member of the Faculty of Theology, University of Oxford, UK: www.oxfordanimalethics.com. **Religion and Animals; Religion and Animals: Animal Theology; Religion and Animals: Christianity; Religion and Animals: Reverence for Life; Religion and Animals: Saints; Religion and Animals: Theodicy; Religion and Animals: Theos Rights; Sentientism**

Randall Lockwood, Senior Vice President, Anti-Cruelty Field Services The American Society for the Prevention of Cruelty to Animals: www.aspca.org; www.aspcapro.org. **Cruelty to Animals and Human Violence; Cruelty to Animals: Prosecuting Anti-Cruelty Laws**

Robert Long, Research Ecologist, Western Transportation Institute, Montana State University: www.wti.montana.edu. **Field Studies: Ethics and Noninvasive Wildlife Research** (by Paula MacKay and Robert Long)

Vonne Lund, Senior Researcher, National Veterinary Institute, Oslo, Norway: http:// www.vetinst.no/eng/. **Fish** (by Cecilie M. Mejdell and Vonne Lund)

William S. Lynn, Assistant Visiting Professor, Environmental Studies, Williams College, Founder and Senior Ethics Advisor, Practical Ethics: www.williams.edu; www.practi calethics.net. **Animal Studies; Practical Ethics and Human-Animal Relationships; Urban Wildlife** (by John Hadidian, Camilla H. Fox and William S. Lynn)

Paula MacKay, Research Associate, Western Transportation Institute, Montana State University: www.wti.montana.edu. **Field Studies: Ethics and Noninvasive Wildlife Research** (by Paula MacKay and Robert Long)

Koen Margodt, Independent Scholar and Author, Philosophy and Moral Sciences, Ghent University, Belgium. **Affective Ethology; The Great Ape Project; Great Apes and Language Research**

Jim Mason, Author, *An Unnatural Order;* co-author with Peter Singer, *The Ethics of What We Eat*. **Dominionism; Misothery; Religion and Animals: Disensoulment**

Cynthia McFadden, Diagnostic Imaging, Austin Diagnostic Clinic, Austin, Texas. **Sports and Animals**

Paul McGreevy, Associate Professor, Veterinary Science, University of Sydney, NSW, Australia: http://www.vetsci.usyd.edu.au/about/staff/pmcgreevy.shtml; www.equita tionscience.com. **Pet Renting**

Cecilie M. Mejdell, Senior Scientist, Section for Domestic Animal Health and Welfare, National Veterinary Institute, Oslo, Norway: http://www.vetinst.no/eng/. **Fish** (by Cecilie M. Mejdell and Vonne Lond)

Michael Mendl, Professor of Animal Behaviour and Welfare, University of Bristol, UK: http://www.vetschool.bris.ac.uk/research/abw/ http://www.vetschool.bris.ac.uk/staff/staff_member.html?person_code=015672. **Animal Welfare: Assessment**

Lieve Meers, Guest Professor, Faculty of Biosciences and Landscape Architecture, University College Ghent, Belgium: http://biot.hoGhent.be/studeren/opleidingen/postgraduaatdoelstellingen.cfm http://www.ethology.uGhent.be/. **Blood Sports** (by William Ellery Samuels, Lieve Meers, Debbie Coultis, and Simona Normando)

Slavoljub Milekic, Professor of Cognitive Science and Digital Design, University of the Arts, Philadelphia, Pennsylvania: www.uarts.edu/faculty/smilekic. **Disneyfication**

Brian J. Miller, Wind River Ranch Foundation, Waltrous, New Mexico: http://www.windriverranch.org/. **Captive Breeding Ethics** (by Richard P. Reading and Brian J. Miller)

Heather Moore, Senior Writer, People for the Ethical Treatment of Animals (PETA), Norfolk, VA: www.peta.org. **Veganism**

David B. Morton, Professor Emeritus, Biomedical Science and Ethics, University of Birmingham, UK. **Distress in Animals; Docking; Pain, Suffering, and Behavior; Polyism; Sizeism**

Nina Natelson, Director, Concern for Helping Animals in Israel, Alexandria, Virginia: http://www.chai-online.org/. **Israel: Animal Protection**

James Lindemann Nelson, Professor of Philosophy, Michigan State University. East Lansing: www.msu.edu/~phl/faculty/profs/nelson.htm. **Xenograft**

Ruth Newberry, Center for the Study of Animal Well-being, Washington State University, Pullman, Washington: http://www.ansci.wsu.edu/People/newberry/faculty.asp; http://www.vetmed.wsu.edu/research_vcapp/newberry.asp. **Chickens**

Ingrid Newkirk, President, People for the Ethical Treatment of Animals: PETA.org. **People for the Ethical Treatment of Animals (PETA)**

Jim Nollman, Interspecies.com, Friday Harbor, Washington: http://interspecies.com. **Field Studies: Ethics of Communication Research with Wild Animals**

Simona Normando, Researcher/Collaborator, Dipartimento di Scienze Sperimentali Veterinarie, Università degli Studi di Padova, Italy: http://www.sperivet.unipd.it/. **Blood Sports** (by William Ellery Samuels, Lieve Meers, Debbie Coultis, and Simona Normando)

F. Barbara Orlans, The Kennedy Institute of Ethics (retired), Georgetown University, Washington, DC. **Dogs**

Wayne Pacelle, President & CEO, The Humane Society of the United States: www.hsus.org. **Animal Protection: On the Future of Activism**

Andrew Page, Senior Director, Wildlife Abuse Campaign, The Humane Society of the United States: www.humanesociety.org; www.humanesociety.org/wildlifeabuse. **Wildlife Abuse**

Elizabeth S. Paul, Department of Clinical Veterinary Science, University of Bristol, UK. **Empathy with Animals**

Norm Phelps, Independent Scholar, Funkstown, Maryland. **Religion, History, and the Animal Protection Movement**

Evelyn Pluhar, Professor of Philosophy, The Pennsylvania State University, Fayette Campus, Union Town, Pennsylvania: http://www.personal.psu.edu/exp5. **Marginal Cases**

Núria Querol i Viñas, MD, Founder of GEVHA (Group for the Study of Violence Towards Humans and Animals), Barcelona, Spain: www.gevha.com www.altarriba.org http://aiudaweb.googlepages.com/ www.proda.es. **Bullfighting**

Richard P. Reading, Director of Conservation Biology, Denver Zoological Foundation, Denver, Colorado. **Captive Breeding Ethics** (by Richard P. Reading and Brian Miller)

Craig Redmond, Campaigns Director, The Captive Animals' Protection Society (CAPS), Manchester, UK: www.captiveanimals.org; www.irishcircuses.org. **Entertainment and Amusement: Circuses, Rodeos, and Zoos** (by Craig Redmond and Garry Sheen)

Tom Regan, Professor Emeritus, Philosophy, North Carolina State University, Raleigh: http://www.lib.ncsu.edu/exhibits/regan/; http://www.tomregan-animalrights.com; http://www.youtube.com/watch?v=ADhNch30Img; http://cultureandanimals.org. **Animal Rights**

Harriet Ritvo, Arthur J. Conner Professor of History, Massachusetts Institute of Technology, Cambridge, Massachusetts: http://web.mit.edu/hnritvo/www/ritvo.htm. **Royal Society for the Prevention of Cruelty to Animals (RSPCA): History**

Jill Robinson, Founder & CEO, Animals Asia Foundation: www.animalsasia.org. **China: Moon Bears and the Bear Bile Industry**

Bernard E. Rollin, Colorado State University, University Distinguished Professor, Professor of Philosophy, Professor of Animal Sciences, Professor of Biomedical Sciences, and University Bioethicist. **Genetic Engineering; Law and Animals, United States; Teleology and Telos; Veterinary Medicine and Ethics**

Holmes Rolston, III, Professor Emeritus, Philosophy, Colorado State University: http://lamar.colostate.edu/~rolston/. **Endangered Species and Ethical Perspectives; Wild Animals and Ethical Perspectives**

Terry L. Root, Senior Fellow/University Faculty, Woods Institute for the Environment, Stanford University: http://terryroot.stanford.edu. **Global Warming and Animals**

Nicole Rosmarino, Wildlife Program Director, WildEarth Guardians: www.wildearth guardians.org. **Endangered Species Act**

Liz Clancy Ross, Federal Policy Advisor, Animal Welfare Institute: www.awionline.org; www.compassionindex.org. **Horse Slaughter** (with Chris Heyde)

Andrew Rowan, Executive Vice President and CEO, Humane International, The Humane Society of the United States. **Pain, Invertebrates; The Silver Spring Monkeys**

Lilly-Marlene Russow, Professor Emerita, Department of Philosophy, Purdue University, West Lafayette, Indiana. **Institutional Animal Care and Use Committees (IACUCS)**

Allen Rutberg, Research Assistant Professor, Tufts Center for Animals and Public Policy, Cummings School of Veterinary Medicine, North Grafton, Massachusetts: http://www.tufts.edu/vet/cfa/home.html. **Wildlife Contraception**

Richard Ryder, RSPCA, UK: www.richardryder.co.uk; http://www.rspca.org.uk/. **Painism; Royal Society for the Prevention of Cruelty to Animals (RSPCA) Reform Group; Speciesism**

Joyce E. Salisbury, Professor Emerita, University of Wisconsin—Green Bay: www.uwgb.edu/salisbuj. **Bestiality: History of Attitudes**

William Ellery Samuels, Director of Assessment, Department of Education, College of Staten Island, City University of New York: http://wesamuels.net/; http://www.pan-inc.org/. **Blood Sports** (by William Ellery Samuels, Lieve Meers, Debbie Coultis, and Simona Normando)

Clinton R. Sanders, Professor, Department of Sociology, University of Connecticut: http://www.sociology.uconn.edu/faculty/sanders.html. **Deviance and Animals**

Constantine Sandis, Senior Lecturer in Philosophy, Oxford Brookes University and New York University in London, UK: http://www.brookes.ac.uk/schools/education/staffinfo/sandis.html!; http://www.nyu.edu/global/london/academics/staff_list/dr_constantine_sandis.htm; http://oxfordbrookes.academia.edu/ConstantineSandis. **Animals in Space**

Lisa M. Savage, Department of Psychology, Behavioral Neuroscience Program, State University of New York, Binghamton, New York. **Native Americans and Early Uses of Animals in Medicine and Research**

James Serpell, Professor of Humane Ethics and Animal Welfare, University of Pennsylvania: http://www.vet.upenn.edu/FacultyandDepartments/Faculty/tabid/362/Default.aspx?faculty_id=6361798; http://www.penncias.org/. **Companion Animals**

Niall Shanks, Curtis D. Gridley Distinguished Professor of History and Philosophy of Science, Wichita State University, Wichita, Kansas. **Krogh Principle** (by Hugh LaFollette and Niall Shanks); **Utilitarianism and Assessment of Animal Experimentation** (by Hugh LaFollette and Niall Shanks)

Kenneth J. Shapiro, Animals & Society Institute, Ann Arbor Michigan: www.animalsandsociety.org. **Objectification of Animals; Scholarship and Advocacy; Student Attitudes toward Animals; Student Objections to Dissection**

Katrina Sharman, Corporate Counsel, Voiceless—The Fund for Animals, Australia: www.voiceless.org.au. **Law and Animals: Australia**

Garry Sheen, Development Director, The Captive Animals' Protection Society, Manchester, UK: www.captiveanimals.org; www.irishcircuses.org. **Entertainment and Amusement: Circuses, Rodeos, and Zoos** (by Craig Redmond and Garry Sheen)

Jo-Ann Shelton, Department of Classics, University of California, Santa Barbara. **Exotic Species**

Mark Simmonds, International Director of Science, WDCS, the Whale and Dolphin Conservation Society, Chippenham, Wiltshire, UK: www.wdcs.org. **Whales and Dolphins: Solitary Dolphin Welfare**

Peter Singer, Ira W. DeCamp Professor of Bioethics, Princeton University, Princeton, New Jersey: www.princeton.edu/~psinger. **Utilitarianism**

Richard Sosis, Associate Professor, Department of Anthropology, University of Connecticut: http://www.anth.uconn.edu/faculty/sosis/. **Signals and Rituals of Humans and Animals**

Marek Spinka, Senior Researcher, Department of Ethology, Institute of Animal Science, Prague: http://www.fhs.cuni.cz/etologie/spinka_main_eng.htm; http://www.vuzv.cz/. **Pigs**

Gary Steiner, John Howard Harris Professor of Philosophy, Bucknell University, Lewisburg, Pennsylvania. **Cosmic Justice**

Ron Swaisgood, Associate Director of Conservation and Research, San Diego Zoo, San Diego, California: http://zooconservation.org; http://cres.sandiegozoo.org/staff/div_applied_cons.html; http://www.eeb.ucla.edu/indivfaculty.php?FacultyKey=2854. **Enrichment and Well-Being for Zoo Animals**

David Sztybel, Department of Sociology, Brock University, St. Catharines, Canada: http://sztybel.tripod.com/home.html. **Animal Welfare and Animal Rights, A Comparison; Religion and Animals: Jainism**

Lee Theisen-Watt, President and Founder, APES: http://www.a-p-e-s.org/. **Sanctuaries, Ethics of Keeping Chimpanzees in** (by Lee Theisen-Watt and Chance French)

Mary Thurston, Independent Scholar, Anthropology and Animal History: www.animal image.com. **War and Animals**

Jacky Turner, Writer on Animal Welfare, England. **Animal Reproduction, Human Control**

Bernard Unti, Senior Policy Advisor, Special Assistant to the CEO, The Humane Society of the United States: www.hsus.org. **Humane Education Movement in Schools**

Gavin Van Horn, Brown Junior Visiting Scholar, Environmental Studies, Southwestern University, Georgetown, Texas: www.religionandnature.com. **Wolves and Ethical Perspectives**

Paul Waldau, Religion and Animals Institute: www.religionandanimals.org/index.html; www.paulwaldau.com. **Religion and Animals: Buddhism; Religion and Animals: Hinduism; Religion and Animals: Judaism and Animal Sacrifice; Speciesism: Ethics, Law, and Policy**

Yvette Watt, Associate Lecturer in Painting, University of Tasmania, Hobart: www. yvettewatt.com.au. **Art, Animals, and Ethics**

John Webster, Professor Emeritus, Veterinary Science, University of Bristol, UK: www. vetschool.bris.ac.uk/animalwelfare/bwapteam. **Sentience and Animal Protection**

Zoe Weil, President, Institute for Humane Education: http://HumaneEducation.org; http://zoeweil.com. **Humane Education**

Jack Weir, Professor Philosophy, Morehead State University, Morehead, Kentucky: http://www.moreheadstate.edu/eflp/index.aspx?id=6381. **Virtue Ethics**

Françoise Wemelsfelder, Senior scientist, Sustainable Livestock Systems, Scottish Agricultural College, Edinburgh, UK. **Animal Subjectivity**

Hal Whitehead, Professor, Department of Biology, Dalhousie University Halifax, Nova Scotia, Canada. **Whales and Dolphins: Culture and Human Interactions**

Fred Whoriskey, Vice President, Research and Environment, Atlantic Salmon Federation, St. Andrews, NB, Canada: www.asf.ca. **Fishing as Sport**

Erin Williams, Communications Director, Factory Farming Campaign, The Humane Society of the United States: www.whyanimalsmatter.com; www.humanesociety.org. **Factory Farms**

Nathan J. Winograd, Director, No Kill Advocacy Center: www.nathanwinograd.com; www.nokilladvocacycenter.org. **Shelters, No-Kill**

Mary W. Wood, Reference Librarian, University of California, Davis, Center for Animal Alternatives Information, Davis, California. **Resources on Animal Rights and Animal Welfare: Searching the Literature Online**

Xu Ping, Manager of Conservation Education, Chengdu Research Base of Giant Panda Breeding, Chengdu, China: http://www.panda.org.cn/english/index.htm. **Humane Education, Animal Welfare, and Conservation**

R. Lee Zasloff, Adjunct Professor, American River College, Sacramento, California. **Cats**

Stephen Zawistowski, Executive Vice President, National Programs and Science Advisor, The American Society for the Prevention of Cruelty to Animals: www.aspca.org. **The American Society for the Prevention of Cruelty to Animals (ASPCA); Humane Education**

Joanne Zurlo, Institute for Laboratory Animal Research, The National Academies: www.nationalacademies.org/ilar. **Alternatives to Animal Experiments: Reduction, Refinement, and Replacement** (by Joanne Zurlo and Alan M. Goldberg)

Index